Farm to Table & Beyond

Grades 5 or 6

Pamela A. Koch
Angela Calabrese Barton
Isobel R. Contento

Published by Teachers College Columbia University

and the National Gardening Association
1100 Dorset St., South Burlington, VT 05403 • *www.garden.org* • *www.kidsgardening.org*

L*I*FE LINKING FOOD AND THE ENVIRONMENT
AN INQUIRY-BASED SCIENCE AND NUTRITION PROGRAM

LINKING FOOD AND THE ENVIRONMENT

AN INQUIRY-BASED SCIENCE AND NUTRITION PROGRAM

Linking Food and the Environment (LiFE) is a collaboration of the Science Education and Nutrition Education programs at Teachers College Columbia University. It was established in 1996 with the vision of promoting scientific habits of mind through thoughtful, inquiry-based activities that integrate the study of food, food systems, and environmental and personal health. *Farm to Table & Beyond* is one module of the LiFE Curriculum Series. The mission of all the LiFE modules is to increase scientific conceptual understandings in life science; improve attitudes toward science; improve attitudes toward personal health and nature; and promote behavior changes in relation to personal and ecological health. For information about LiFE, please visit *www.tc.edu/life*.

TEACHERS COLLEGE
COLUMBIA UNIVERSITY

Teachers College Columbia University, Center for Food & Environment, 525 West 120th Street, Box 137, New York, NY 10027

The National Gardening Association (NGA) is a nonprofit organization established in 1972. Its mission is to promote home, school, and community gardening as a means to renew and sustain the essential connection between people, plants, and the environment. NGA's programs and initiatives are targeted at five areas: plant-based education, health and wellness, environmental stewardship, community development, and responsible home gardening. For more information on NGA and its programs, please visit *www.garden.org* and *www.kidsgardening.org* or call (800) 538-7476.

Department of Health and Human Services • National Institutes of Health
Supported by a Science Education Partnership Award (SEPA) from the National Center for Research Resources
This publication was made possible by a Science Education Partnership Award (SEPA), grant number R25 RR12374, from the National Center for Research Resources (NCRR), a component of the National Institutes of Health (NIH). Its contents are solely the responsibility of the authors and do not necessarily represent the official views of the NCRR or NIH.

Additional funding for publication was provided by the W. K. Kellogg Foundation for Rethinking Food, Health, and the Environment: Making Learning Connections, a joint project of the Center for Ecoliteracy and Teachers College Columbia University. For more information, please visit *www.rethinkingschoollunch-life.org*.

ISBN 978-0-915873-50-0

Library of Congress Control Number: 2008932243

Printed in Canada.

Linking Food and the Environment Project Team

PRINCIPAL INVESTIGATOR
Isobel R. Contento, PhD, Mary Swartz Rose Professor of Nutrition Education, Teachers College Columbia University

CO-PRINCIPAL INVESTIGATOR
Angela Calabrese Barton, PhD, Associate Professor of Science Education, Michigan State University

AUTHOR AND PROJECT DIRECTOR
Pamela A. Koch, EdD, RD, Executive Director, Center for Food & Environment, Teachers College Columbia University

Farm to Table & Beyond Team

Author
Pamela A. Koch, EdD, RD

Lesson Development, Implementation, and Evaluation
Emily Bielecki, BS
Tracy Cullen, MS, RD
Marcia Dadds, MS, RD
Sumi Hagiwara, PhD
Stacia Helfand, MS, EdM, RD
Toby Jane Hindin, EdD
Ana De Lourdes Islas, EdD, RD
Meredith Smart, MS
Michelle Trudeau, MA
Karen Wadsworth, EdD

Teacher Contributors
Steven Broder, MBA, MS, P.S. 50
Nicholas Graham, EdM, IS 528
Kathy Ortiz, Bilingual Bicultural Mini School

Advisors
Ronald DeMeersman, PhD, Human Physiology
Joan D. Gussow, EdD, Food Systems
Toni Liquori, EdD, Food Systems and Implementation Issues
Patricia Zybert, PhD, Statistician

Field-Test Partners
Center for Ecoliteracy, Berkeley, CA, Zenobia Barlow, Executive Director
Environmental Education Council of Marin, San Rafael, CA, Sandy Wallenstein, Executive Director
The Food Trust, Philadelphia, PA, Sandra Sherman, EdD
Lawrence Hall of Science, Berkeley, CA, Katherine Barrett, PhD
University of Missouri, St. Louis, MO, William Kyle, PhD
University of Texas, Austin, TX, Julie Loft, PhD

Production Team

Editorial Director
Margo Crabtree

Senior Editors
Loaiza Ortiz
Blakely Tsurusaki, Technical Review

Associate Editor
Rachel Bartlett

Copy Editor
Kate Norris

Proofreader
Patricia Egan

Original Design for LiFE Curriculum Series
Lisa Cicchetti

Design, Art Direction, and Page Layout
Alison Watt, National Gardening Association

Illustrations
Anne Faust

Photography
Tyler/Barlow, Center for Ecoliteracy

To all of the children who participate in LiFE
and to the future of our food supply.

Contents

UNIT 1: BECOMING FOOD SCIENTISTS

UNIT 2: INTERACTING PARTS

UNIT 3: FOOD PROCESSING

UNIT 4: ENVIRONMENTAL EFFECTS

UNIT 5: WASTE

UNIT 6: MAKING CHOICES

Acknowledgments

The seed for LiFE was planted in the 1970s when Joan Gussow, EdD, the Mary Swartz Rose Professor Emeritus of Nutrition and Education, brought the perspective of food-system study to Teachers College Columbia University. LiFE began with the simple goal of developing an inquiry-based science curriculum that would educate children about food systems, food choice, and personal health. Because of the ideas, thoughts, and dedication of so many, the LiFE Curriculum Series has grown and expanded to include the interplay of biology, personal behavior, and the present food system and technological environment, which encourage overconsumption and sedentary behavior. It is our hope that the LiFE Curriculum Series will enhance students' personal motivation and competence to use their science understandings to reflect upon and purposefully act upon their world with the aim of transforming themselves and the conditions of their lives.

Many people have been involved in the development of the *Farm to Table & Beyond* module of the LiFE Curriculum Series, and we are indebted to them all. To those who have reviewed our materials, offered suggestions and insights, and shared experiences and challenges, we thank you. To the many educators and students who tested lessons and offered feedback, we appreciate your enthusiasm for *Farm to Table & Beyond* and your valuable contributions. To our field-test-school principals, teachers, and students, we thank you for working with us to make *Farm to Table & Beyond* what it is today. We learned so much from you. To our families and friends who have lived with this project for many years, we are deeply grateful for your warm support and endless patience.

1997–2007 Field-Test Partners

New York City Department of Education, New York, NY: Community School Districts 3, 4, 6, 10, and 11

2001–2004 Field-Test Partners

Center for Ecoliteracy, Berkeley, CA: Berkeley Unified School District

Food Trust, Philadelphia, PA: Philadelphia School District

Lawrence Hall of Science, Berkeley, CA: Hayward Unified School District and Mt. Diablo School District

University of Missouri, St. Louis, MO: Maplewood-Richmond Heights School District and Normandy School District

University of Texas, Austin, TX: Austin Independent School District

Introduction

Welcome! You and your students are about to embark on an exciting adventure — learning science through the study of our farm-to-table system, from understanding what our complex and highly technological food system is, to the interactions and feedback among the subsystems of the larger food system, to the trade-offs and environmental effects of the food-system choices we make as a society.

Human impact on the natural world is expected to increase as human populations grow and as science and technology develop ever more sophisticated ways to manage the natural world to meet human desires more effectively. Today's children, as tomorrow's adults, need a solid understanding of science concepts and skills to engage in scientific discussions and to participate in public debate about important issues that involve science and technology. During their lifetime, today's children will be called upon to make many decisions about their personal health, including how to choose foods that will lead to both nutritional and environmental well-being.

Children are naturally curious. They are investigators and problem-solvers, attempting to understand the natural and designed worlds. They already are "doing science." They may not always be aware of that fact, since "science" is often thought of as content that is abstract from everyday life. But science can be made personally meaningful … and it is in LiFE! We hope this module of the LiFE Curriculum Series brings enjoyment, learning, and growth to you and your students.

Isobel Contento
Angela Calabrese Barton
Pamela Koch

Goals

Students who participate in LiFE will:

- **increase their knowledge and conceptual understandings** about how the biological world works and how it interacts with the human-designed world;

- **develop skills in scientific inquiry** about the natural and designed world; use evidence to justify statements; use both logical reasoning and imagination; and be able to explain, to predict, and to identify and limit bias;

- **expand their ways of thinking or habits of mind** to include curiosity, flexibility, open-mindedness, informed skepticism, creativity, and critical thinking;

- **improve their attitude toward the processes of science** through enjoyable activities in a domain that is meaningful and familiar to them — food;

- **improve their attitude toward the natural environment** by developing an appreciation of nature's complexity, diversity, change, and constancy; respect for the natural environment; and concern for the impact of human food systems on the environment;

- **improve their attitude toward their personal health** through an understanding of the impact of food on health and an appreciation of healthful eating habits;

- **appreciate the connectedness** of science, technology, the natural environment, and everyday life in ways that are life-changing for the students themselves, for society, and for the natural environment;

- **increase confidence and commitment** to apply the above conceptual understandings, skills, attitudes, and ways of thinking (habits of mind) to personal decisions and public debate of issues related to food systems, health, and the natural environment.

Making Science Real, Meaningful, and Successful

Children construct meanings about their environment through their explorations of the world around them all the time. Using their senses, they make observations and use those to make predictions about how things work. They use their developing understanding to build a complex framework for how the world works.

Yet children's explanations of the natural world are incomplete and sometimes even scientifically unsound. For example, from a scientific perspective, we know that living things are distinguished from nonliving things in their ability to carry on the following life processes: movement, metabolism, growth, responsiveness to environmental stimuli, and reproduction. However, many children believe that objects are living if they move or grow. For example, the sun, wind, and clouds are seen as living because they move. Fires are seen as living because they consume wood, move, require air, reproduce (sparks cause other fires), and give off waste (smoke). [1,2]

As teachers, we need to continually remind ourselves that children build their personal scientific ideas over many years of explorations. Therefore, it is difficult, at best, for science teachers to help children "change" their understandings with one or two lessons. Children need time — as a whole class, in small groups, and individually — to think through new forms of evidence, new explanations, and new ideas alongside their preexisting ideas. For new ideas to take hold, they must not only make sense to students, but they should also "fit" the complex framework that children have created outside of the classroom.

Since the early 1970s, research on learning has shown us how important it is that we begin science instruction from the standpoint of students' experiences. David Ausubel[3] emphasized this by distinguishing between "meaningful learning" and "rote learning." For meaningful learning to occur, the learner must be able to relate new knowledge to relevant existing concepts in his or her cognitive structure. As the National Science Education Standards remind us, "In the same way that scientists develop their knowledge and understanding as they seek answers to questions about the natural world, students develop an understanding of the natural world when they are actively engaged in scientific inquiry — alone and with others."[4]

Despite advances in our understanding of how children learn and its importance for how we teach, elementary and middle school science has not followed suit. Science is often taught as if students do not already have their own ideas about how the world works or as if their out-of-school experiences could be easily displaced by school knowledge. Yet we know, from our own experiences and from the literature on misconceptions, that children come to school with certain beliefs about how things happen and that these beliefs are tenacious. Unless, as teachers, we draw upon these experiences and connect what we are teaching to the students' worlds, we will make little headway with meaningful learning, and children's incomplete ideas will persist. Indeed, science-education research shows that many students complete elementary, middle, and even high school with strongly held misconceptions about core science concepts, like the flow of matter and energy in ecosystems (such as the food-making process or the release of energy from food), even though these core ideas are commonly taught and are covered in most middle school curricular materials.[5]

Why is this? There are many factors that shape how and why students are not learning science for deep conceptual understanding, including instruction, the curriculum, and the knowledge base of the teacher. However, recent studies have found how important curricular materials are. The American Association for the Advancement of Science (AAAS) has shown that "while better curriculum materials alone are unlikely to improve student learning, high-quality curriculum materials can positively influence student learning directly and through their influence on teachers."[6] Other studies have found that new teachers rely heavily on curriculum materials. In addition, teachers' pedagogical decisions are influenced by curriculum materials. [7,8,9] After all, nearly 90 percent of science teachers use curricular materials 95 percent of the time. [10]

The AAAS Project 2061 has put forth a research-based framework for evaluating how well curricular materials support teachers and students in meaningful teaching and learning.[11] Its framework is partly structured around the following:

- How well does the **content** of the curriculum align with the big ideas of science?

- How well does the **presentation of ideas** within the curriculum support a teacher's instructional approach in ways that help teachers take account of students' prior ideas?

- Are the suggestions for **assessment** aimed at specific benchmarks and standards that are likely to reveal what students actually know (as opposed to rote memorization of these goals)?

In the design of the LiFE Curriculum Series, these three criteria are central. The topics covered in this module, *Farm to Table & Beyond*, are about systems and interacting parts, issues in technology, and the effects of our global food system on the environment. Information is presented for teachers to use in questioning students about their current understanding of the topics of study. This allows students to combine their current thinking with what they are learning to develop new knowledge constructs. Our assessments include activity sheets that challenge the students to collect thorough, accurate data and to carefully interpret these data, along with writing assignments in their LiFE Logs, to reflect on what they learned while expanding their thinking.

Common Misconceptions

The driving question for *Farm to Table & Beyond* is *What is the system that gets food from farm to table, and how does this system affect the environment?* There are several important misconceptions connected to this question to be aware of.

Students may believe that:

- *Farmers take the food they grow directly to supermarkets where it is sold.* Students may know that food is grown on farms but may have little understanding of what happens to food between the farm and the store where it is sold. Nor do students realize the impact that food-supply chains have on the environment. Foods can go through many steps, including production, processing, preservation, packaging, and transportation, before being sold at a grocery store. At each of these steps, there is an impact on the environment.[12, 13]

- *What we choose to eat has little or no impact on the environment.* Students may know that turning off lights or fixing leaking faucets can have a positive impact on the environment. They may even know that recycling or reusing food packaging is good for the environment. However, they may not be aware that the steps that food goes through between the farm and the table, such as transporting, processing, and packaging food, creates waste. This waste can have a negative impact on the environment, and choosing foods that create less waste can help the environment.

- *The ingredients used in processed foods originate in factories.* Students may know that foods are made in factories, but may not realize that most of the ingredients in even highly processed foods originated from plants and animals. A quick read of an ingredients list often offers a series of very long, complicated names that may confirm the idea that they are "chemicals" that come from factories. However, a careful analysis can help students figure out that the ingredients originated from plants or animals.

- *Microorganisms are not living things.* Students may not believe that mold is alive. Students often think about mold on food as they do rust on metal objects. It is important to help students understand that mold and other microorganisms grow on food, just as other living things grow, when the right conditions are present.

- *The greenhouse effect is bad for the environment.* Students may believe that the terms "global warming" and "greenhouse effect" refer to the same phenomenon. They may also use the terms "greenhouse gas," "greenhouse-gas emission," and "pollution" interchangeably. If students do not make a distinction between "greenhouse gas" and "emission," they may think of greenhouse gases as "bad" rather than as essential in making Earth habitable. Additionally, students who think "greenhouse gases" and "pollution" are synonymous may think that all types of pollution enhance the greenhouse effect.[14] This leads to the belief that the greenhouse effect is exclusively an environmental problem rather than a natural phenomenon that is essential to life on earth. The *enhanced* greenhouse effect is caused by emissions of man-made greenhouse gases that trap more terrestrial radiation near the ground, causing increases in the earth's temperature and climatic changes. It is important to distinguish between the naturally occurring greenhouse effect, the enhanced greenhouse effect, and global warming.[15]

- *Global warming is caused by the hole in the ozone layer.* Students, and adults,[16] have a tendency to confuse global warming with ozone depletion, or to attribute the former to the latter. Students may believe that ultraviolet radiation passes into the Earth's atmosphere through the hole in the ozone layer, becomes trapped by greenhouse gases, and causes the Earth's temperature to rise, causing global warming. It is important to stress the distinct roles of greenhouse gases, which trap terrestrial radiation, and the ozone layer, which is an essential shield in the upper atmosphere that protects living things from harmful ultraviolet radiation.[17, 18] Global warming is caused by the increased amount of greenhouse gases, such as carbon dioxide.

- *Oil comes from dinosaurs.* Perhaps due to the use of dinosaurs as symbols for prehistoric eras, students often believe that petroleum originated from decayed dinosaurs. Remind students that petroleum results from the transformation of the remains of ancient marine organisms over millions of years and that petroleum is a nonrenewable resource — humans are using petroleum at a much faster rate than the environment can replace it.[19]

- *Petroleum exists in large pools or puddles underground.* Students often have misconceptions about the configuration of petroleum underground, largely due to misrepresentations in television and print. Most petroleum exists interstitially between grains of sand and other rocks.[20]

- *A system of objects must be doing something in order to be a system; a system that loses a part of itself is still the same system.* Students may not believe that most things are made up of different parts and an object may not work if its parts are missing. It's important to help students understand that a system can perform functions that the single parts cannot perform on their own.[21]

[1] Driver, Squires, Rushworth, & Wood-Robinson, 1994
[2] Kyle & Shymansky, 1989
[3] Ausubel, 1968
[4] NSES, 1996, p. 27
[5] Stern & Roseman, 2004
[6] Stern & Roseman, 2004, p. 539
[7] Ball & Cohen, 1996
[8] Ball & Feiman-Nemser, 1988
[9] Grossman & Thompson, 2004
[10] Renner, Abraham, Grzybowski, & Marek, 1990
[11] Kesidou & Roseman, 2002

[12] Calabrese Barton, Koch, Contento, & Hagiwara, 2005
[13] Tsurusaki & Anderson, 2007
[14] Rebich & Gautier, 2005
[15] Koulaidis & Christidou, 1999
[16] Kempton, 1997
[17] Koulaidis & Christidou, 1999
[18] Meadows and Wiesenmayer, 1999
[19] Rule, 2005
[20] Rule, 2005
[21] AAAS, 1993, p. 355

Getting Acquainted with *Farm to Table & Beyond*

Throughout this module, you and your students will investigate the question *What is the system that gets food from farm to table, and how does this system affect the environment?* This question is more complicated than it may seem. You and your students will learn more about the complex nature of our food system in this module, and begin to explore the impact of human activity on the natural world.

As you begin your investigations, you'll find yourself immersed in a world of interconnected systems. You will be exploring the subsystems of our immense, technological farm-to-table system — packaging, processing, transportation, and the byproducts, including waste, that are produced by this system. This module is an interweaving of learning about science, technology, and society. You and your students will come to understand that technology is not innately good or bad. Typically, the effects produced by technology are hard to estimate accurately and likely to have different values for different people at different times. In this module you and your students will create a time line of the food system with milestones in packaging, transportation, food processing, energy, and garbage. Additionally, student readings will enhance your studies with a deeper sense of the history of the food system. Overall, we hope students come away with a fuller appreciation of the enormous effects that our ability as humans to invent tools and processes have on the lives of other living things, on natural systems, and on the state of our collective environment.

As part of a complex biological system, humans have responsibilities to the other parts of our system. To insure our own survival, we must respect the wide diversity of living creatures — both large and small — and the physical components that all life depends on for growth and sustainability. Our planet has limited resources that cannot be carelessly wasted or polluted without running the risk of jeopardizing life on Earth. By investigating this module's question, your students will become informed citizens, prepared to make choices that will help support a sustainable future for generations to come.

If you have access to a school garden, we encourage you to engage your students in growing their own food. Such gardens can also serve as living laboratories where students can set up investigations and monitor them over time. Gardens are a great place to monitor the effects of changing weather conditions on ecosystems, decomposition, the seed-to-table life cycle of garden plants, and much more. The National Gardening Association's Web site (*www.kidsgardening.org*) is an excellent resource for both ideas and materials.

Overview

This module consists of six units, each with its own driving question.

Unit 1

Becoming Food Scientists introduces LiFE and explores the question *What is a food scientist?* Investigating corn motivates students to study food. Studying grapes introduces students to LiFE's QuESTA Learning Cycle. Assessing what students already know about our farm-to-table food system and how this system affects the environment offers a baseline to track student growth throughout this module.

Unit 2

Interacting Parts introduces students to thinking about the farm-to-table system in terms of its parts and how the various parts relate to and interact with each other to create a "whole" through the question *What is the system that gets food from farm to table?* Students investigate processing, packaging, and transportation as subsystems within our larger food system. They explore different kinds of packaging and investigate modes of transportation. They begin long-term observations of what happens to fresh foods that are left out in their classroom, to gain an understanding of why we preserve food. Students tie together what they have learned by mapping out the relationships and interactions of the parts to build a new conceptual understanding of the whole.

Unit 3

Food Processing asks students to dive deeper into exploring the changes food undergoes in our food system with the question *What happens to food as it moves from farm to table?* Students continue their long-term observations of fresh foods left out in the classroom and develop theories to explain the changes they observe. They learn about several methods of preserving food through understanding what microorganisms need to grow and methods to prevent their growth. Students experience being food processors by creating cornmeal, whole-wheat flour, and butter, then using these ingredients to make pancakes from scratch. Through continuing with pancakes and the various ways they are available in our marketplace, they examine differing degrees of processing that food can undergo. They culminate their studies by becoming food preservers, making pickles. The unit ends with the students synthesizing what they have learned about the science and technology of food processing.

Unit 4
Environmental Effects leads students through a critical examination of the effects of our dependence on such a highly technological food system through an exploration of the question *What are the environmental effects of our farm-to-table system?* Students investigate what is needed to sustain the natural environment and how we, as humans, use natural resources to create "goods" that will meet our needs. The students revisit transporting, packaging, and processing food to layer onto what they learned an understanding of the trade-offs and impacts of our farm-to-table system. Following this, they examine the importance of fossil fuels in our society and how our dependence on fossil fuels affects the environment, specifically exploring these fuels' role in increasing greenhouse-gas emissions and global warming. Students examine how by-products of our food system can contaminate the air, water, and soil that we depend on to produce food.

Unit 5
Waste allows students to conduct research that is a practical application of what they have learned by doing waste-analysis investigations. Through this they explore the question *How can we reduce the food-related waste that we produce?* Students conduct studies of how much food and food-packaging waste are produced by themselves, their families, and their communities. They learn about reducing, recycling, reusing, and composting as ways to decrease waste production. They develop and carry out a research plan to investigate how much food-related waste is generated by the school cafeteria. Finally, they analyze their personal food habits.

Unit 6
Making Choices brings the module to a close and encourages students to make informed decisions and changes in their lives by asking, *How can we use the science we learned to make ecologically sound food-system choices?* In this unit, students put on a *Farm to Table & Beyond* expo to share what they have learned about the global food system. As a post-assessment, students revisit the food-system posters they made in Lesson 4. Finally, they develop their own food choice guidelines based on what they've learned throughout the module.

Promoting Inquiry
Teaching science as inquiry makes science a process of doing and thinking instead of learning a set of predetermined facts. This changes your role as teacher. Instead of being a source of science facts, you are a partner with your students as you seek answers or explanations. It means turning students' questions back on them. If a student asks, *"Why does food rot?"* respond with, *"Well, I'm not sure. How shall we find out?"* This sends a powerful message — that knowing how to find an answer is as important as knowing the answer.

Ask open-ended questions that promote reflection and further questions. "How" and "why" questions work well: *"Why do you think that is?"* or *"How would we find out?"* Ask questions that encourage critical thinking, like: *"What evidence or observation leads you to that conclusion?"* Help your students develop theories and bring closure to their explorations and experiments by asking: *"How would you explain your results?"* and *"What theories can you think of to explain this?"*

QuESTA Learning Cycle
How students learn is as important as what they learn in the LiFE Curriculum Series. The questions that drive the modules and units in LiFE challenge students to explore, question, investigate, analyze, synthesize, and act. LiFE's five-phase learning cycle, QuESTA, guides students through this process.

 **QUESTIONING**

Students explore their prior knowledge and experiences related to the area of study and develop and refine meaningful questions to guide further inquiry. They also share their current conceptions about the topic so that any misconceptions can be addressed.

EXPERIMENTING

Students plan and conduct experiments to answer the questions within the area of study. Thus, students identify problems, state hypotheses, select methods, display results, and draw conclusions from these experiments to further their knowledge.

SEARCHING

Students seek out other information already known about their topic through readings provided in the lessons, researching in the library or on the computer, and interviewing people.

THEORIZING

Through thoughtful reflection and synthesis of what they have learned in the previous phases, students develop their own theories and constructs about how the world works. Students gain skills that enable them to articulate theories, give evidence to support their arguments, and appropriately challenge the theories of others.

APPLYING TO LIFE

Students apply the new constructs and processes they learned through the unit to decisions and actions they make each day. Students develop new questions to continue their exploration in the area of study. This phase of QuESTA is also an opportunity for you and your students to extend the LiFE activities. For example, in Lesson 2, as students learn about grapes, you may wish to have them investigate what climate grapes grow in, where the grapes they buy in their local market are grown, and how far the grapes have to travel to reach the market. Look for ideas for going further on the LiFE Web site (*www.tc.edu/life*).

Using QuESTA

The activities that focus on questioning, experimenting, and searching are engaging and often easy to implement in the classroom. Activities that call for students to theorize and apply to life help students refine their abilities to construct explanations and theories about what they have learned from their exploring and experimenting and to apply their learning to their

daily lives. Pay special attention to the theorizing and application activities in the lessons. These activities will help you meet some important and challenging standards. The National Science Education Standards and the American Association for the Advancement of Science Project 2061 Benchmarks suggest that in addition to making observations, and designing and conducting investigations, students should:

- use logical reasoning and critical thinking to link evidence with explanations;

- use communication skills to describe observations, summarize results, articulate theories and constructs about how the world works, consider alternative explanations, and challenge the explanations proposed by others;

- apply scientific constructs and processes to everyday decisions and actions.

Assessment Strategies

Authentic assessment tasks provide students with opportunities to construct meaning from what they have learned. The LiFE Curriculum Series offers different assessment strategies to help you track your students' progress. Many of these are integrated into the lessons.

Pre-Assessment

Lesson 4 serves as this module's pre-assessment. Students answer the Module Question *What is the system that gets food from farm to table, and how does this system affect the environment?* As students respond to this question, remind them that they will not be graded on their answers. Encourage students to write down what they know and think now.

Post-Assessment

In Lesson 30, students revisit the Module Question, look at their responses to the question in the pre-assessment, and reflect on what they have learned. Make this post-assessment an exciting academic challenge for your students. As a teacher you not only want your students to *know* the content taught; you want them to be able to *use* their knowledge and skills in the real world.

Ongoing Assessment

Throughout the module, students have multiple opportunities to participate in full-class discussions, work and discuss materials in small groups, and present their work to the class. These interactions offer oppor-

QUESTA PHASES AND ASSOCIATED TERMS

This table includes terms for each phase of QuESTA. We developed these lists to help you and your students understand and differentiate among the types of action or activities appropriate for each phase. These terms are used throughout the teacher and student materials.

QUESTIONING	EXPERIMENTING	SEARCHING	THEORIZING	APPLYING TO LIFE
assess	check	discover	analyze	apply
consider	conduct	explore	build theories	carry out
contemplate	create research questions	find out	compare	embark
inquire	design experiments	gain knowledge about	conclude	employ
mull over	determine	learn about	construct knowledge	implement
ponder	display data	look into	contrast	put into action
question	evaluate	research	create ideas	undertake
speculate	examine	search	debate	use
think about	experiment	seek	deliberate	utilize
wonder	gather data		discuss	
	hypothesize		envision	
	identify variables		explain	
	inspect		imagine	
	investigate		infer	
	manipulate		realize	
	observe		reason	
	predict		recognize	
	probe		reflect	
	prove		summarize	
	solve		think through	
	study			
	test			

tunities to assess how students are thinking about the topics being studied, their level of sophistication in what they are thinking and saying, and their ability to engage in discussions, debates, and scientific arguments with their peers. These ongoing assessments may be particularly helpful for students who are challenged by writing and public speaking. In each lesson, students write in their LiFE Logs, reflecting on what they have learned. This reflective writing gives students the freedom to express in their own words what they are learning in class. Often the LiFE Log assignment will be an answer to an open-ended question, which will help you assess how students have internalized what they learned in the lessons, how they made meaning of new concepts, and how they brought earlier ideas to bear on new understandings.

How to Use this Book

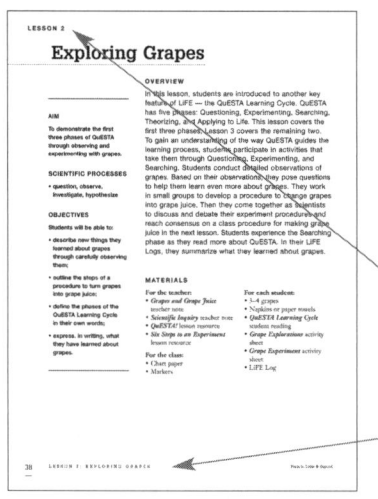

Each module in the LiFE Curriculum Series includes lesson plans, teacher materials, and student pages. In this way, everything you need to teach each lesson is all in one place. We recommend reviewing these materials as you prep for the lesson.

Each page is designed so, at a glance, you will know where you are.

Header: This indicates the type of page. For example, lesson plan, sample conversation, experiment sheet, teacher note, lesson resource, student reading, or activity sheet.

Footer: The lesson number, title, and page number are shown at the bottom of each page.

Units: The module's unit titles are listed in order down the side of the page. The name of the unit you are working with is shown in bold type.

LESSON FORMAT

Each lesson contains an activity and supplemental materials that support the activity. These materials include any background information, teaching suggestions, illustrations, student readings, and student activity sheets that are needed to teach the lesson. The lessons appear in a specific order in the module. This reflects a logical storyline designed to keep students engaged. However, we encourage you to think about your students' needs and abilities and to adjust the order of the lessons accordingly. For example, we found that students were motivated to conduct the waste studies in Unit 5 after they had learned about the environmental effects in Unit 4.

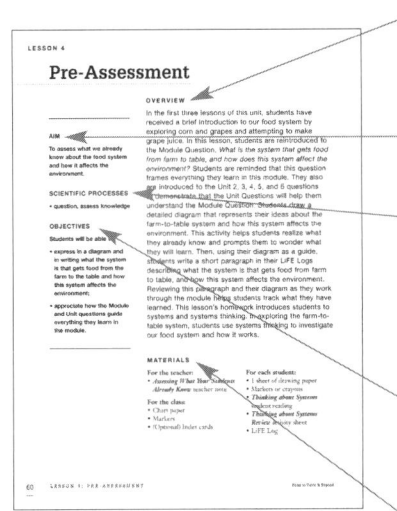

Overview: This section provides a description of the lesson and details both what the students do and what they learn during the lesson.

Aim: This summarizes the main idea of the lesson.

Scientific Processes: These terms are related to specific phases of the QuESTA Learning Cycle. They indicate which QuESTA phases are emphasized in the lesson and the skills that students will be using as they complete the activity.

Objectives: These highlight what you can expect your students to know and be able to do at the end of the lesson.

Materials: This list includes materials that are commonly found in the classroom, such as chart paper, construction or drawing paper, markers, crayons, hand lenses, scissors, and rulers. It also includes LiFE Logs (composition notebooks that each student uses to record observations and for reflective writing). Materials listed in ***bold italics*** are teacher and student pages provided in this book. Some lessons with class demonstrations or experiments call for materials you may need to bring in. Look for the **Supply List** lesson resource that itemizes the supplies and equipment that you will need.

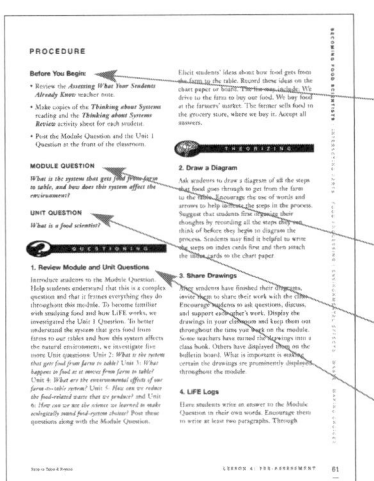

Before You Begin: All activities require advance preparation. You may need to make copies of reproducibles, gather materials, post the Module and Unit questions, or review the teacher note, sample conversation, or student readings.

Module Question: Each lesson lists the Module Question.

Unit Question: Each lesson lists the Unit Question.

Lesson Procedure: The lesson procedure provides step-by-step information to complete the lesson. QuESTA icons used throughout the procedure indicate which phase of the QuESTA Learning Cycle is being emphasized. Each lesson begins by engaging students in the concept, and presents an opportunity for you to check for student understanding, review the Module Question, and introduce or review the Unit Question.

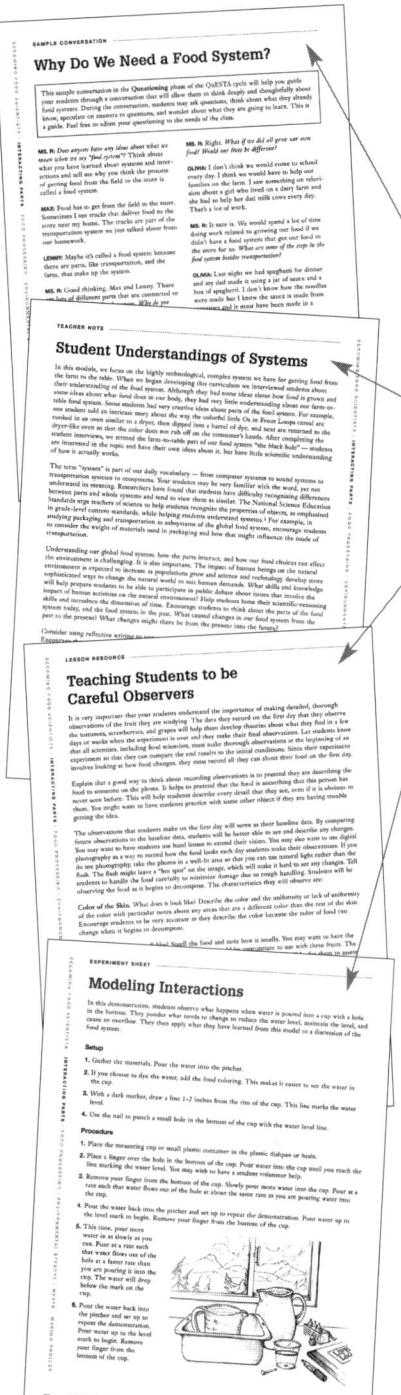

SUPPLEMENTAL TEACHER MATERIAL

Teacher pages, such as the teacher note and lesson resource, also are distinguished by the absence of a QuESTA icon.

Sample Conversations demonstrate the level of thinking and discussion recommended for students to reflect on what they have learned in order to develop new theories from the activity.

Teacher Notes include helpful background information. Look here for information about student misconceptions and background information to help you as you prepare the lesson.

Lesson Resources provide information that supports the activity. The pages include detailed supply lists for experiments and demonstrations and reference material that you can use to guide the lesson's activities.

Experiment Sheets provide a detailed description of the setup, procedure, and questions to guide class discussion.

STUDENT PAGES

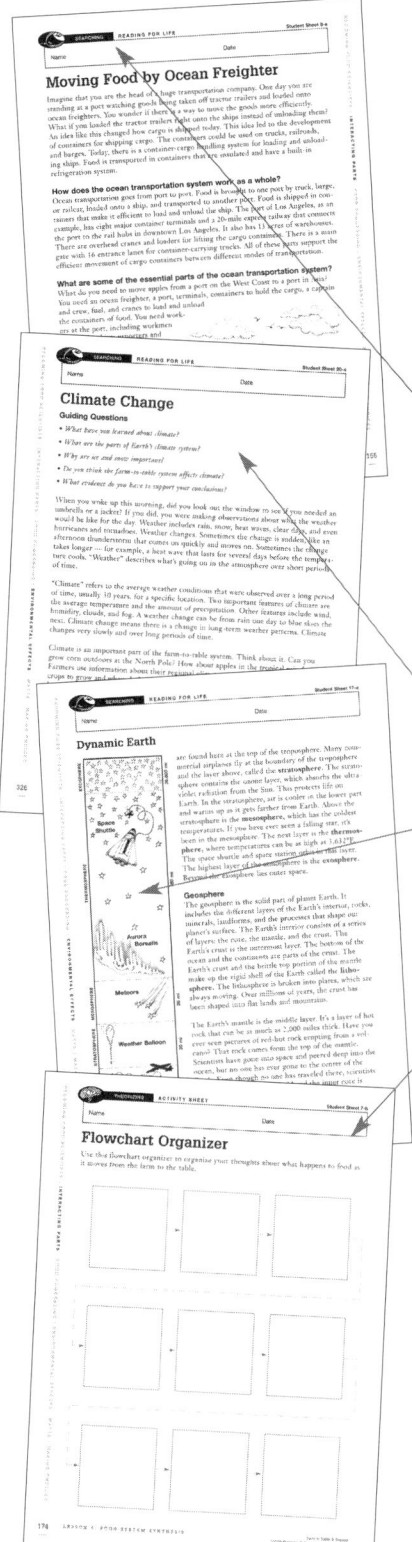

Each module in the LiFE Curriculum Series includes reproducible student pages: readings, activity sheets, and take-home recipes.

All student pages include a QuESTA icon indicating the phase of the QuESTA Learning Cycle that corresponds with the activity students are engaged in. For example, the Searching icon appears on the Reading for Life pages. All student pages are designed to be copied for use as handouts.

Reading for Life pages introduce students to each lesson's topic. These readings provide background information, explanation of concepts, and connections to current scientific research.

These readings have a Searching QuESTA icon to denote that these pages provide students with interesting information that other scientists have already learned about the study topic.

Guiding questions help students organize the new information they are learning.

Illustrations, both editorial and science, are included on many of the student pages. The often-humorous editorial artwork makes the pages more engaging to a student audience. The science illustrations reinforce the science explanations in the text. These illustrations provide additional information to enhance the science learning and help communicate the reading's ideas.

Activity Sheets. These pages help students focus their learning and organize data collection and analysis. They provide questions to guide student thinking and help students reflect on what they have learned. Completed activity sheets can help you assess student learning as you progress through each unit.

Some activity sheets contain graphic organizers. These visual tools help students organize information, synthesize it, and communicate what they have learned to others. Brace maps helps students understand relationships between a whole and its parts, while flowcharts provide a way for students to sketch out step-by-step processes. Flowcharts also help students see relationships between the different steps.

Graphic organizers promote active learning and can be used in all phases of the QuESTA Learning Cycle: from brainstorming to summarizing what students have learned. Use them to access students' prior knowledge before introducing a new concept. They are also helpful with activities that require using critical-thinking skills. Once students have completed their graphic organizers, be sure to allow enough time for them to look at what they have created so they can begin to see patterns and relationships that they may not have seen before.

Making the Most of *Farm to Table & Beyond*

Social Studies

During the development of Linking Food and the Environment, we worked with some teachers who taught both science and history–social science. These teachers pointed out some obvious, and not so obvious, connections across the curriculum. Here are some of the connections that we made. Communication is an important science skill. The student readings and writing activities in LiFE are a natural link to language arts. For example, in *Farm to Table & Beyond*, students undertake a research project that requires working with different electronic sources, evaluating information, and presenting this information to their peers. As scientists collecting data for analysis, students work with numbers, graphs, and charts. As they analyze their data, they discuss why the way in which the question was asked might have influenced the results.

Food, an important part of culture, is a thread that weaves throughout history. Think about how studying any culture, time, and place connects to food. Economics and food are interconnected through trade. Early explorers drove the global spread of plants and animals. Global trade brought new food crops to all the continents. Technological innovations, like the ability to keep food from spoiling, made it possible to ship food from coast to coast and overseas. Food also connects to government through policies and laws. One of the LiFE pilot teachers, Ms. Wright, who taught sixth graders science and social studies found that once her students had LiFE in science, they demanded each social studies class begin with a study of agricultural practices. This included climate, what could be grown where, as well as information about the food system and the politics of food. Ms. Wright found food studies to be the hook that motivated her students in social studies. We developed the time line woven throughout this module as one way to highlight the connections.

Food System Time Line

Throughout this module, we have included lesson resources that highlight milestones in the development of today's global food system and student readings that discuss some of the key events in narrative form. Look for milestones in the history of packaging (p. 118), transportation (pp. 139–140), food processing (pp. 230–231), energy (p. 313), and garbage (p. 376) You may want to add other milestones that are particularly relevant to your local community — for example, when the first grocery store opened or when your farmers' market was established.

Use these milestones to create a time line that students can add to as they work through the module. Time lines help students see how one development or discovery may have resulted in something else happening. For example, refrigerated railroad cars made it possible to ship perishable foods from coast to coast. Time lines are visual records that will help your students make connections and begin to understand complex relationships.

If you include all of the milestones we list, your time line may stretch around your classroom. If at all possible, start the time line with Lesson 7 and leave it up as you move through the module. Teachers who are experienced in using time lines suggest making it easy to access — otherwise you may not use it. Here are some suggestions for different ways to set up your time line:

- **Single Line.** Visually, this approach makes it clear that the time line is a continuing process and not individual sections. Paper on a roll, like freezer paper or newsprint, makes it easy to stretch the line around the room. Add to the line as you introduce new milestones. Have students record the milestones on index cards and post them on the time line. Use a pencil to mark out time periods in case you want to modify the scale later on.

- **Sentence Strips.** Some teachers use sentence strips and post them around the room. Each strip represents a time period (1800–1850, etc.). If the sentence strips are made of card stock, it is possible after they are taped together to fold the time line up accordion-style to store it. More sentence strips or index cards are used to record the milestone and posted above or below the appropriate time period.

Learning Stations

We suggest using learning stations in Unit 3. We believe it is important for all students to have hands-on experience and be able to "mess around" as they explore how the world works. Many of the lessons in Unit 3 involve grinding, shaking, mixing, and cooking. The more students are directly involved with labor-intensive activities, such as grinding grain by hand, the more likely it is that they will appreciate what motivated people to develop more efficient forms of technology to accomplish these tasks.

At the same time, we recognize the challenges of managing a classroom full of eager students, multiple materials lists, and, often, limited resources. We have found that using learning stations is an effective way for small groups of students to move through the series of activities and still have a hands-on experience. Using a learning-center approach is also a way to save time and maximize the use of limited resources — you use one setup that students rotate through.

Jigsaw Method

In Lesson 8 and Lesson 17, students learn about five different kinds of transportation used in our food system. Timewise it is not realistic for students to learn about each type of transportation. However, using the jigsaw method, small groups of students become experts on each form of transportation and report back to their "home" group. Using this method, students are in charge of their own learning. It helps students develop a depth of knowledge that would not be possible if the students had to learn all of the material about all of the kinds of transportation on their own. This method also helps you assess student understanding. When students report back to the home group, you can assess a student's understanding of a concept, as well as discover any misunderstandings.

Food and Food Systems in the News

As you and your students begin your *Farm to Table & Beyond* studies, look in the media for articles that relate to what you are studying in class. Newspapers, the business section, in particular, are a rich source of stories related to new products and processes that are entering the market. Stories related to the environment are often news stories and are found on the front page or in the front section. Quite often these stories are disaster stories, such as container ships leaking oil; or severe weather conditions shutting down major high-ways used by trucks carrying freight; or severe weather causing crop failure. A newspaper's food section is a great way to get students to look at food and food systems from the point of view of personal taste and preferences. It's also a place to find information about local and seasonal foods that are available and different ways to prepare them.

Look for ways to connect informational materials, such as newspaper and magazine stories, to what students are learning in their language-arts classes. Engage students in a discussion of ways that the format, graphics, or illustrations help make the information accessible. Use these materials to help students gain experience learning to distinguish fact from opinion in text.

Consider keeping a scrapbook of food system–related stories and reviewing them at the end of the module. You may find that your students have a different perspective on stories when they apply what they have learned.

Connecting what students are learning in the classroom to real-world stories that are covered by the media is a great way to help students realize that the science they are learning is relevant to public debate about important issues that involve science and technology.

Science Standards Matrices

NATIONAL SCIENCE EDUCATION STANDARDS to be met by the end of 8th grade*	*FARM TO TABLE & BEYOND* UNITS					
	UNIT 1	UNIT 2	UNIT 3	UNIT 4	UNIT 5	UNIT 6
A. SCIENCE AS INQUIRY						
1. Abilities necessary to do scientific inquiry	X	X	X	X	X	X
2. Understandings about scientific inquiry	X	X	X	X	X	X
B. PHYSICAL SCIENCE						
1. Properties and changes of properties in matter			X	X	X	X
2. Motions and forces						
3. Transfer of energy						
C. LIFE SCIENCE						
1. Structure and Function in Living Systems						
2. Reproduction and Heredity						
3. Regulation and Behavior						
4. Populations and Ecosystems			X			
D. EARTH AND SPACE SCIENCE						
1. Structure of the Earth System				X		X
2. Earth in the Solar System				X		X
E. SCIENCE AND TECHNOLOGY						
1. Abilities of Technological Design	X	X	X	X	X	X
2. Understandings about Science and Technology	X	X	X	X	X	X
F. SCIENCE IN PERSONAL AND SOCIAL PERSPECTIVES						
1. Personal Health						
2. Populations, Resources, and Environments						
3. Natural Hazards						
4. Risks and Benefits		X	X	X	X	X
5. Science and Technology in Society	X	X	X	X	X	X
G. HISTORY AND NATURE OF SCIENCE						
1. Science as a Human Endeavor	X	X	X	X	X	X
2. Nature of Science	X	X	X	X	X	X
3. History of Science		X	X	X	X	X

*National Research Council, Center for Science, Mathematics, and Engineering Education, National Science Education Standards (1996) **KEY: X** Addresses Standard

BENCHMARKS FOR SCIENCE LITERACY
to be met by the end of 5th grade*

BENCHMARKS FOR SCIENCE LITERACY to be met by the end of 5th grade*	UNIT 1	UNIT 2	UNIT 3	UNIT 4	UNIT 5	UNIT 6
1. THE NATURE OF SCIENCE						
A. The Scientific World	X	X	X	X	X	X
B. Scientific Inquiry	X	X	X	X	X	X
C. The Scientific Enterprise	X	X	X	X	X	X
3. THE NATURE OF TECHNOLOGY						
A. Technology and Science	X	X	X	X	X	X
B. Design and Systems	X	X	X	X	X	X
C. Issues in Technology	X	X	X	X	X	X
4. THE PHYSICAL SETTING						
B. The Earth				X		X
5. THE LIVING ENVIRONMENT						
A. Diversity of Life						
B. Heredity						
C. Cells						
D. Interdependence of Life		X	X	X	X	X
E. Flow of Matter and Energy		X	X	X	X	X
F. Evolution of Life						
7. HUMAN SOCIETY						
A. Cultural Effects on Behavior						
B. Group Behavior						
C. Social Change						
D. Social Trade-Offs						
E. Political and Economic Systems						
F. Social Conflict						
G. Global Interdependence		X	X	X	X	X
8. THE DESIGNED WORLD						
A. Agriculture	X	X	X	X	X	X
B. Materials and Manufacturing		X	X	X	X	X
C. Energy Sources and Use		X		X	X	
D. Communication						
E. Information Processing						
F. Health Technology						
11. COMMON THEMES						
A. Systems	X	X	X	X	X	X
B. Models		X				
C. Constancy and Change		X	X	X	X	X
D. Scale						
12. HABITS OF MIND						
A. Values and Attitudes	X	X	X	X	X	X
B. Computation and Estimation						
C. Manipulation and Observation	X	X	X	X	X	X
D. Communication Skills	X	X	X	X	X	X
E. Critical-Response Skills	X	X	X	X	X	X

*Benchmarks for Science Literacy by the American Association for the Advancement of Science (AAAS), Project 2061.
www.project2061.org/publications/bsl/online/bolintro.htm **KEY: X** Addresses Benchmark

BENCHMARKS FOR SCIENCE LITERACY
to be met by the end of 8th grade*

FARM TO TABLE & BEYOND UNITS

BENCHMARKS	UNIT 1	UNIT 2	UNIT 3	UNIT 4	UNIT 5	UNIT 6
1. THE NATURE OF SCIENCE						
A. The Scientific World	X	X	X	X	X	X
B. Scientific Inquiry	X	X	X	X	X	X
C. The Scientific Enterprise	X	X	X	X	X	X
3. THE NATURE OF TECHNOLOGY						
A. Technology and Science	X	X	X	X	X	X
B. Design and Systems	X	X	X	X	X	X
C. Issues in Technology	X	X	X	X	X	X
4. THE PHYSICAL SETTING						
B. The Earth				X		X
5. THE LIVING ENVIRONMENT						
A. Diversity of Life						
B. Heredity						
C. Cells						
D. Interdependence of Life		X	X	X	X	X
E. Flow of Matter and Energy						
F. Evolution of Life						
7. HUMAN SOCIETY						
A. Cultural Effects on Behavior						
B. Group Behavior						
C. Social Change						
D. Social Trade-Offs						
E. Political and Economic Systems						
F. Social Conflict						
G. Global Interdependence						
8. THE DESIGNED WORLD						
A. Agriculture	X	X	X	X	X	X
B. Materials and Manufacturing	X	X	X	X	X	X
C. Energy Sources and Use						
D. Communication						
E. Information Processing						
F. Health Technology						
11. COMMON THEMES						
A. Systems	X	X	X	X	X	X
B. Models		X				
C. Constancy and Change						
D. Scale						
12. HABITS OF MIND						
A. Values and Attitudes	X	X	X	X	X	X
B. Computation and Estimation						
C. Manipulation and Observation						
D. Communication Skills						
E. Critical-Response Skills	X	X	X	X	X	X

*Benchmarks for Science Literacy by American Association for the Advancement of Science (AAAS), Project 2061.
www.project2061.org/publications/bsl/online/bolintro.htm **KEY: X** Addresses Benchmark

Project Planner

PROJECTS	Food-System Diagram	Food Observation Experiment	Farm to Table & Beyond Time Line	Food Packaging Research Project	Analyzing Waste	Farm to Table & Beyond Expo
Lesson 1		Order materials				
Lesson 2						
Lesson 3						
Lesson 4	Pre-assessment: draw diagram					
Lesson 5		Collect baseline data				
Lesson 6			Begin time line; add packaging events	Begin packaging project		Review expo information, p. 27; introduce expo to class
Lesson 7						
Lesson 8		Collect midline data	Add transportation events		Review Lesson 25; order recycling information	
Lesson 9	Redraw diagram					
Lesson 10						
Lesson 11		Collect final data				
Lesson 12			Add food processing events	Check progress of research projects		
Lesson 13						
Lesson 14						
Lesson 15						
Lesson 16						
Lesson 17			Add energy events			
Lesson 18						
Lesson 19				Present packaging research to class		
Lesson 20						
Lesson 21						
Lesson 22			Add garbage events	Present research at expo		
Lesson 23						
Lesson 24						
Lesson 25						
Lesson 26					Begin waste inventory	
Lesson 27					Analyze waste	
Lesson 28					Take action	
Lesson 29						
Lesson 30	Post-assessment: draw diagram					Present expo

Kitchen Chemistry

In *Farm to Table & Beyond*, students learn about different compounds that are in foods naturally or are used during food preservation and processing. The lessons in Unit 3, in particular, provide an opportunity for students to have hands-on experience with applied chemistry. Your students probably already know that the chemical name for water is H_2O, but they may not be familiar with the chemical names for other substances. Cooking provides a real-world opportunity to introduce your students to some chemicals commonly found in the kitchen and to experience chemistry in action.

This table lists the compounds that are included in the activities in Unit 3.

COMPOUND	FORMULA	COMMERCIAL EQUIVALENT	SOURCE	USE / PURPOSE
Acetic acid	CH_3COOH	Vinegar (5%)	Grocery store	Preservation and sour flavor in non-fermented pickling
Amylose	$(C_6H_{10}O_5)_n$	Cornstarch	Grocery store	Binder, thickener, or drying agent
Sodium bicarbonate + potassium bitartrate + amylose	$NaHCO_3$ + $KHC_4H_5O_6$ + $(C_6H_{10}O_5)_n$	Baking powder*	Grocery store	Leavening agent containing both alkaline and acid
Fructose	$C_6H_{12}O_6$	Fruit sugar	Grocery store	Makes fruit sweet
Glucose	$C_6H_{12}O_6$	Dextrose	Drug store	Sweetener
Potassium bitartrate or potassium hydrogen tartrate	$KHC_4H_5O_6$	Cream of tartar	Grocery store	Acidic agent that has many culinary purposes, one of which is to cause leavening when used with alkaline agent
Sodium bicarbonate or sodium hydrogen carbonate	$NaHCO_3$	Baking soda	Grocery store	Alkaline agent that causes leavening when used with acidic agent
Sodium chloride	$NaCl$	Plain table salt (not iodized)	Grocery store	Brine for pickling
Sucrose	$C_{12}H_{22}O_{11}$	Table sugar	Grocery store	Sweetener
Water	H_2O	Water	Ground water, precipitation	Provides moisture

*There are several formulations for baking powder. All contain an alkali (usually sodium bicarbonate), an acid in the form of salt crystals (often potassium bitartrate), and a starch to keep it dry (such as amylose).

Planning a *Farm to Table & Beyond* Expo

In Unit 6, we encourage you to hold an expo or conference, which is a great way for students to be able to share what they have learned. You can start small, holding it in your classroom and inviting grade-level peers, or hold a school-wide expo and invite the entire student body and parents. Either way, your students will be the food-systems experts ready to share information with their local community. Once you make the decision to hold an expo, the next step is to pick a date and decide on a location. Will it be during the school day, after school, or on a weekend?

One way to organize the presentations is to follow the major topics covered in the module: food scientists, food packaging, food transportation, food processing, kitchen chemistry, environmental effects, waste management, and making choices. Next, identify leaders in your class and assign them to help you organize. Plan to have a publicity committee and a program committee that will oversee the presentations and exhibits. Make sure you have a cleanup committee as well. Make it clear that they are not the only ones who will be cleaning up, but they will be the ones to develop a cleanup plan. The program committee can have several teams, with each team responsible for a different topic area. Your role will be conference coordinator. Have each committee leader work with a team so the entire class is involved. Hold regular meetings with the committee leaders to make sure the work is progressing smoothly. You may want to use the time line as a focal point during the expo.

For each of the topics above, students will have researched information, produced posters or graphic organizers, or conducted an experiment and written up the results. Have the team leaders pull together the student-produced materials and look for ways to display this information or use it to create new materials to display during the expo. Ask each team to pull together a list of ideas of what they would like to present during the expo. Encourage students to use illustrations, graphs, and diagrams with their presentations. Consider having them invite members of the community, such as local farmers, your school food-service staff, members of local environmental groups, or the media, and others to participate or contribute to your expo. For example, students may want to invite the food-service staff to help run taste tests or kitchen chemistry demonstrations. Include a farmer's

market to highlight local food. Your community may have an environmental group that helps residents set up worm composting or collects materials for recycling. Local farmers or representatives from your farmers' market may have materials about seasonal availability of local crops or farm-trails maps to distribute.

Have the publicity committee come up with a name for the expo and create a banner to display at the entrance of the room. You may wish to have the committee think of several names and submit them to a class vote. The publicity committee is also responsible for designing informational materials to go out to your intended audience. Make sure you work closely with them to oversee the text and graphics. Consider inviting your colleagues to share their expertise with different expo committees. Invite the language-arts teacher to oversee the publicity. Ask the technology teacher to help students create graphs and diagrams on the computer.

When you are working on the program, be sure to sketch out where each presenter will be located. Make sure each presenter or presenting group has enough space to display its materials. Don't crowd the presenters together.

Here's a checklist that you can use to help you get started. As you begin to develop the specifics of your expo, you'll want to add to this list.

- Pick a date, time, and location and get school approval.
- Set up committees.
- Draft an expo agenda that includes ideas for presentations and exhibits.
- Decide on presenters you would like to invite.
- Send out invitations to outside presenters.
- Send out early announcements about the expo.
- Create a place in the classroom where students can store the materials they create for the expo.
- Invite members of the local media to attend the expo.
- Have a digital camera available for taking photographs during the expo.

Once the expo is over, make sure you set aside time to discuss the experience with students. Encourage them to share both what they got out of the experience and what they think attendees got from the experience. Be sure to congratulate students on a job well done.

UNIT 1

Becoming Food Scientists

Corn Investigations

AIM

To explore corn and speculate on how corn is changed into oil, syrup, and flour.

SCIENTIFIC PROCESSES

• observe, investigate, speculate

OBJECTIVES

Students will be able to:

• describe how corn gets turned into corn oil, corn syrup, and corn flour;

• recognize science as a process of inquiry;

• communicate what they think it will be like to study science through learning about food;

• use their LiFE Logs to record questions, observations, data, and conclusions.

OVERVIEW

In this lesson, students are first introduced to the Module and Unit 1 questions to help them begin to understand how questions and questioning frame inquiry-based science. Second, students are introduced to the investigative aspect of science through thinking about corn, and how it can get turned into three very different products: corn oil, corn syrup, and corn flour. Students do not need to understand the exact details of what happens but rather realize that questioning, observing, speculating, and creating new ideas can help them learn about how the world works. This lesson's sample conversation and teacher note offer suggestions to guide students through this process. Finally, students are introduced to the communicative aspect of science through being introduced to LiFE Logs, a notebook in which students will record questions, observations, data, and conclusions. Through periodically reviewing what they have written in their logs, students can track their own learning.

MATERIALS

For the teacher:
• *Thinking about Corn Oil* sample conversation
• *Corn-Based Food Products* teacher note

For the class:
• 1 ear of corn, fresh, frozen, or dried (Indian corn)
• 1 bottle corn oil
• 1 bottle corn syrup
• 1 bag corn flour
• *Corn Investigations* experiment sheet
• (Optional) Several kernels of popcorn

• (Optional) 1 sharp paring knife
• (Optional) Hand lens
• (Optional) Overhead transparency film
• (Optional) Chart paper
• Module and Unit questions

For each group of 4–6 students:
• 1 paper cup
• *Corn Kernel* lesson resource

For each student:
• LiFE Log (composition notebook)

PROCEDURE

Before You Begin:

- Follow the setup instructions on the *Corn Investigations* experiment sheet.

- Review the *Thinking about Corn Oil* sample conversation and the *Corn-Based Food Products* teacher note.

- Make copies of the *Corn Kernel* lesson resource to distribute to each group of students.

- Post the Module Question and the Unit 1 Question at the front of the classroom.

MODULE QUESTION

What is the system that gets food from farm to table, and how does this system affect the environment?

UNIT QUESTION

What is a food scientist?

 **QUESTIONING**

1. Introduce LiFE

Explain to students that they are about to begin a science program that focuses on the study of food. In LiFE, your students are scientists — a special kind of scientist called a food scientist. *What do you think food scientists might do? What kind of knowledge do they need?* Accept all answers. Record students' ideas on chart paper or on the board.

Food scientists investigate food in lots of different ways. Some study how food is produced. Others might look at how a food, like corn, is changed and combined with other foods to make another kind of food, like cereal, or cake, or pizza. There are food scientists who try to understand how what we eat influences our personal health. And

there are food scientists who investigate how the waste and pollution created through growing, processing, and packaging our food affects our natural environment.

2. Discuss Module and Unit Questions

Post the Module Question and the Unit 1 Question on the board. Invite volunteers to read the questions out loud. Tell students that as LiFE food scientists, they will be investigating answers to questions like these.

Understanding what food scientists do will help students be better prepared for their work as LiFE food scientists. Make sure you give your students time to discuss their ideas about a food scientist's work. As students work through this lesson, check to see how their understanding has changed.

3. Explain and Conduct Corn Investigation

Show students the ear of corn and the ingredients made from corn: corn oil, corn syrup, and corn flour. *How do corn kernels get changed into corn oil, corn syrup, and corn flour?*

Follow the procedure outlined on the *Corn Investigations* experiment sheet. Challenge students to think about different ways that corn gets changed into other products. Create a sense of mystery and intrigue. Make it clear that all ideas and thoughts are welcome. Remind students that they are asking questions and wondering about the products. They are not trying to come up with a correct answer. The sample conversation and teacher note can help you guide your students through this inquiry. Be sure to review the questions on the experiment sheet.

4. Have Groups Share Findings

Encourage a whole-class conversation led by students. You may wish to have each group select a reporter to share the group's thoughts with the class. Remind students that this activity is about thinking, exploring, and learning. It is not about finding the correct answer. As the discussion comes to a close, remind students that as food scientists they are going to be investigating, experimenting, and developing new ideas about lots of topics related to food.

(Optional) Distribute the cut-up popcorn kernels to each group of students. Have students look at the inside of the kernel. The small core part near the bottom is called the germ. There is oil in the germ. The germ is the part that is squeezed to get out the oil. Invite students to try to rub the germ on paper. *Do you see a stain? What does that tell you?*

5. Discuss Science Inquiry

Explain that the corn investigation is an example of a science inquiry. *What do you think "science inquiry" means?*

After several students have shared their ideas, review this definition of **science inquiry:** using your own curiosity about a topic to help you put together what you already know with what you are learning to construct new knowledge that you can use in your daily life.

6. Introduce the LiFE Logs

Throughout *Farm to Table & Beyond,* students will keep a log to record thoughts, observations, data, and conclusions about what they are learning. This allows students to reflect on what they learn and to understand how their thinking grows and changes.

7. LiFE Logs

Have students write a paragraph that responds to the following statement: "What I think it will be like to be a food scientist."

If your students are not accustomed to this type of reflective writing, they may find it challenging. Help students understand that it is fine to sit in front of a blank page for a few moments as they think about what they want to write.

You may wish to brainstorm a list of ideas (as a whole class or individually) to serve as prompts. Students can use this list as they write their paragraphs.

8. Assign Homework

Have students write two questions in their LiFE Logs that reflect what they would like to learn about food.

Have students look at home and select five different kinds of food. Have them look at the ingredients lists to see if they can identify any ingredients made from corn. They can make a simple table, like the one shown below, in their LiFE Logs. Make sure students know how to read an ingredients list. You may wish to demonstrate in class.

Name of the Food	Ingredients from Corn in this Food

Thinking about Corn Oil

This sample conversation in the **Questioning** phase of the QuESTA cycle will help you guide your students through a discussion that will allow them to think deeply and thoughtfully about the corn investigation they are about to conduct. During the conversation, students may ask questions, think about what they already know, speculate on answers to questions, and wonder about what they are going to learn. This is a guide. Feel free to adjust your questioning to the needs of your class.

MS. D: Look closely at the kernels of corn. Now compare that to the oil. *Any ideas about how the corn kernel is changed into corn oil?*

JESSIE: One time I noticed that we had sunflower seed oil at my home. I asked my dad how the sunflower oil gets made from the seeds because I didn't see any oil in the seeds. He said the sunflower seeds are squeezed very hard and the oil comes out. Maybe if we squeeze the corn kernels very hard oil will come out.

MS. D: That's an interesting comparison to sunflower seeds. We can try squeezing one of our corn kernels. *What do you think would happen?*

ALEX: The corn kernel seems watery to me. I don't think water and oil are the same. I think if we squeeze a corn kernel, we will get wet, juicy stuff out, not oil. *How can we get oil from corn?*

MS. D: Good thinking. Let's start by figuring out how we can tell if we get water or oil from the corn. *Does anyone know how we can tell if what we get from squeezing a corn kernel is wet, juicy stuff or oil?*

ROSANNA: When I get pizza my mom always says that the stuff that gets on the paper plate and stains it is oil. If we squeeze the stuff that comes out of a corn kernel onto a paper plate and it's oil, it will stain the paper plate.

MS. D: Good idea. Let's try it. We don't have a paper plate here, but we can squeeze the kernel's juice onto a piece of paper. Let's see if what comes out stains the paper. (Do this simple experiment.) We'll have to let it sit for several minutes to dry.

This is a good start to what we often will do in the LiFE Curriculum Series. We'll start with a question and do some investigations that will help us develop theories about what the answer to the question might be. Our class discussions will help us come up with ideas. Talking through our ideas, even if your idea seems silly at first, is much more important than knowing the right answer. Scientists do this type of thinking and discussing all the time. Don't worry if you don't completely understand how corn flour, corn oil, and corn syrup are made from corn. You all did some excellent thinking about how corn gets changed. I think doing science this way is fun. I hope you do, too.

Continue the conversation to include a discussion of how corn kernels are made into corn syrup and corn flour.

Corn-Based Food Products

What's the first thing that comes to mind when you think of corn? Probably it's corn on the cob, fresh from the field. Yet most of the corn in the American diet is not fresh, whole corn. It's corn that has been refined and processed into its component parts and used in the highly processed food products we buy at the supermarket. Although corn has been refined for more than 150 years, the number of corn products used as ingredients in our food has increased tremendously in the last 50 years. If you look at the ingredient list on the most highly processed foods, you'll find that one or more ingredients are corn-based. In this lesson we're focusing on three corn products: corn flour, corn oil, and corn syrup.

Corn flour is made either from the entire corn kernel or from a kernel that has been degermed. Flour made from the whole kernel has a richer, more robust flavor, but it is more likely to turn rancid because of the oil found in the germ. Commercially, corn flour, also called cornmeal, is made from the mechanical grinding of dried kernels of white or yellow corn and is available in fine, coarse, and stone-ground varieties. You can make your own corn flour by grinding dried corn in a coffee grinder or a hand grinder.

Corn oil is made from the germ of the corn kernel. Before corn can be degermed it is steeped in 50°F water for about 35 hours. This adds moisture to the corn and releases the starch. Large corn refiners have bins that can hold up to 3,000 bushels of corn! After it's steeped, the corn is coarsely ground so that the germ breaks free from the rest of the kernel. A centrifuge spins out the low-density germs. These germs are pumped onto screens and thoroughly washed to remove any remaining starch. Next, they are pressed to release the oil. Chemical solvents are applied to extract any remaining oil. Filtering and further refining removes free fatty acids and phospholipids. The final product is high in polyunsaturated fat — healthy fat — and low in saturated and trans fats — less healthy fats.

Over the past few decades, the corn product that has experienced the greatest increase in use is corn syrup. In 1966 the average American consumed about 14 pounds of corn sweeteners a year. By 2004 this amount had increased more than five-fold. High fructose corn syrup (HFCS) accounted for much of the increase. Introduced into the food supply in 1968, HFCS consumed by Americans has steadily increased to the current level of 80 pounds per person per year. If you take a look at food labels, you'll find HFCS in products ranging from salad dressing to chewing gum. Corn syrup or corn sweeteners are made from cornstarch. After corn is degermed, what remains is the gluten, fiber, and starch. A series of steps removes the starch, yielding cornstarch that is more than 99% pure. Some cornstarch is sold directly as starch, but the vast majority is further processed to make corn syrup. To make syrup, the cornstarch is treated with enzymes to break the starch down into sugar. HFCS is the most refined of the corn sweeteners and has been processed until there is virtually no starch — it's all sugar.

To help students begin to develop an awareness of how many different kinds of food contain corn products, they will have a homework assignment to look for processed foods on their kitchen shelves. When you give the homework assignment, reinforce the fact that students are applying what they have learned in the classroom to their daily lives.

Corn Kernel

The **seed coat** is mostly fiber and protects the kernel.

The **air cavity** is the space between the seed coat and the endosperm.

The **endosperm** is mostly starch. It provides food for the young corn plant.

The **germ** contains most of the oil found in a corn kernel. When the kernel is planted, the germ develops into a corn plant.

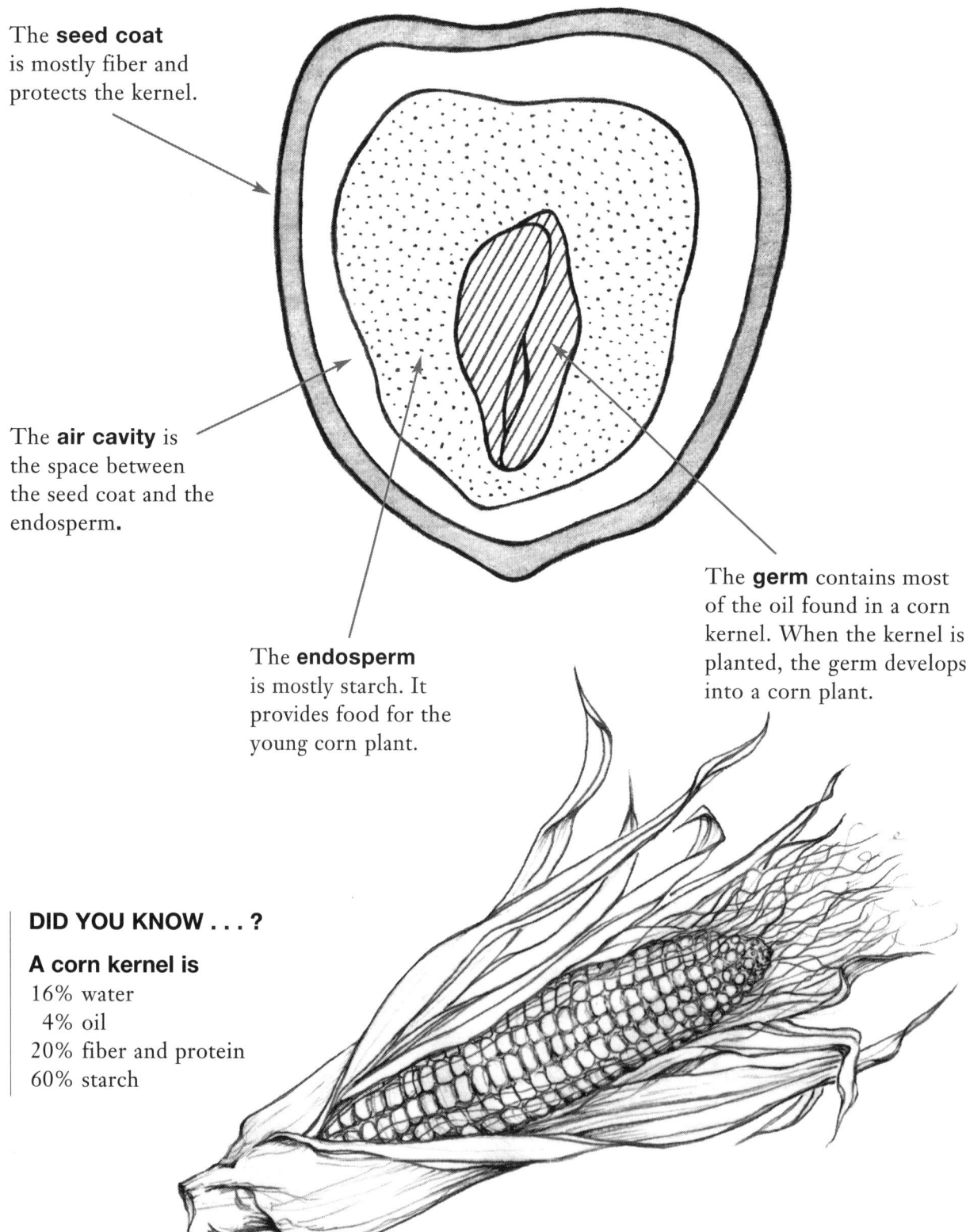

DID YOU KNOW . . . ?

A corn kernel is
16% water
 4% oil
20% fiber and protein
60% starch

Corn Investigations

Students observe three different corn products and try to determine how corn is changed into these products.

Setup

1. Prepare enough paper cups of corn oil, corn flour, and corn syrup so that each group of 4–6 students gets one paper cup of corn oil, corn flour, or corn syrup.

2. Make an overhead transparency of the ***Corn Kernel*** lesson resource or make one photocopy for each student group.

3. (Optional) If you choose to examine the popcorn endosperm and germ, use the sharp paring knife and cut several popcorn kernels in half.

Procedure

1. Have students work in groups of 4–6. Assign each group one product (corn oil, corn syrup, or corn flour) to investigate.

2. Give each student group several kernels from the ear of corn to help them with their investigation.

3. Encourage students to be creative in their thoughts and discussion. This experiment is not about figuring out exactly how these products are made, but rather to have an enjoyable experience as a food scientist.

4. Have groups share their findings. Remind students that all thoughts and ideas are welcome.

5. Encourage questions, discussion, and debates during these presentations.

Questions

1. *How does your product look similar to a kernel of corn? How is it different?*

2. *What could you do to the corn kernels that might make them become more like your product?* List all the steps you can that might happen in this process.

3. *Do you think the whole corn kernel is used to make your product or only part of the kernel? What part?*

4. *Does the picture of the corn kernel with the parts labeled help you at all?*

5. *Is your product similar to any other food product? If it is, does this help you think about how your product was made?*

Exploring Grapes

AIM

To demonstrate the first three phases of QuESTA through observing and experimenting with grapes.

SCIENTIFIC PROCESSES

- **question, observe, investigate, hypothesize**

OBJECTIVES

Students will be able to:

- **describe new things they learned about grapes through carefully observing them;**

- **outline the steps of a procedure to turn grapes into grape juice;**

- **define the phases of the QuESTA Learning Cycle in their own words;**

- **express, in writing, what they have learned about grapes.**

OVERVIEW

In this lesson, students are introduced to another key feature of LiFE — the QuESTA Learning Cycle. QuESTA has five phases: Questioning, Experimenting, Searching, Theorizing, and Applying to Life. This lesson covers the first three phases; Lesson 3 covers the remaining two. To gain an understanding of the way QuESTA guides the learning process, students participate in activities that take them through Questioning, Experimenting, and Searching. Students conduct detailed observations of grapes. Based on their observations, they pose questions to help them learn even more about grapes. They work in small groups to develop a procedure to change grapes into grape juice. Then they come together as scientists to discuss and debate their experiment procedures and reach consensus on a class procedure for making grape juice in the next lesson. Students experience the Searching phase as they read more about QuESTA. In their LiFE Logs, they summarize what they learned about grapes.

MATERIALS

For the teacher:
- *Grapes and Grape Juice* teacher note
- *Scientific Inquiry* teacher note
- *QuESTA!* lesson resource
- *Six Steps to an Experiment* lesson resource

For the class:
- Chart paper
- Markers

For each student:
- 3–4 grapes (with seeds)
- Napkins or paper towels
- *QuESTA Learning Cycle* student reading
- *Grape Explorations* activity sheet
- *Grape Experiment* activity sheet
- LiFE Log

PROCEDURE

Before You Begin:

- Review the background information in the *Grapes and Grape Juice* and *Scientific Inquiry* teacher notes.

- Detach and wash grapes, 3–4 per student

- Make copies of the *QuESTA Learning Cycle* student reading and the *Grape Explorations* and *Grape Experiment* activity sheets for each student.

- Review the questions on the *Grape Explorations* and *Grape Experiment* activity sheets.

- Post the *Six Steps to an Experiment* lesson resource at the front of the class.

- Review the *QuESTA!* lesson resource.

- If you have not already done so, post the Module Question and Unit 1 Question at the front of the classroom.

MODULE QUESTION

What is the system that gets food from farm to table, and how does this system affect the environment?

UNIT QUESTION

What is a food scientist?

 **QUESTIONING**

1. Review Module and Unit Questions

Explain that in this lesson students will investigate grapes. As food scientists, students will explore ways to turn grapes into grape juice.

2. Explore Grapes

Pass out 3–4 grapes to each student. Have students share what they already know about grapes. Record their ideas on the board. Ask students to think about some ways that food scientists might investigate grapes. Brainstorm different ways to make observations about the grapes. Encourage students to break the grapes in half and to use all of their senses, including taste, to learn more about this fruit. Challenge students to discover something they did not know about grapes before. Have students record their findings on the *Grape Explorations* activity sheet.

Ask volunteers to share their observations. Accept all answers. Encourage discussion, questions, and debate among the students.

 **EXPERIMENTING**

3. Develop Experiment Procedure

Have students work in groups of 3–4 to develop an experimental procedure to answer the question: *How is grape juice made from grapes?* Students in each group can record the methods and materials on their *Grape Explorations* activity sheet. Encourage students to include as much detail as possible.

4. Have Groups Share Their Methods

Have each group write its method on the board or on chart paper. Invite groups to present their methods to the class. Encourage student groups to ask questions of their classmates. Explore similarities and differences among the groups' methods. *What steps did all the groups include? What steps did only some groups include?*

What steps seem to be very important to turn grapes into grape juice? Do you think any steps listed are not important? Why? Most likely, all the groups included smashing the grapes as part of their method. Investigate each group's grape-smashing method. Be sure to discuss how to handle the seeds and skins. *How would you guess the grape juice might turn out using these different mashing methods? What ideas do you have for handling the seeds and skin of the grapes?*

5. Plan Whole-Class Experimental Procedure

Consider all of the groups' experimental procedures. Then, as a whole class, develop one method to use in Lesson 3 when you will make grape juice from grapes. This is an excellent opportunity to model how to plan a complete, detailed procedure for an experiment. Write your methods on chart paper so they can be posted when you do Lesson 3. If you wish to have students record the class experiment, they can copy it into their LiFE Logs. See the ***Grapes and Grape Juice*** teacher note for a description of how grape juice is made.

Make a materials list and determine where you will get the supplies. You may wish to have students bring in some of the kitchen supplies that are needed. Modify methods, as necessary, based on available materials.

Create a class hypothesis for the experiment. Or, you can have the students write their own hypotheses in their LiFE Logs.

Alternative method: Instead of creating one class experiment, let each group try the process as designed. This way the class can compare results and each group can learn what works and does not work in their experimental procedure. However, this will take more materials and more time.

6. Introduce QuESTA Learning Cycle

Explain that LiFE uses the QuESTA Learning Cycle to help students understand the process of learning. *Have you ever thought about how you learn? What might be the steps involved in learning something new? Can you share with the class an example of how you learned something?*

Distribute the **QuESTA Learning Cycle** student reading. Have students read it individually or as a class. Discuss each phase and ask students to define each one using their own words. Encourage questions and discussion.

7. LiFE Logs

Have students write a paragraph that begins with: "Today I learned some new things about grapes. Here's what I learned…" Ask students to include a description of the kind of observations they made. Prompt them to describe what they learned by using each of their senses. Remind them to include any questions they still have about grapes.

Grapes and Grape Juice

Did you know that grapes are the most commonly grown fruit in the world? This fruit and its juice have been popular for a long time. Archaeologists have found evidence of fermented grape juice in pottery jars dated to Neolithic times, about 5400 to 5000 B.C. And even though wild grapes didn't grow in ancient Egypt, artwork on tomb walls depicts the wine-making process. Extracting juice from grapes has a long history!

What's the first thing you think of when you think about making grape juice? For most people, smashing grapes — whether with feet or food processor — is what first comes to mind. Undoubtedly, this will also be the first thing that comes to your students' minds as well. You can count on hearing some rather creative ideas for smashing the fruit! While smashing grapes is part of the process, a very important step occurs after the smashing. This is the step when the grape pulp is heated. The heating process brings out the sugar in the grapes and incorporates elements from the skins. If purple grapes are used, the juice takes on the purple color of the skin.

Grape-Juice Recipe

Most grape-juice recipes recommend mashing the grapes with a potato masher, then heating the grapes to boiling. Continue to mash the grapes during heating. After the grape pulp boils for about 3 minutes, reduce the heat and simmer for 20 minutes. The final step is straining the juice through a fine strainer or cheesecloth to remove any remaining skin fragments and seeds. If possible, try heating your grapes when your class makes grape juice. Even if you don't do it in class, perhaps you, or one of your students, could try adding the heating step at home and bringing in the juice for everyone to try. Commercial grape juice is heated at least one more time during the pasteurization process before it reaches store shelves.

Most commercial grape juice is made from Concord grapes, which have a deep purple skin but light green insides. Concord grapes are highly perishable and can be hard to find in stores. If they grow in your region, you may be able to find them in farmers' markets. Another popular variety to look for is Niagara.

Scientific Inquiry

With this lesson, LiFE introduces students to the scientific method. Your students will gain valuable experience recording and describing each step in the grape juice–making procedure.

It's important for students to understand that different kinds of investigations are used to answer different kinds of questions. Sometimes observation might be appropriate. *What happens to grapes if we leave them out on our desks?* Sometimes a question can be answered through research. *Are all grapes purple?* Sometimes a thought analysis can answer a question. *Could animals live if there were no plants?* Other times, a controlled experiment can be used to answer the question. *Does sugar affect the growth of yeast?* In an experiment, scientists need to control the variables — the factors that can affect the results. For example, water and temperature might also affect yeast growth. To test just the effect of sugar, a scientist would control the other factors (temperature and water).

While students are introduced to controlled experiments in this lesson, they do not conduct one until later in the module. You may choose to save your discussion of controls and variables until then.

QuESTA!

How students learn is as important as what they learn. The Module and Unit questions ask students to think hard by challenging them to explore, question, investigate, analyze, synthesize, and act. QuESTA is a five-phase cycle that guides students through this process. Although the phases are presented linearly, they are dynamic. Once you get acquainted with QuESTA, you and your students will flow among the phases. Here are some sample questions to help you guide your students' learning.

QUESTIONING

- What do I already know about the topic?
- What don't I know about the topic but would like to learn?
- What am I curious about?
- How might I find answers to my questions?
- What if…?

EXPERIMENTING

- How can I set up my experiment?
- What are the steps in my experiment?
- What materials do I need for my experiment?
- What do I think will happen?
- Did my experiment work as well as I thought it would? Is there anything I would like to change about it?
- What data do I have?
- What are the results of my experiment?

SEARCHING

- What can I learn from reading or talking to people?
- Where can I find out more information?
- What do scientists already know about this topic?
- How can I find out if my results are accurate?
- How can I tell the difference between a fact and an opinion?

THEORIZING

- What have I learned?
- What evidence do I have to support my conclusion?
- Have my ideas changed?
- What are some different ways that I can analyze what I have learned from questioning, experimenting, and searching?
- What conclusions can I draw?
- Has my thinking about this topic changed? Why or why not?

APPLYING TO LIFE

- How can I use what I have learned?
- How can I remember to think about what I have learned as I do my daily activities?
- What can I teach my family and friends?
- What new questions do I have about the topic now that I am using this new knowledge in the real world?

Six Steps to an Experiment

Step 1: Develop an experiment question. This is the question your experiment is trying to answer. To develop a question, think about what you want to learn.

Ask Yourself: What question do I want to answer with my experiment?

Step 2: Decide on a **hypothesis.** A hypothesis is a prediction, or a guess, about the results of the experiment.

Ask Yourself: What do I think will happen?

Step 3: Develop your **experimental design.** Write down the steps you will do to conduct the experiment. These are called your methods. If you are conducting an experiment to make comparisons, identify the **control group** and the **experimental group.** With the experimental group, be sure to identify the **variable** (the thing that will change). Include a list of the materials you will need for your experiment.

Ask Yourself: What are all the steps in my experiment? What materials do I need to conduct the experiment?

Step 4: Do the experiment following your methods.

Ask Yourself: Is my experiment working as planned? Does it need to be changed? How could I change it to make it better?

Step 5: Record your data in a table, a chart, or in your LiFE Log.

Ask Yourself: What are the results? How should I record my data? Should I use a chart, a graph, a table, or write a paragraph to describe my results?

Step 6: Examine your results and think about what you have learned. Use your answers to this question to make your conclusions.

Ask Yourself: What did my experiment teach me, and how can I use this?

LESSON 2: EXPLORING GRAPES

Name	Date

QuESTA Learning Cycle

You have been learning science since you were born. Have you ever thought about how you learn? You will in the LiFE Curriculum Series!

In the LiFE program, you will use the QuESTA Learning Cycle. When you first begin to study a topic, you will question what you already know about the topic. For example, what do you already know about grapes? After you think about what you already know, you ask yourself what **questions** you still have about grapes.

How can you find answers to the questions you still have? One way is to conduct a scientific test called an **experiment.** When you conduct an experiment, you may make new discoveries or develop new ideas as you try to find answers to your questions. You may also **search** for more information by reading or talking to people.

What do you do with all of this new information you've learned from questioning, doing experiments, and searching? You can use it to develop new **theories,** or explanations of how things work.

You may be asking yourself how you can use what you have learned. Try using it in your daily life and sharing it with your family. For example, the next time you have a glass of grape juice, think about how it was made. You may even want to teach your family and friends what you've learned. When you do this, you'll be **applying to life** what you have learned.

Once you start applying what you learned, you may find you have more questions. This starts the learning cycle all over again.

| QUESTIONING | EXPERIMENTING | SEARCHING | THEORIZING | APPLYING TO LIFE |

Name	Date

Grape Explorations

Answer the following questions.

1. What do you already know about grapes?

_____

_____

_____

_____

_____

_____

2. Look at your grapes very closely and record some of your new observations.

_____

_____

_____

_____

_____

_____

_____

(continued on next page)

Name   Date

Grape Explorations

3. Make a drawing of the grape and what you observe.

4. Based on what you already know about grapes and what you have just learned from your observations, write down some ideas about how you think grapes are turned into grape juice.

_____

_____

_____

_____

_____

Name

Date

Grape Experiment

Answer the following questions.

Research Question: *How can we make grape juice from grapes?*

1. Working with your group, list all of the steps that are needed to make grape juice from grapes. Try to include every step from start to finish. These are called your methods.

_____

_____

_____

_____

_____

_____

_____

_____

_____

_____

_____

(continued on next page)

Name   Date

Grape Experiment

2. List everything you will need for your grape juice–making procedure. Think through the entire process and try not to leave anything out.

_____

_____

_____

_____

_____

_____

3. What do you think your results will be?

_____

_____

_____

_____

_____

Making Grape Juice

AIM

To continue to learn about grapes and QuESTA through making grape juice and synthesizing new ideas about grapes.

SCIENTIFIC PROCESSES

- experiment, gather data, infer, theorize, apply

OBJECTIVES

Students will be able to:

- describe their experiences of making grape juice from grapes;

- develop theories on how commercial grape juice is made, based on what they have learned about grapes;

- express in writing in their LiFE Logs what it will be like to learn using QuESTA.

OVERVIEW

Students learn more about the QuESTA Learning Cycle. They continue with the Experimenting phase by conducting their class experiment for making grape juice, doing the experiment and variations on it as many times as they can. After the experiment they move to the Theorizing phase through speculating on how store-bought grape juice is made. They speculate on how much effort it took to make a small amount of juice and come up with ideas about how factories might make very large amounts of grape juice. This gives them the opportunity to process what they have learned in order to synthesize new knowledge constructs, something they will do throughout the LiFE Curriculum Series. Finally, they Apply the lesson to life by discussing what they will think, feel, and do differently in the future, based on what they have learned about grapes and grape juice.

MATERIALS

For the teacher:
- *Grapes and Grape Juice* teacher note (p. 41)

For the class:
- 3 bunches of grapes
- Chart paper with experiment recorded (from Lesson 2)
- Materials as listed for class experiment (from Lesson 2)

For each student:
- *Turning Grapes into Juice* activity sheet
- *Grape Juice Theories* activity sheet
- *Applying What I Have Learned* activity sheet
- LiFE Log

PROCEDURE

Before You Begin:

- Review the *Grapes and Grape Juice* teacher note.

- Complete any setup needed for your experiment.

- Post the experiment recorded on chart paper in Lesson 2.

- Make enough copies of the *Turning Grapes into Juice, Grape Juice Theories,* and *Applying What I Have Learned* activity sheets for each student.

- If you have not already done so, post the Module Question and Unit 1 Question at the front of the classroom.

MODULE QUESTION

What is the system that gets food from farm to table, and how does this system affect the environment?

UNIT QUESTION

What is a food scientist?

1. Review Module and Unit Questions

Explain that in this lesson the class is going to conduct an experiment. Conducting experiments is one way that food scientists learn more about food.

2. Conduct Grape Juice Experiment

As the experiment is being done, help the students see how you are following the methods just as they were written up in the last session. If you have the supplies and the time, conduct the experiment a second time to compare with the results from the first time. You may do the experiment in the exact same way or make modifications based on what you learned the first time. Explain to the students that scientists often do experiments more than once to determine if their results come out the same each time. If the results are different, scientists carefully look over what they did to determine why.

3. Record Results

Have students describe the results of the experiment on the *Turning Grapes into Juice* activity sheet. Review the questions on the activity sheet with students. Check for understanding. Have students answer the questions.

4. Develop Grape Juice Theories

From what you learned in your experiment, how do you think the grape juice you buy in the store is made? Discuss as a whole class, then have students write their answers on the *Grape Juice Theories* activity sheet.

Encourage students to think about the factories that make grape juice in very large quantities. *How might what happens in a factory be similar to what we did in class? How might it be different? What do you think factories do with the seeds and skins of the grapes? How might the equipment used in a factory be like the equipment we used in class? How might it be different?* Use the *Grapes and Grape Juice* teacher note to guide you through this discussion.

Remind students that the theorizing phase of QuESTA is particularly important. It is their opportunity to put together everything they have learned and to gain new knowledge about what they are studying that they will be able to use now and in the future.

5. Share Written Answers

Encourage questions, discussion, and debate among the students. Continue the conversation until you feel most students have developed a new understanding about making grape juice.

6. Discuss Applications

Ask students to describe how their thoughts and actions related to grapes and grape juice have changed as a result of this lesson. Have students complete the *Applying What I Have Learned* activity sheet.

Explain that in this final phase of the learning cycle, students will think about how they will apply what they learned to their daily decisions.

7. LiFE Logs

Have students define, in their own words, the five phases of QuESTA (Questioning, Experimenting, Searching, Theorizing, and Applying to Life) and describe what they think it will be like to learn using the QuESTA Learning Cycle.

Name	Date

Turning Grapes into Juice

Use these questions to help you describe the results of your experiment.

1. Does what you made look like grape juice?

_____

_____

_____

2. How is the juice you made similar to grape juice you have had before? How is it different?

_____

_____

_____

_____

3. Is the juice thick or thin? What is the texture?

_____

_____

_____

(continued on next page)

BECOMING FOOD SCIENTISTS : INTERACTING PARTS : FOOD PROCESSING : ENVIRONMENTAL EFFECTS : WASTE : MAKING CHOICES

Name

Date

Turning Grapes into Juice

4. Does the juice have skin or seeds in it?

_____

_____

_____

_____

5. How does it smell? How does it taste?

_____

_____

_____

_____

6. Compare your results with your hypothesis. What was similar? What was different?

_____

_____

_____

_____

(continued on next page)

LESSON 3: MAKING GRAPE JUICE

Farm to Table & Beyond
©2008 Teachers College Columbia University

Name   Date

Turning Grapes into Juice

7. What worked about your experiment? Why?

_____

_____

_____

_____

8. What didn't work about your experiment? Why?

_____

_____

_____

_____

9. What would you do differently next time? Why?

_____

_____

_____

_____

Name   Date

Grape Juice Theories

Now that you have made grape juice, you're ready to start developing some theories about how the grape juice you buy in the store is made. Remember how you made grape juice in class. To develop your theories, you will need to think about what you have learned, what evidence you have to support your theory, and how your ideas about grape juice may have changed. Maybe you have some new, or different, ideas about grape juice. Think about what you know about grapes and grape juice. Now think about what you learned by making the juice.

1. How do you think the grape juice you buy in the store is made?

_____

_____

_____

_____

_____

_____

_____

_____

(continued on next page)

Name   Date

Grape Juice Theories

2. What evidence do you have for your ideas?

_____

_____

_____

_____

_____

_____

3. How have your ideas changed?

_____

_____

_____

_____

_____

_____

_____

Name   Date

Applying What I Have Learned

Now that you've completed your experiment and developed theories about grape juice, it's time to think about how you can use what you have learned in your life. Maybe you want to teach other people how to make grape juice. Perhaps you will think about the kind of grapes that were used to make the juice you drink, where the juice was made, and how it was made. Maybe you'll discover that as you think about making grape juice, you have even more questions that you'd like to investigate. There are no right or wrong answers. This is a time for you to put into action what you have learned in this lesson.

1. How will you use what you learned about grapes and grape juice in the future?

_____

_____

_____

_____

_____

_____

_____

_____

(continued on next page)

Name

Date

Applying What I Have Learned

2. Based on what you have learned about making grape juice in class, do you have any new questions about the grape juice you buy in the store?

_____

_____

_____

_____

_____

_____

3. How can you apply what you have learned to your daily life? Is there anything that you will do differently?

_____

_____

_____

_____

_____

_____

BECOMING FOOD SCIENTISTS : INTERACTING PARTS : FOOD PROCESSING : ENVIRONMENTAL EFFECTS : WASTE : MAKING CHOICES

Pre-Assessment

AIM

To assess what we already know about the food system and how it affects the environment.

SCIENTIFIC PROCESSES

- **question, assess knowledge**

OBJECTIVES

Students will be able to:

- **express in a diagram and in writing their current understandings of what the system is that gets food from the farm to the table and how this system affects the environment;**

- **appreciate how the Module and Unit questions guide everything they learn in the module.**

OVERVIEW

In the first three lessons of this unit, students have received a brief introduction to our food system by exploring corn and grapes and attempting to make grape juice. In this lesson, students are reintroduced to the Module Question, *What is the system that gets food from farm to table, and how does this system affect the environment?* Students are reminded that this question frames everything they learn in this module. They also are introduced to the Unit 2, 3, 4, 5, and 6 questions to demonstrate that the Unit Questions will help them understand the Module Question. Students draw a detailed diagram that represents their ideas about the farm-to-table system and how this system affects the environment. This activity helps students realize what they already know and prompts them to wonder what they will learn. Then, using their diagram as a guide, students write a short paragraph in their LiFE Logs describing what the system is that gets food from farm to table, and how this system affects the environment. Reviewing this paragraph and their diagram as they work through the module helps students track what they have learned. This lesson's homework introduces students to systems and systems thinking. In exploring the farm-to-table system, students use systems thinking to investigate our food system and how it works.

MATERIALS

For the teacher:
- *Assessing What Your Students Already Know* teacher note

For the class:
- Chart paper
- Markers
- (Optional) Index cards

For each student:
- 1 sheet of drawing paper
- Markers or crayons
- *Thinking about Systems* student reading
- *Thinking about Systems Review* activity sheet
- LiFE Log

PROCEDURE

Before You Begin:

- Review the *Assessing What Your Students Already Know* teacher note.

- Make copies of the *Thinking about Systems* reading and the *Thinking about Systems Review* activity sheet for each student.

- Post the Module Question and the Unit 1 Question at the front of the classroom.

MODULE QUESTION

What is the system that gets food from farm to table, and how does this system affect the environment?

UNIT QUESTION

What is a food scientist?

 **QUESTIONING**

1. Review Module and Unit Questions

Introduce students to the Module Question. Help students understand that this is a complex question and that it frames everything they do throughout this module. To become familiar with studying food and how LiFE works, we investigated the Unit 1 Question. To better understand the system that gets food from farms to our tables and how this system affects the natural environment, we investigate five more Unit questions: Unit 2: *What is the system that gets food from farm to table?* Unit 3: *What happens to food as it moves from farm to table?* Unit 4: *What are the environmental effects of our farm-to-table system?* Unit 5: *How can we reduce the food-related waste that we produce?* and Unit 6: *How can we use the science we learned to make ecologically sound food-system choices?* Post these questions along with the Module Question.

Elicit students' ideas about how food gets from the farm to the table. Record these ideas on the chart paper or board. The list may include: We drive to the farm to buy our food. We buy food at the farmers' market. The farmer sells food to the grocery store, where we buy it. Accept all answers.

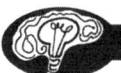 **THEORIZING**

2. Draw a Diagram

Ask students to draw a diagram of all the steps that food goes through to get from the farm to the table. Encourage the use of words and arrows to help indicate the steps in the process. Suggest that students first organize their thoughts by recording all the steps they can think of before they begin to diagram the process. Students may find it helpful to write the steps on index cards first and then attach the index cards to the chart paper.

3. Share Drawings

After students have finished their diagrams, invite them to share their work with the class. Encourage students to ask questions, discuss, and support each other's work. Display the drawings in your classroom and keep them out throughout the time you work on the module. Some teachers have turned the drawings into a class book. Others have displayed them on the bulletin board. What is important is making certain the drawings are prominently displayed throughout the module.

4. LiFE Logs

Have students write an answer to the Module Question in their own words. Encourage them to write at least two paragraphs. Through recording what they already know about this question, students will be able to assess how their thinking changes as they go through the module lessons. This process of assessing their own understanding and knowledge can be a stimulating academic challenge. Be sure to spend time discussing this with students and brainstorming ideas they have in order to track their learning. With each lesson, be sure to provide time for students to share what they have learned with their classmates.

5. Assign Homework

Hand out the *Thinking about Systems* student reading and the *Thinking about Systems Review* activity sheet. Draw students' attention to the guiding questions at the beginning of the student reading. Invite student volunteers to read them out loud. Tell students to think about these questions as they read. Remind students to answer the questions on the activity sheet after they have read *Thinking about Systems*.

Assessing What Your Students Already Know

The first unit of this module, Becoming Food Scientists, familiarized students with the study of food, inquiry-based science, and the phases of the QuESTA Learning Cycle. Units 2–6 are about the system that gets food from the farm to the table and how this system affects the natural environment. This lesson is an authentic assessment that allows your students to record their current thinking about the Module Question and to track how their thinking changes as they complete the lessons in this module. We have found that when students fully understand and explore the Module Question at the beginning of the module, they are more fully engaged throughout the entire module. Please make it clear to students that what they draw in their diagram and what they write in their LiFE Logs offers an understanding of what they know right now. By doing a thoughtful, reflective job with this today, they can chart what they learn throughout the module, since they will draw another diagram and write an answer to the Module Question again in Lesson 30 at the end of the module. This lesson also offers an opportunity to set up how each of the Unit questions builds to a great understanding of the Module Question.

Name

Date

Thinking about Systems

Guiding Questions

- *What is a system?*
- *Is it possible for parts of a system to also be systems?*
- *What does it mean to think about things as systems?*

What do a scooter, a computer, a chicken, a school, a garden, and a human all have in common? Each one is an example of a **system.** Think about what we mean by the word "system." What is a system? Basically, a system is made up of parts that interact with each other to function as a whole. For example, take a look at the scooter. A kick scooter has wheels, brakes, a frame, and handlebars. All these parts work together. Remove one part — let's say a wheel — and the scooter will no longer work the same way. Without the wheel, the scooter system has one part that is missing, so the system does not work.

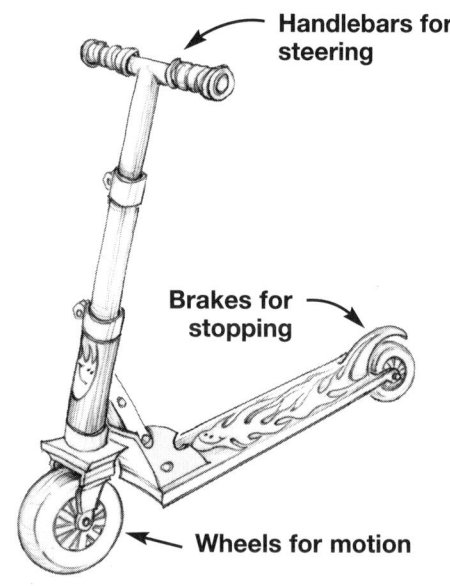

Handlebars for steering

Brakes for stopping

Wheels for motion

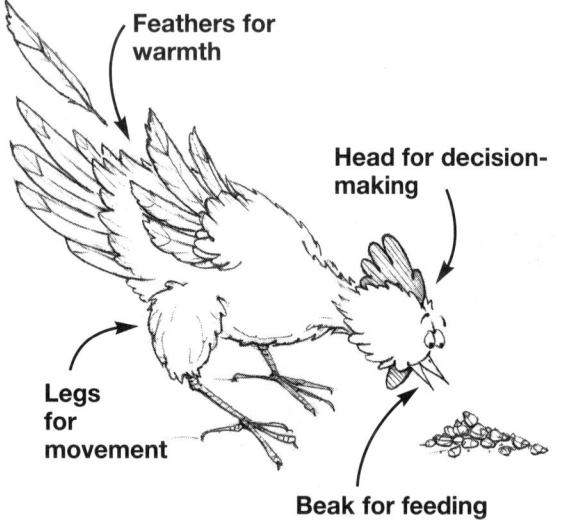

Feathers for warmth

Head for decision-making

Legs for movement

Beak for feeding

Can the parts of a system also be systems? Think about a chicken. It's a system. A chicken is made up of parts like a digestive system and a circulatory system. These parts are **subsystems** of the whole system, the chicken. That means they are systems with parts that interact. For example, the chicken's digestive system has parts that interact and make up the digestive system. All the parts of the chicken interact and make a whole system, the chicken. If you removed the chicken's head and put in the scooter's handlebars, the chicken would no longer be a whole system that works like a chicken.

You are an example of a system, too. You have lots of interacting parts that make up you, a human system. Cells, lungs, veins, arteries, the digestive system, the skeletal system, and the nervous system are just a few of the parts that make it possible for you to function as a whole. What would happen if you broke a leg, part of your skeletal

(continued on next page)

LESSON 4: PRE-ASSESSMENT

Name   Date

Thinking about Systems

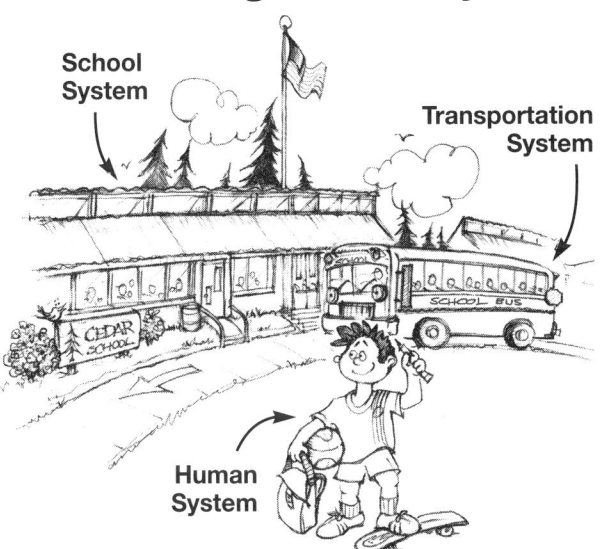

School System

Transportation System

Human System

system? The broken bone, one of your parts, can't interact with your other parts. You can't move as well as you did before you broke your leg. When the bone heals, you can run and jump again. All the parts interact once more.

You, a human system, also are part of a school system. You are a subsystem of your class at school, which is another subsystem of the whole school. You are also a subsystem of your family, and your family members are also parts of your family system.

The fruits and vegetables that we eat were once part of a garden or farm system. The garden or farm is a kind of **ecosystem,** which means it is a system of interacting living organisms and their environment. This ecosystem includes the humans who live there and work in the fields, insects, birds, worms, vegetable plants, fruit trees, rabbits, bacteria, soil, sun, water, and more. Each of these parts is also a subsystem and they all interact. Think about the farmer who uses irrigation to bring water to the fields. The farmer is interacting with the water system and the plants.

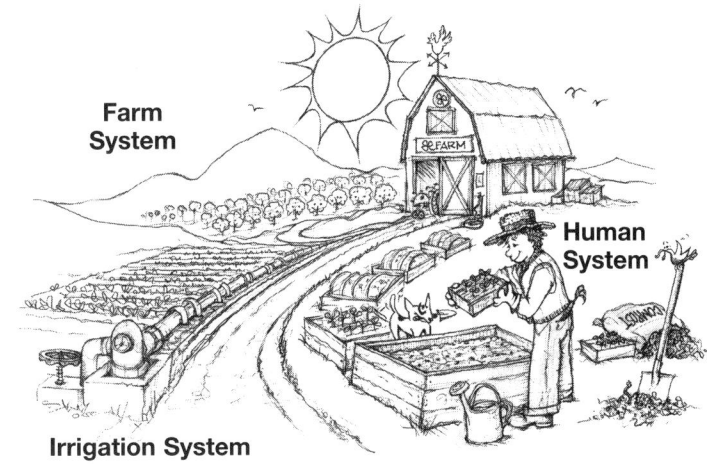

Farm System

Human System

Irrigation System

Thinking about systems means thinking about the whole system, the parts of the system, and how the parts are connected and interact. It means looking for all the ways that each part relates to the other parts. It means trying to discover what systems are connected to one another and learning about the interactions. In your investigations of the food system, you will be looking for the subsystems that are part of the whole food system. You will also be looking for all the ways that the subsystems interact to make the whole system "work." The food system we have today is very complex. As you begin to unravel all the connections and look at all the parts, you may be surprised to discover that the jar of jelly you bought at your local store is part of a food system that includes subsystems from around the world.

Name

Date

Thinking about Systems Review

1. What is a system?

_____

_____

_____

_____

2. What does it mean when someone says a system can be very different from its parts? (*Hint:* Think about the scooter wheel and the whole scooter. Can a wheel take a person from place to place? How about a scooter?)

_____

_____

_____

_____

_____

3. Is it possible for part of a system to also be described as a system? Explain your answer.

_____

_____

_____

_____

_____

(continued on next page)

LESSON 4: PRE-ASSESSMENT

Name   Date

Thinking about Systems Review

4. Study the illustration. Write down all of the parts of the farm system that you can see. Think about the different ways they are connected to one another. Record at least three interactions.

PARTS OF THE FARM SYSTEM

[Example] Trees and Sun

INTERACTION

[Example] Tree leaves get energy from the sun.

_____   _____

_____   _____

_____   _____

_____   _____

_____   _____

(continued on next page)

Name   Date

Thinking about Systems Review

5. You are a systems scientist. Your assignment is to think about as many different connections and interactions as you can in your home-to-school transportation system. Think about what you have learned about systems in your reading. Next think about how you get to school from your home. List all the systems that you can think of that are part of your home-to-school transportation system. (Hint: Start with you, a human system, and end with your school, another system. Do you interact with other humans on your way to school? Do you ride a bike or take a bus or do you walk? Does the bus use fuel? Where does the fuel come from? Do you travel along a highway system or sidewalks?)

SYSTEM   **INTERACTION**

[Example] Me   *[Example]* I ride a bus to school

_____   _____

_____   _____

_____   _____

_____   _____

_____   _____

_____   _____

Farm to Table & Beyond

Interacting Parts

From Field to Store

AIM

To begin to investigate the system that gets food from field to store.

SCIENTIFIC PROCESSES

- question, investigate, build theories

OBJECTIVES

Students will be able to:

- explain what a system is;

- identify parts of a system and describe how they interact;

- discuss the steps a type of food goes through from field to store;

- identify processing, packaging, and transportation as steps in the food system.

OVERVIEW

In this lesson, students continue their investigations of systems and interactions. The lesson starts with a review of the Lesson 4 homework, which you can use as an opportunity to check for student understanding of systems. Students share their ideas about the home-to-school transportation system that they participate in each day. This discussion is followed by a class demonstration that introduces students to interactions and feedback. In introducing the Unit 2 Question, *What is the system that gets food from farm to table?,* students review what they have learned and then brainstorm the parts of a food system that get food from the farm to the store. As students ponder systems and interacting parts, encourage them to think about the energy that is used in our food system. With this activity, students begin in earnest their exploration of the Module Question: *What is the system that gets food from farm to table, and how does this system affect the environment?*

MATERIALS

For the teacher:
- *Why Do We Need a Food System?* sample conversation
- *Student Understandings of Systems* teacher note
- *Apples to Applesauce Concept Map* lesson resource
- *Modeling Interactions* experiment sheet
- Chart paper
- Markers

For the class:
- Materials from the *From Field to Store Supply List* lesson resource

For each student:
- *Exploring Our Food System* student reading
- *Thinking about Systems* student reading (pp. 64–65)
- *Thinking about Systems Review* activity sheet (pp. 66–68)
- LiFE Log

PROCEDURE

Before You Begin:

- Review the *Why Do We Need a Food System?* sample conversation, *Student Understandings of Systems* teacher note, and the *Apples to Applesauce Concept Map* lesson resource.

- Remind students to bring in their homework from Lesson 4. Review the *Thinking about Systems* student reading and the *Thinking about Systems Review* activity sheet.

- Review the *Modeling Interactions* experiment sheet.

- Gather the materials listed on the *From Field to Store Supply List* lesson resource and set up the *Modeling Interactions* demonstration.

- Make copies of the *Exploring Our Food System* student reading to distribute to students.

- If you have not already done so, post the Module Question and the Unit 2 Question at the front of the classroom.

MODULE QUESTION

What is the system that gets food from farm to table, and how does this system affect the environment?

UNIT QUESTION

What is the system that gets food from farm to table?

 **QUESTIONING**

1. Review Student Understandings of Systems

Ask students to take out their completed *Thinking about Systems Review* activity sheet, their homework from Lesson 4. In a whole-class discussion, elicit students' ideas about systems. You may wish to use the *Student Understandings of Systems* teacher note and the guiding questions from the student reading to direct the discussion. *What does the word "system" describe?* (Something made up of parts that interact to make a whole.) *Can a system be different from its parts? What is an example of a whole system that is different from its parts? Can part of a system also be a system? What are some examples of parts of a system that also are systems?* Use this homework review as a way to gauge student understanding of systems and interactions. Pay particular attention to student understanding and discussion of the interconnections. Urge them to think beyond naming the parts of the system to how these systems interrelate and what the connections are.

Invite students to share their ideas about the home-to-school transportation system that they are part of. List the parts on chart paper. Accept all answers. Ask student volunteers to describe or draw on the chart paper different interactions that take place between the parts. For example, if students ride a bus to school, the bus, a mechanical system, and human systems interact when students get on the bus and ride it to school. Another human, the bus driver, interacts with the bus by applying the brake system to stop the bus to pick up the students; the bus driver takes the students to school, part of the school system, which also includes the students, teachers, administration, and other staff. The bus runs on fuel, which can be traced back to the environment. The bus also travels along a network of highways and is part of the school system's transportation system, and so forth.

 **EXPERIMENTING**

2. Discuss Interactions

Use the *Modeling Interactions* experiment sheet to conduct the water level demonstration. If you have not already done so, gather the materials listed on the *From Field to Store Supply List* lesson resource and set up the *Modeling Interactions* demonstration.

Explain that interactions are an important source of information in studying systems. Tell students that you are going to model a system with interactions and feedback. The feedback is the information that you will get based on the action you take. The feedback will tell you how the system is working. Students may be familiar with feedback from their own experiences, such as feedback from their teachers on their schoolwork, or feedback from coaches. Invite students to talk about what they do when they get feedback. Tell them that feedback in a system is similar. Information from the system is passed back through the system. Follow this discussion with the demonstration.

Do the first demonstration. Pour the water into the cup at a steady rate to maintain the water level. *What is the action that is taking place?* (Pouring water into the cup as water flows out of the bottom of the cup.) *What information do we get back from this action?* (We pour water in at the same rate that it flows out the bottom.) *What can we do to reduce the water level?* (Pour water in at a slower rate than it flows out.) Repeat the demonstration, pouring the water at a slower rate. *What information did we get this time?* (We did not pour fast enough.) *What changed?* (The water level dropped.)

Do the demonstration again, this time pouring water faster so that the cup overflows. *What happened?* (We poured water into the cup faster than it flowed out of the cup.)

 **THEORIZING**

3. Discuss Systems

Encourage students to think through what they observed in the demonstration. Imagine this model represents a food system. If the cup represents a farmer's stand at a market, and the water is a crop, like raspberries, the water in the cup would represent the raspberries that the farmer is selling. *What would the water flowing out represent?* (Customers buying raspberries.) *What would the water flowing into the cup represent?* (The berries the farmer grows for the market.) Continue the discussion. Point out that if there aren't as many customers, it's like more water flowing in than water flowing out. There are berries left at the end of the day. If there are more customers than there are berries, some customers go away empty-handed. With this feedback (not enough berries for the customers), the farmer might bring more berries the next week. If the customers don't buy them all, the farmer might sell them at a reduced price at the end of the day or might have to throw them away or compost them. *Can you see why the farmer might want the system to be balanced?* Bring the discussion to a close.

4. Introduce the Unit 2 Question

Remind students that in their *Farm to Table & Beyond* studies, they have been investigating the Module Question. In Unit 2, the class will be investigating the system that gets food from farm to table.

Invite students to reflect on what they have learned about systems and to brainstorm the parts of a food system that get food from the farm to the store. Consider referring to a food that most students are familiar with, such as applesauce. Record their answers on chart paper. The *Why Do We Need a Food System?* sample conversation and *Apples to Applesauce Concept Map* lesson resource can help you guide your students through this discussion of the parts of the food system. You may wish to prompt student thinking with questions such as: *What ingredients are in applesauce? Where are apples grown?* (Orchard, farm, trees.) *What happens when the apples are ripe?* (Someone picks them.) *What happens after the apples are picked? How do they become applesauce?* (Someone makes the applesauce; apples are made into applesauce in a factory.) *How does the applesauce get to a place where you can buy it?* (It's put in jars or cans and then into boxes, shipped, a truck takes it to a store.) *What happens when the applesauce gets to the store?* (Someone puts it out for people to buy.) *Can you think of any feedback that someone might get in this food system?* (People want more applesauce than the store has and the store sells out; the apples are not the right ones to use for applesauce and it doesn't turn out well; there are not enough apples to make applesauce.)

5. Make Connections

Review student responses. Point out that by describing all these parts, students have started to describe a food system. Ask if anyone would like to change any of the answers or add to them. Record any changes or additions. Next, elicit student ideas about the connections between the parts of the whole food system that they outlined. *Is the natural environment connected in any way? Where do you see connections to the natural environment?* (Water, trees, soil, air.) *Are humans connected to this food system? Where do you see the connections? Can you think of any points in this food system where energy is used?* (Fuel for tractors, human energy to pick the apples, energy to run the machines in the factory, energy for lights in the factory, fuel for the truck, energy in the stores.)

Discuss how the parts of the applesauce system interact. *What would happen if half of the apple harvesters were out sick one day? Do you think that the harvesters being out sick would affect the total number of apples that were picked and placed on the truck that day?* (If there weren't as many people picking the apples, there would be fewer apples on the truck.) *If a factory bought machines that could make twice the amount of applesauce, do you think that would affect the rest of the apples-to-applesauce system? What kind of changes do you predict would take place?* (It would create a demand for more apples. There would be more applesauce available in the marketplace.) *Do you think it would have an affect on you, the applesauce consumer?* (There might be a change in the price of applesauce.)

6. LiFE Logs

Ask students to describe the farm-to-table system in their own words. Tell them to include where they think there may be interactions that would provide feedback. Have students record any questions they still have about food systems, including interactions and feedback. Invite students to share their descriptions with the class. Engage students in a discussion of why we need a food system. Use the sample conversation as a guide. As you close the lesson, record the questions students still have about food systems on chart paper and post it at the front of the classroom.

7. Homework

Give each student a copy of the *Exploring Our Food System* student reading to read after he or she completes this lesson.

Why Do We Need a Food System?

This sample conversation in the **Questioning** phase of the QuESTA cycle will help you guide your students through a conversation that will allow them to think deeply and thoughtfully about food systems. During the conversation, students may ask questions, think about what they already know, speculate on answers to questions, and wonder about what they are going to learn. This is a guide. Feel free to adjust your questioning to the needs of the class.

MS. R: *Does anyone have any ideas about what we mean when we say "food system"?* Think about what you have learned about systems and interactions and tell me why you think the process of getting food from the field to the store is called a food system.

MAX: Food has to get from the field to the store. Sometimes I see trucks that deliver food to the store near my home. The trucks are part of the transportation system we just talked about from our homework.

LENNY: Maybe it's called a food system because there are parts, like transportation, and the farm, that make up the system.

MS. R: Good thinking, Max and Lenny. There are lots of different parts that are connected to make up the whole food system. *Why do you think we need a food system?*

SOFIA: My grandmother told me that she grew up on a farm in California and that back then there were lots of people who were farmers. But I don't think there are as many farmers now. I don't know anyone who is a farmer.

MS. R: You're right, Sofia. There are not as many farmers today as there were when your grandmother was a little girl. *So, if we have fewer farmers, what does that mean? How do we get our food?*

SOFIA: It means that most people don't grow their own food, so we have to get it from the store.

MS. R: Right. *What if we did all grow our own food? Would our lives be different?*

OLIVIA: I don't think we would come to school every day. I think we would have to help our families on the farm. I saw something on television about a girl who lived on a dairy farm and she had to help her dad milk cows every day. That's a lot of work.

MS. R: It sure is. We would spend a lot of time doing work related to growing our food if we didn't have a food system that got our food to the store for us. *What are some of the steps in the food system besides transportation?*

OLIVIA: Last night we had spaghetti for dinner and my dad made it using a jar of sauce and a box of spaghetti. I don't know how the noodles were made but I know the sauce is made from tomatoes and it must have been made in a factory somewhere.

MS. R: You bring up an important idea, Olivia. In the food system we have today, we can buy foods that are already prepared in some way. This is called processing. Tomatoes from a farm were processed into sauce. It is convenient for us to buy foods that already are prepared, like the tomato sauce. The sauce is convenient because we don't have to spend time making it from scratch. So we have more time to spend doing other things.

Continue the conversation to include a discussion of packaging as part of the food system.

Student Understandings of Systems

In this module, we focus on the highly technological, complex system we have for getting food from the farm to the table. When we began developing this curriculum we interviewed students about their understanding of the food system. Although they had some ideas about how food is grown and some ideas about what food does in our body, they had very little understanding about our farm-to-table food system. Some students had very creative ideas about parts of the food system. For example, one student told an intricate story about the way the colorful little Os in Froot Loops cereal are cooked in an oven similar to a dryer, then dipped into a barrel of dye, and next are returned to the dryer-like oven so that the color does not rub off on the consumer's hands. After completing the student interviews, we termed the farm-to-table part of our food system "the black hole" — students are interested in the topic and have their own ideas about it, but have little scientific understanding of how it actually works.

The term "system" is part of our daily vocabulary — from computer systems to sound systems to transportation systems to ecosystems. Your students may be very familiar with the word, yet not understand its meaning. Researchers have found that students have difficulty recognizing differences between parts and whole systems and tend to view them as similar. The National Science Education Standards urge teachers of science to help students recognize the properties of objects, as emphasized in grade-level content standards, while helping students understand systems.[1] For example, in studying packaging and transportation as subsystems of the global food system, encourage students to consider the weight of materials used in packaging and how that might influence the mode of transportation.

Understanding our global food system, how the parts interact, and how our food choices can affect the environment is challenging. It is also important. The impact of human beings on the natural environment is expected to increase as populations grow and science and technology develop more sophisticated ways to change the natural world to suit human demands. What skills and knowledge will help prepare students to be able to participate in public debate about issues that involve the impact of human activities on the natural environment? Help students hone their scientific-reasoning skills and introduce the dimension of time. Encourage students to think about the parts of the food system today, and the food system in the past. What caused changes in our food system from the past to the present? What changes might there be from the present into the future?

Consider using reflective writing to uncover students' thinking about systems and interactions. Encourage them to use diagrams to demonstrate the flow of information and feedback loops. Use the time-line resources included in the module to help students develop a historical perspective and to spur them on to think of creative solutions for the future. After all, your students will be the scientists, engineers, designers, and consumers who will be making the decisions that influence the food system in years to come.

---

[1] National Science Education Standards, 1996, p. 116

Apples to Applesauce Concept Map

List the terms and phrases that students suggest during the brainstorming session. Terms might include: apples (or food); orchard, farm, trees; someone picks the apples; trucks take the apples to the factory; someone makes applesauce, applesauce is made at a factory; it's put in jars or cans and then into boxes; it gets shipped, trucks take it to the store; someone puts the applesauce on the shelf.

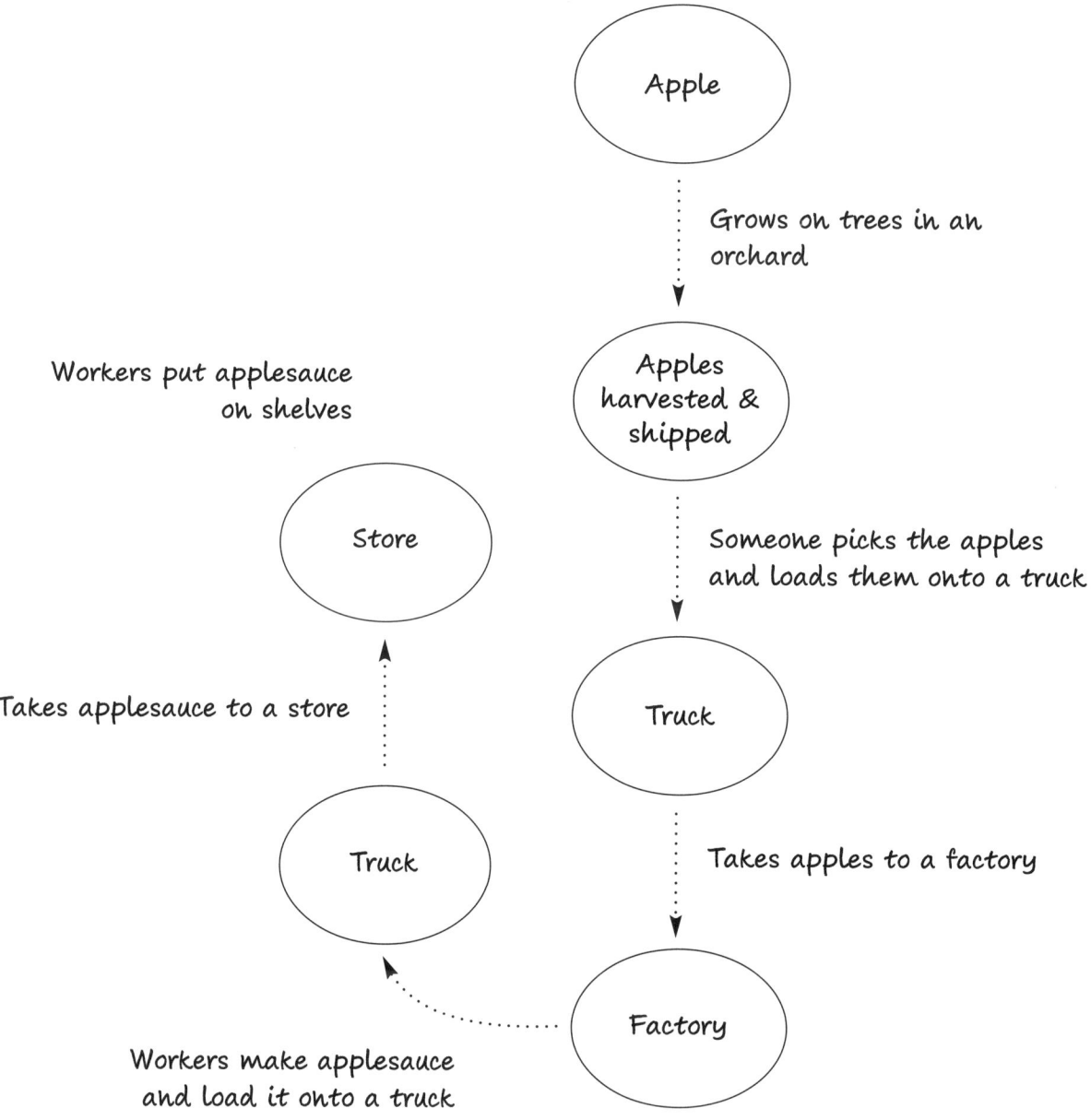

From Field to Store Supply List

MODELING INTERACTIONS (pp. 78–79)

Supplies

For the class
- 3 cups of water
- Pitcher (1 quart or larger)
- (Optional) Food coloring
- (Optional) Spoon to stir food coloring in water
- 1 clear plastic cup, no larger than 8 ounces
- 1 nail (to punch a hole in the bottom of the cup)
- 1 two-cup measuring cup or small plastic container to catch the water
- Plastic dishpan or basin
- Paper towels (to clean up any spills)

Modeling Interactions

In this demonstration, students observe what happens when water is poured into a cup with a hole in the bottom. They ponder what needs to change to reduce the water level, maintain the level, and cause an overflow. They then apply what they have learned from this model to a discussion of the food system.

Setup

1. Gather the materials. Pour the water into the pitcher.

2. If you choose to dye the water, add the food coloring. This makes it easier to see the water in the cup.

3. With a dark marker, draw a line 1–2 inches from the rim of the cup. This line marks the water level.

4. Use the nail to punch a small hole in the bottom of the cup with the water level line.

Procedure

1. Place the measuring cup or small plastic container in the plastic dishpan or basin.

2. Place a finger over the hole in the bottom of the cup. Pour water into the cup until you reach the line marking the water level. You may wish to have a student volunteer help.

3. Remove your finger from the bottom of the cup. Slowly pour more water into the cup. Pour at a rate such that water flows out of the hole at about the same rate as you are pouring water into the cup.

4. Pour the water back into the pitcher and set up to repeat the demonstration. Pour water up to the level mark to begin. Remove your finger from the bottom of the cup.

5. This time, pour more water in as slowly as you can. Pour at a rate such that water flows out of the hole at a faster rate than you are pouring it into the cup. The water will drop below the mark on the cup.

6. Pour the water back into the pitcher and set up to repeat the demonstration. Pour water up to the level mark to begin. Remove your finger from the bottom of the cup.

7. This time, pour more water in as fast as you can. Pour at a rate such that water flows into the cup at a faster rate than it flows out of the hole. The cup will overflow.

Questions

1. *What action is taking place? What happened the first time we poured water into the cup? What information did we get?*

2. *What action can we take to reduce the water level while still pouring water into the cup?*

3. *What evidence do you have to support your views?*

4. *Based on this feedback, if we don't want the water level to drop, what should we do?*

5. *What evidence do you have to support your views?*

6. *What action can we take to have the water overflow as we pour water into the cup?*

7. *What evidence do you have to support your views?*

Name	Date

Exploring Our Food System

Guiding Questions

- *What are the parts of a food system?*
- *Who produces food?*
- *Why do we package food?*
- *How does processing change food?*
- *How is food transported?*

The next time you sit down to eat spaghetti with tomato sauce, stop for a minute and think about how this food gets to your table. The sauce's journey begins in a field where the summer sun is hot and the tomato plants are covered with tomatoes. A farmer grows these tomatoes for others to eat. This step of the **food system** is called **food production.** As soon as the tomatoes are ripe, someone picks them, puts them in boxes, and prepares them to ship from the farm to the next step in the food system. People load the boxes onto trucks and the truck drivers drive off. During the trip to the factory, the drivers stop to buy fuel and eat a meal. After resting, they continue the drive along highways and across bridges until they reach a factory. This step of the food system includes **transportation.**

At the factory, people sort the tomatoes and keep all the ones that are perfect for sauce. The damaged and rotting tomatoes are thrown away. The tomatoes for sauce are moved from one part of the factory to another, where they are processed into sauce. They are washed and heated, and the skins are removed and thrown away. They are cooked at high temperatures. Water, salt, and herbs are added to the tomatoes. **Food processing** means that food, like the tomato, is changed in some way.

When the sauce is ready, it goes to the next step of the food system, called **packaging.** The can that tomato sauce comes in is one kind of packaging. Sometimes packaging is made from cardboard, plastic, or glass. Packaging protects the food. It keeps food from drying out, getting moist, breaking into small pieces, and spilling. It can also protect food from microorganisms. The packaging usually includes a label, so you know what kind of food is inside. Making the labels and attaching them is another step in the food system. It can include taking a photograph of the food, or drawing it, designing a label, printing the label, and pasting it onto the package. Packaging also helps advertise the food, which can increase the sales of the food product. Once the packaging is complete,

(continued on next page)

Name   Date

Exploring Our Food System

the tomato-sauce cans are put into boxes or crates and are sealed, ready to be shipped to the next stop on this journey.

The transportation system connects with the food system again. People load the boxes of tomato sauce onto trucks, or into railroad cars, or onto boats, or even onto planes. The tomato sauce is on the move and heading to its next destination. Once the tomato sauce arrives, it might be stored in a warehouse. From the warehouse, boxes of tomato sauce are moved directly to your neighborhood market. At your market, workers unload the trucks, open the cases, and put the cans of tomato sauce on the shelves.

The next time you enjoy spaghetti with tomato sauce, think about all the steps that it takes to bring those farm-fresh tomatoes to your table.

Why Package Food?

AIM

To further student understanding of the food system and interactions between natural and human-designed systems.

SCIENTIFIC PROCESSES

- **question, search, observe, theorize**

OBJECTIVES

Students will be able to:

- **identify packaging as a subsystem of the food system;**

- **discuss parts of the packaging system and how they are connected;**

- **list different kinds of packaging;**

- **describe the purpose of different kinds of packaging.**

OVERVIEW

This lesson has two parts. First, students begin to explore one part of the food system: food packaging. This part of the food system is a system, too, and includes many other parts. Students investigate different kinds of packaging and discover some of the interconnections between human-designed systems, such as factories, and natural systems, such as forests. Challenge students to think about why the food we purchase in markets comes in packages. Encourage them to consider both the advantages and disadvantages of packaging, including its impact on the natural environment. The second part of this lesson sets up a long-term observation that helps students build understandings about why we need to preserve food. Students observe what happens to strawberries, tomatoes, and grapes when they are left out at room temperature for several weeks. They record their observations and gather data that help them with their explorations of food processing in Unit 3.

MATERIALS

For the teacher:
- *Food-Change Experiment* teacher note
- *Investigating Cereal Packaging* lesson resource
- *Teaching Students to be Careful Observers* lesson resource
- *Food Change* experiment sheet
- Chart paper
- Markers
- (Optional) Digital camera

For each group:
- Materials from *Why Package Food? Supply List* lesson resource
- Chart paper
- Markers

For each student:
- *Trade-offs* student reading
- *Exploring Our Food System* student reading (pp. 80–81)
- *Packaging Analysis* activity sheet
- *Food-Change Observations* activity sheet
- *Food-Change Monitoring* activity sheet
- *Food-Change Final Data* activity sheet
- LiFE Log
- (Optional) Hand lenses

PROCEDURE

Before You Begin:

- Purchase the materials listed on the *Why Package Food? Supply List* lesson resource.

- Review the *Food-Change Experiment* teacher note and the experiment sheet.

- Review the *Investigating Cereal Packaging* lesson resource. You may wish to write the questions from the lesson resource on chart paper to post at the front of the classroom.

- Review the *Teaching Students to be Careful Observers* lesson resource.

- Review the *Trade-offs* student reading and the *Food-Change Observations, Food-Change Monitoring, Food-Change Final Data,* and *Packaging Analysis* activity sheets. Make copies for each student. Please note that you do not need to give each student a copy of all of the fruit outlines that are part of the food-change observations activity sheets. Students will be observing one fruit and only need copies of the outline of the fruit they are studying.

- If you have not already done so, post the Module Question and Unit 2 Question at the front of the classroom.

MODULE QUESTION

What is the system that gets food from farm to table, and how does this system affect the environment?

UNIT QUESTION

What is the system that gets food from farm to table?

QUESTIONING

1. Review Food System

If you have not already read the *Exploring Our Food System* student reading from Lesson 5, read it as a class before going on with this lesson. Invite students to discuss new information they learned about food systems. Draw attention to the packaging part of the food system. Ask students if they have any new questions, particularly about food packaging. Record them on chart paper and post it at the front of the classroom.

2. Review Module and Unit Questions

Remind students of the Module Question and that the Unit 2 Question explores the parts of the food system. Tell students that they will use systems thinking to explore in more detail some of the connections between the parts of the packaging system, which is part of the food system.

SEARCHING

3. Investigate Packaging

Hold up the box of cereal. Tell students that with this lesson, they are going to think like product designers. Use the *Investigating Cereal Packaging* lesson resource to guide the class discussion.

4. Analyze Packaging

Have students work in small groups. Distribute the *Packaging Analysis* activity sheet and one of the food items to each group (tomato sauce in a can, fresh tomato, and so forth). Make sure each group has one food item. Tell students that they are going to analyze the food's packaging. Remind them of the class discussion on cereal packaging. Have each group nominate a recorder. The recorder will take notes on the group's analysis on the activity sheet. Walk among the student groups. If students are struggling with the analysis, you may wish to review the ***Investigating Cereal Packaging*** lesson resource with the groups.

5. Present Analysis

Explain that an important part of being a scientist is sharing work with other scientists. Allow each student group about five minutes to present its food-packaging analysis. Have student groups select one person in the group to be the presenter. Post each group's analysis at the front of the room.

Invite the class to ask each presenting group clarifying questions or to debate any of the information that is presented. Once all of the groups have presented, have a class discussion about food packaging. Tie packaging back to the food system. Elicit from students their ideas about packaging as a subsystem of the whole food system. *Do you see any subsystems within the packaging part of the food system?* (Manufacturing; the system to get the materials for packaging; trees; computer systems for designing.) *What interactions do you see at this part of the food system?* (Humans harvesting trees for paper; factory workers making the cans; food producers interacting with package designers.) *What would happen if the packaging system that was designed did not work? What if it failed?* (Products might be ruined; consumers might not buy the products.) Monitor the length and depth of this discussion to gauge how well your students understand systems and interactions.

6. Fruit Observations

Review the ***Food Change*** experiment sheet. While students are analyzing the food packaging, set up the food-change experiment.

Tell students they are going to be food scientists. Explain that through the process of observing

the food, students will begin to learn about food changes and why we need to preserve food. *Why do we preserve food?* (To make it last longer.) *What are some ways that we preserve food?* Have students think back to some of the food packaging they analyzed. *Was any of that food preserved? How was it preserved?* (Canned, chemicals added.)

7. Collect Baseline Data

Explain that this food study involves thoroughly observing what happens to fresh food that is left out at room temperature. Follow the ***Food Change*** experiment sheet to begin the observations. Have students record their observations on their ***Food-Change Observations*** activity sheet. Give students in each food group the recording sheet and the appropriate fruit drawing. When students have completed their observations, have them place the food in the food-observation storage box for later study. Remind students to wash their hands after they handle the fruit. Make sure students have easy access to their observation sheets and the ***Food-Change Monitoring*** and ***Food-Change Final Data*** activity sheets. Students will need them each time they study the food.

8. LiFE Logs

Have students write two paragraphs about the food-packaging system. Ask them to point out two parts within the food-packaging system and interactions between these parts.

9. Assign Homework

Distribute the ***Trade-offs*** student reading. Have students read this before beginning Lesson 7.

Food-Change Experiment

This experiment is a long-term observation of food and what happens to food when it is left out at room temperature. Students collect baseline data and refer back to their observations later as they continue to study the food. The ongoing series of observations helps students develop their own understanding of why we need to preserve foods.

In this lesson you will be letting food decompose. You may find that some students feel this is wasteful. Help students understand by explaining that as food scientists they need to know what makes food decompose in order to learn how to make food last longer. With this knowledge, students will be better able to keep food from decomposing in their own homes in the future. Remind them that they will be applying what they have learned in science class to their daily lives.

As students observe the foods change over the next few weeks, they will learn that when food is left out at room temperature, microorganisms grow on it. From here they will learn that these microorganisms need warmth, air, and water to grow. Thus, in order to reduce the growth of microorganisms, we must take away the conditions they need to grow. In lessons following this experiment, students learn that refrigeration works as a short-term way to preserve food, and canning (taking away air), freezing (taking away warmth), and drying (taking away water) are longer-term ways of preserving food.

Students make their initial observations during this lesson. Midway observations and final observations happen in future lessons (see the Project Planner, p. 25). Depending on the temperature and other conditions in your classroom, the initial physical condition of the fruits, and how quickly you and your class are moving through the module, your students may or may not observe noticeable changes in the fruits when you get to Lesson 8 or 10. If you find that you are moving quickly through the lessons you may wish to extend the experiment and have students make midway and final observations later on. If you are moving through the lessons more slowly and the fruits are decomposing quickly, you may choose to have students make their observations before the time suggested in the lesson plans.

If you and your students have completed the composting activities in the *Growing Food* module of the LiFE Curriculum Series, remind students of what they learned about the decomposition cycle. Challenge them to describe how the decomposition cycle is part of the food system.

Why Package Food? Supply List

INVESTIGATING CEREAL PACKAGING (p. 87)

Supplies

For the class
- 1 box breakfast-cereal flakes

FOOD CHANGE (p. 89)

Supplies

For each group of 3 students
- 1 strawberry
- 1 tomato
- 1 grape
- 3 paper towels
- 1 food-observation storage box (mini clear-plastic aquaria/terraria)*

For the class
- Indoor/outdoor min-max thermometer
- (Optional) Psychrometer kit to measure humidity

* The food-observation storage boxes are available as small insect cages from pet stores and online: *www.lllreptile.com/store/catalog*; *www.amazon.com*. We have had success with mini Kritter Keepers — although in this case, they are used to keep critters out. You will use these boxes to store the fresh food that students are observing in the food-change investigations. Check the National Gardening Association Web site (*www.kidsgardening.org*) for indoor/outdoor thermometers, and Carolina Biological Supply (*www.carolina.com*) if you choose to record the humidity.

PACKAGING ANALYSIS (pp. 92–94)

Supplies

For the class
- 1 can tomato sauce
- 1 tomato from a farmers market or garden (no label, no packaging)
- 1 glass jar tomato sauce/pasta sauce
- 1 plastic produce bag with several tomatoes
- 1 can tomato juice (or vegetable juice with tomato as an ingredient)
- 1 small box of cherry tomatoes
- (Optional) 1 tube of tomato paste

Investigating Cereal Packaging

What does packaging do? While a box of cereal may look like a simple box, it actually was designed to meet specific requirements and to communicate information to a consumer. Use this sheet to guide a class discussion about packaging.

1. *What is the product?* Encourage students to think back to the product that was grown. For example, cereals are made from edible grains. The food may be corn, rice, wheat, or some other grain.

2. *What does the packaging do?* Package designers work with very specific design requirements. The packaging has to be strong enough for the product to be shipped without breaking or crushing. It also has to keep out moisture, dust, and other kinds of contamination.

– It keeps the cereal from getting crushed.

– It keeps the cereal dry.

– It protects the cereal from unwanted consumers (insects and so forth).

– It makes it possible to stack the boxes, which helps stores display the product.

– It makes it easy to ship.

– It makes it convenient for consumers (shoppers).

– It has to be flexible enough so that the box can be opened many times without tearing.

3. *What does the packaging communicate to the consumer?* What the package looks like is very important in terms of sales. The kinds of images and information on the package offer clues to the customer the cereal manufacturer wants to attract. For example, popular cartoon characters will attract young children, while terms such as "organic," "healthy," "low sodium," will appeal to adults. Research suggests that at least two-thirds of the decision to purchase a product is made at the point of sale. It stands to reason that the look of the package influences sales.

– It tells the consumer what is inside.

– It conveys important information to the consumer that might help sell the product (organic, low fat, low sodium, grams of fiber).

– It tells the consumer how many calories there are per serving, how many servings per package.

– It tells the consumer what ingredients were added to the grain during processing.

– It gives a nutritional analysis of the product.

– It might give a serving suggestion, such as a photograph of the cereal in a bowl with milk and fruit added.

– It tells the consumer where the food was distributed and sold, but not where the cereal was grown.

– It tells the consumer the weight.

– It might include a sales pitch that tells the consumer why it is important to eat this kind of cereal.

4. *What materials are used in the packaging?* Cereal boxes have flexible cardboard on the exterior and an inner package that helps keep moisture out.

– If the box has printing on it, the materials also include inks.

– The box and inner packaging are sealed with glue. Both are designed so the consumer can open and close them with ease.

Teaching Students to be Careful Observers

It is very important that your students understand the importance of making detailed, thorough observations of the fruit they are studying. The data they record on the first day that they observe the tomatoes, strawberries, and grapes will help them develop theories about what they find in a few days or weeks when the experiment is over and they make their final observations. Let students know that all scientists, including food scientists, must make thorough observations at the beginning of an experiment so that they can compare the end results to the initial conditions. Since their experiment involves looking at how food changes, they must record all they can about their food on the first day.

Explain that a good way to think about recording observations is to pretend they are describing the food to someone on the phone. It helps to pretend that the food is something that this person has never seen before. This will help students describe every detail that they see, even if it is obvious to them. You might want to have students practice with some other object if they are having trouble getting the idea.

The observations that students make on the first day will serve as their baseline data. By comparing future observations to the baseline data, students will be better able to see and describe any changes. You may want to have students use hand lenses to extend their vision. You may also want to use digital photography as a way to record how the food looks each day students make their observations. If you do use photography, take the photos in a well-lit area so that you can use natural light rather than the flash. The flash might leave a "hot spot" on the image, which will make it hard to see any changes. Tell students to handle the food carefully to minimize damage due to rough handling. Students will be observing the food as it begins to decompose. The characteristics they will observe are:

Color of the Skin. What does it look like? Describe the color and the uniformity or lack of uniformity of the color with particular notes about any areas that are a different color than the rest of the skin. Encourage students to be very accurate as they describe the color because the color of food can change when it begins to decompose.

Smell. What does it smell like? Smell the food and note how it smells. You may want to have the class brainstorm a list of descriptive words that would be appropriate to use with these fruits. The more descriptive and accurate students are at the beginning, the easier it will be for them to assess how the smell changes.

Texture. What does it feel like? Describe the texture — how firm or soft the skin is. Take particular care in noting any areas that are firmer or softer than the rest of the fruit. The texture of food can change as the fruit decomposes.

Damage to the Skin. Carefully examine the skin to note the specific location of any small tears or weaknesses in the skin. Encourage students to note this information on the fruit outline. This is very important because the microorganisms involved in decomposition get into the food where the skin is torn or weak.

Food Change

Over a period of time, students will observe what happens to three different kinds of fresh food when the food is left out at room temperature.

Setup

1. Gather materials.

2. If you have not already done so, make copies of the *Food-Change Observations, Food-Change Monitoring,* and *Food-Change Final Data* activity sheets for each student.

3. Assemble the materials for each group: one food-observation storage box, a set of activity sheets for each group member, and one piece of each of the three types of fresh food. Place a paper towel under each piece of fruit. Throughout the activity each group member will observe the same piece of fruit. Label or number the boxes with group names.

4. Set up the thermometer and, if you choose to use it, the psychrometer.

Procedure

1. Divide the class into groups of three. Have each group member choose one type of fruit to observe throughout the experiment.

2. Review the observation and data-collection sheets with students.

3. Have students carefully observe their food and record their observations. Allow time for group members to share their observations with each other.

4. Once students have completed their observations, have each group carefully place their food into their food-observation storage box.

Questions

1. *Did you notice any damage or marks on the fruit?*

2. *Why will you compare future observations to the observations you make today?*

3. *Do you think the fresh fruit will change over time? What evidence do you have to support your answer?*

Name

Date

Trade-offs

Guiding Questions

• *What are constraints?*

• *How does knowing the constraints help someone design a new product?*

• *What are resources?*

• *What are trade-offs?*

Imagine that you are on a hill, surrounded by wild blueberry bushes. You ate as many berries as you could and now you want to bring some home with you to enjoy later on. But, there's a problem. You do not have a container for the blueberries. What do you do?

You can solve this problem by designing a product to hold the blueberries. First, you think about the **constraints,** or limitations, that you have to work around. For example, the blueberries are round, which means they roll, so you need to design something that takes into consideration the fact that blueberries are round and roll. Blueberries are also soft, so you need to design something that will help prevent the berries from being squished. You also know that you love blueberries and you want to bring home as many as you possibly can. Finally, you need to be home by dinner, just a few hours away, and you need to walk at least two miles before you get home. This means the container needs to be big enough to hold lots of berries, but not so big that you can't carry it. You also need to make the container rather quickly so you have time to pick the berries and walk home before dinnertime.

You look around at all the **resources,** or materials, that you might use. There is a vine growing up a tree. The vine has really big leaves. Maybe you could wrap the blueberries in some leaves and make leaf packages. There is a field of tall grass nearby. You think about pulling up the grass and weaving a basket. There is also a log nearby. You break off a long piece of bark. It's curved, like the shape of the log. It might work as a container, too.

The vine leaves, the grass, and the bark all are possibilities, but none of them is the perfect solution. There are **trade-offs** with each one. The vine leaves are big, but it's hard to reach them because they are way up in the tree. You take a leaf, put some berries in the middle and then try to fold it up. Some of the berries get squished and the packet starts to unfold. It would be much easier if you had a string to tie it. You try using the stem to tie it closed, but it doesn't work very well. The trade-offs are that

(continued on next page)

Trade-offs

although the leaf holds the berries, it does not really protect them from getting squished and it doesn't always stay folded. If you use the leaves, the leaf packages might unfold and you would arrive home with squished berries or no berries.

The grass is light and flexible, so you might be able to make a big basket and carry lots of berries home. However, it will take you at least a day to make a basket big enough to carry more than a handful of berries. This is a problem. You can't wait until tomorrow to go home. You need to be home for dinner tonight.

You look back at the piece of bark. What are the trade-offs here? The bark has sloping sides, but both ends are open. You can stuff the ends with grass or moss, but then you will lose some of the space inside the bark. If you don't stuff them, you might lose some or even most of the berries on the trip home.

It's time to make a decision so you can pick the berries and still be home for dinner. You think about the possible solutions, consider the trade-offs, and make a decision. There are too many **risks** with the vine leaves. The leaf packages might fall apart or the berries might get squished. That would mean no blueberries for later. The grass basket seems perfect, except that you can't make a large enough basket in the time you have. You might not even have enough time to finish a small basket. The risk is too great. With less time than you need to make a basket, you could not bring home any berries.

You choose the bark. It is easier to work with than the vine leaves and it's already almost a container. You just need to block the ends. Should you use the vine leaves or the grass? You choose the grass. The leaves are big, but they are not as soft or easy to handle as the grass. If the leaves tear as you are stuffing the bark, you might lose the berries. You pull some grass and use it to line the inside of the bark, blocking both ends. Then you pick your berries, and head for home, dreaming of blueberry pie.

Name   Date

Packaging Analysis

Look carefully at your food packaging. Think about the food that is stored in the packaging and the person who might buy this kind of food. Think about all the steps between the food being harvested and then being sold in a store.

1. Name of product: _____

2. What is the main ingredient? _____

3. How is the food packaged? *Hint:* Is the food in a metal can? Is it in a plastic bottle? Is it in a glass jar?

_____

4. What does the packaging do? *Hint:* Is it easy to stack the packages? Does the package protect the food from damage or from dust or animals? Is the packaging flexible or is it firm and hard to squeeze? Why do you think it has this kind of packaging?

_____

_____

_____

_____

_____

_____

_____

(continued on next page)

Name   Date

Packaging Analysis

5. What does the packaging communicate? *Hint:* Does the packaging have a label? What kind of information is on the label or on the package? Do you think children or adults would be interested in this kind of information? Are there any colors on the packaging?

_____

_____

_____

_____

_____

_____

_____

6. What materials or natural resources were used to make this packaging? *Hint:* Do you see any paper? Is anything printed on the packaging?

_____

_____

_____

_____

(continued on next page)

Name

Date

Packaging Analysis

7. Do you think this is a successful package design? Does the package help "sell" the food? Why or why not? Would you change anything?

_____

_____

_____

_____

_____

_____

8. How does this packaging protect or preserve the food?

_____

_____

_____

_____

9. Can this packaging be recycled?

_____

_____

_____

_____

Name   Date

Food-Change Observations

Look carefully at your food. Describe the food as clearly and with as much detail as you can. Draw as many details as you like on the food outline. Your observations include both your words and the drawing. The observations you make about the food today, at the beginning of your experiment, are called your **baseline data.** The more details you have in your baseline data, the more you will be able to note whether or not the food changes during the experiment. Handle the food with care so you do not damage it.

Type of Food:_____

Date:_____

Room Temperature:_____

Humidity:_____

   (If your class is not recording the humidity, skip this and go on to the next item.)

Color of the Skin
Describe the color. Use adjectives such as bright, dull, or spotted. Describe any areas where the color changes. Draw any areas where there are color changes on the food outline.

_____

_____

_____

Smell
Describe the smell. Try to use as many adjectives as you can. Some examples are: flowery, musty, moldy, fragrant, sweet.

_____

_____

_____

(continued on next page)

Name

Date

Food-Change Observations

Texture

Describe how it feels. Use words like firm or soft, smooth, rough. If there are any soft areas, use a ruler to measure the area.

_____

_____

_____

_____

Damage to the Skin

Study the skin carefully. Look for any spots where there are tears or bruises. Use a ruler to measure them. Draw any damaged spots on the food outline.

_____

_____

_____

_____

Prediction

What do you think will happen to the fresh fruit? What evidence do you have to support your answer?

_____

_____

_____

_____

(continued on next page)

Name   Date

Food-Change Observations

Name   Date

Food-Change Observations

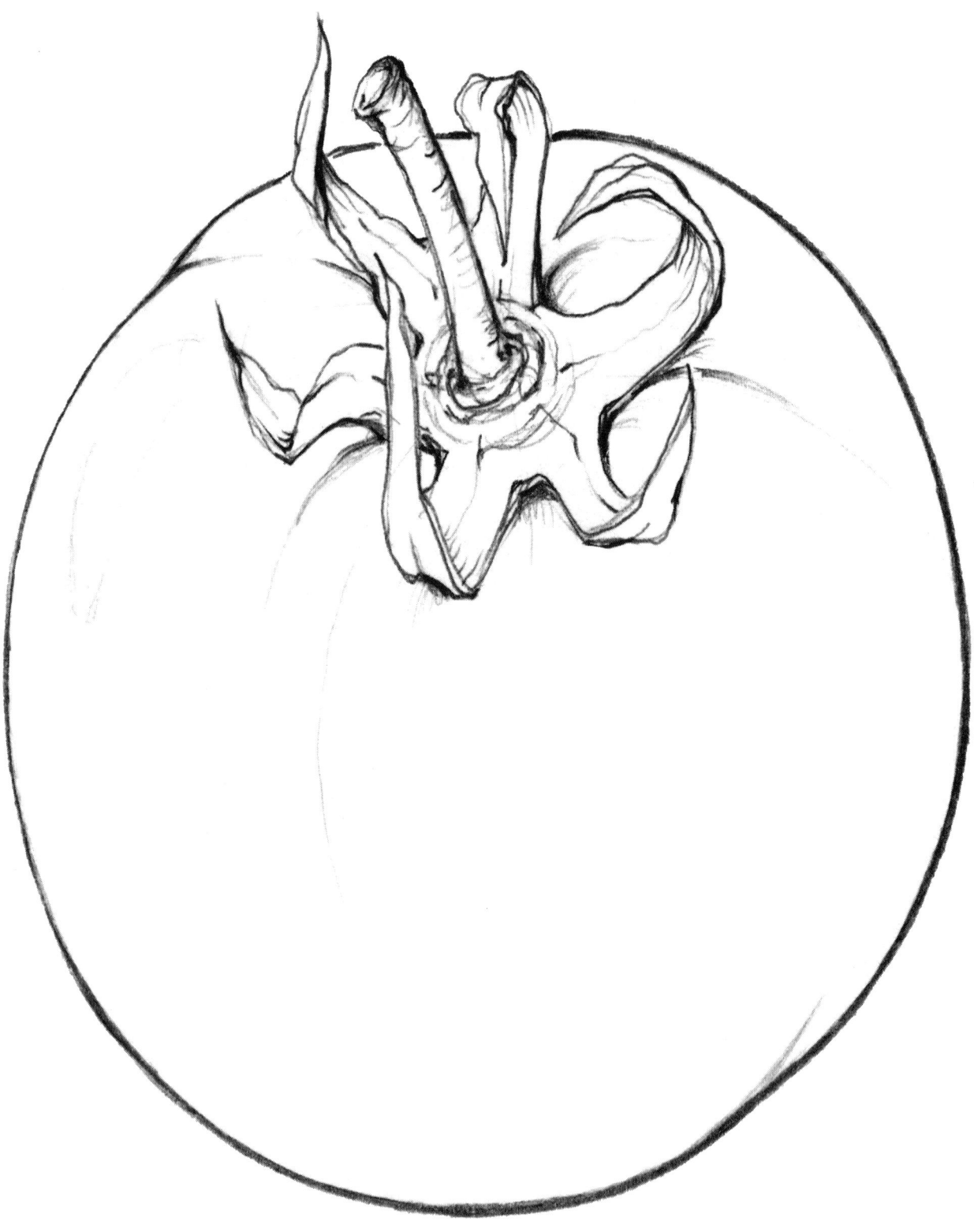

Farm to Table & Beyond
©2008 Teachers College Columbia University

Name   Date

Food-Change Observations

Name   Date

Food-Change Monitoring

Look carefully at your food. Describe the food as clearly and with as much detail as you can. Draw as many details as you like on the food outline. Your observations include both your words and the drawing.

Type of Food: _____

Date: _____

Room Temperature: _____

Humidity: _____

(If your class is not recording the humidity, skip this and go on to the next item.)

Color of the Skin

Describe the color. Use adjectives such as bright, dull, or spotted. Describe any areas where the color changes. Draw any areas where there are color changes on the food outline.

_____

_____

_____

_____

Smell

Describe the smell. Try to use as many adjectives as you can. Some examples are: flowery, musty, moldy, fragrant, sweet.

_____

_____

_____

_____

(continued on next page)

Name

Date

Food-Change Monitoring

Texture

Describe how it feels. Use words like firm or soft, smooth, rough. If there are any soft areas, use a ruler to measure the area.

_____

_____

_____

_____

Damage to the Skin

Study the skin carefully. Look for any spots where there are tears or bruises. Use a ruler to measure them. Draw any damaged spots on the food outline.

_____

_____

_____

_____

(continued on next page)

Name   Date

Food-Change Monitoring

Name

Date

Food-Change Monitoring

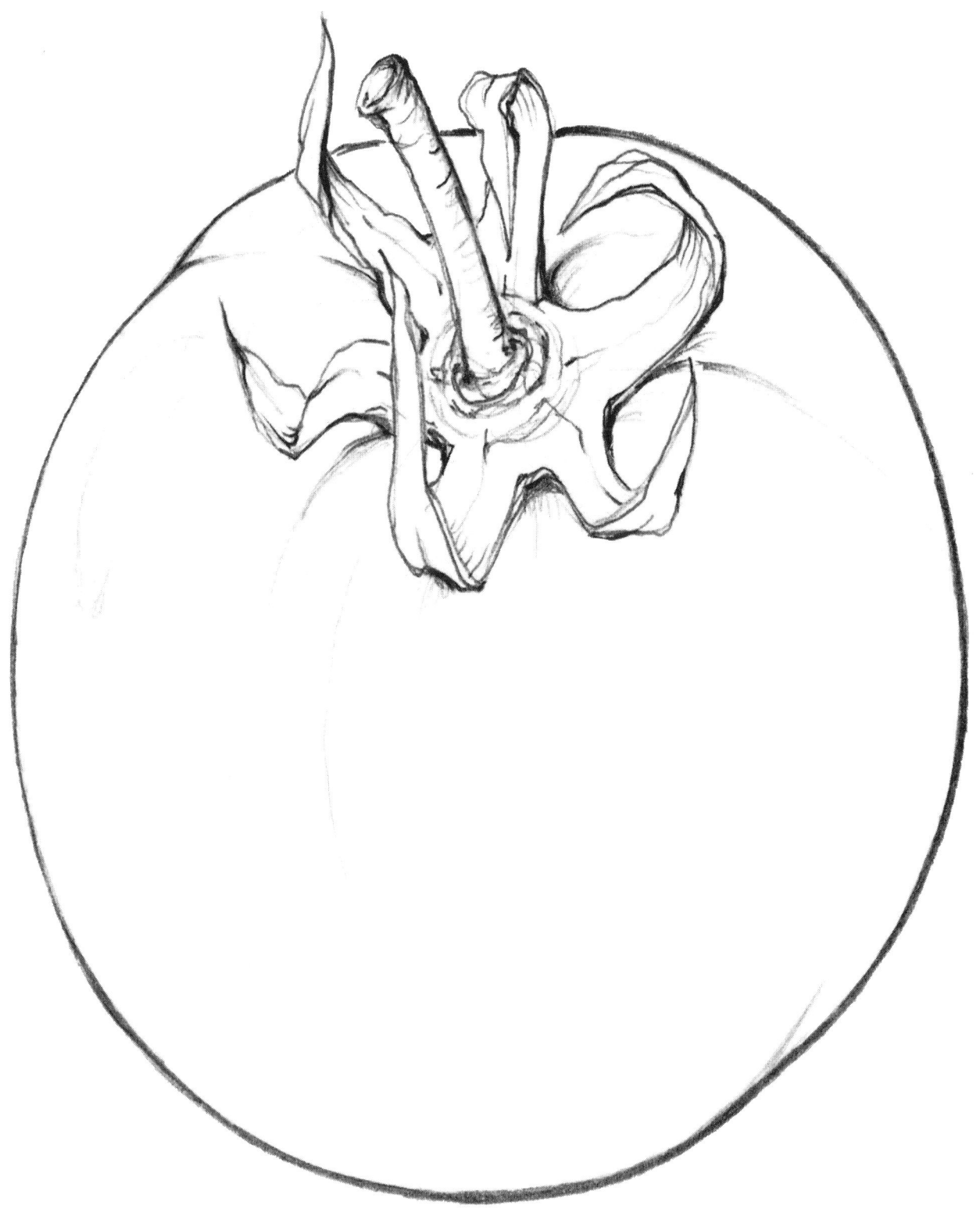

Name   Date

Food-Change Monitoring

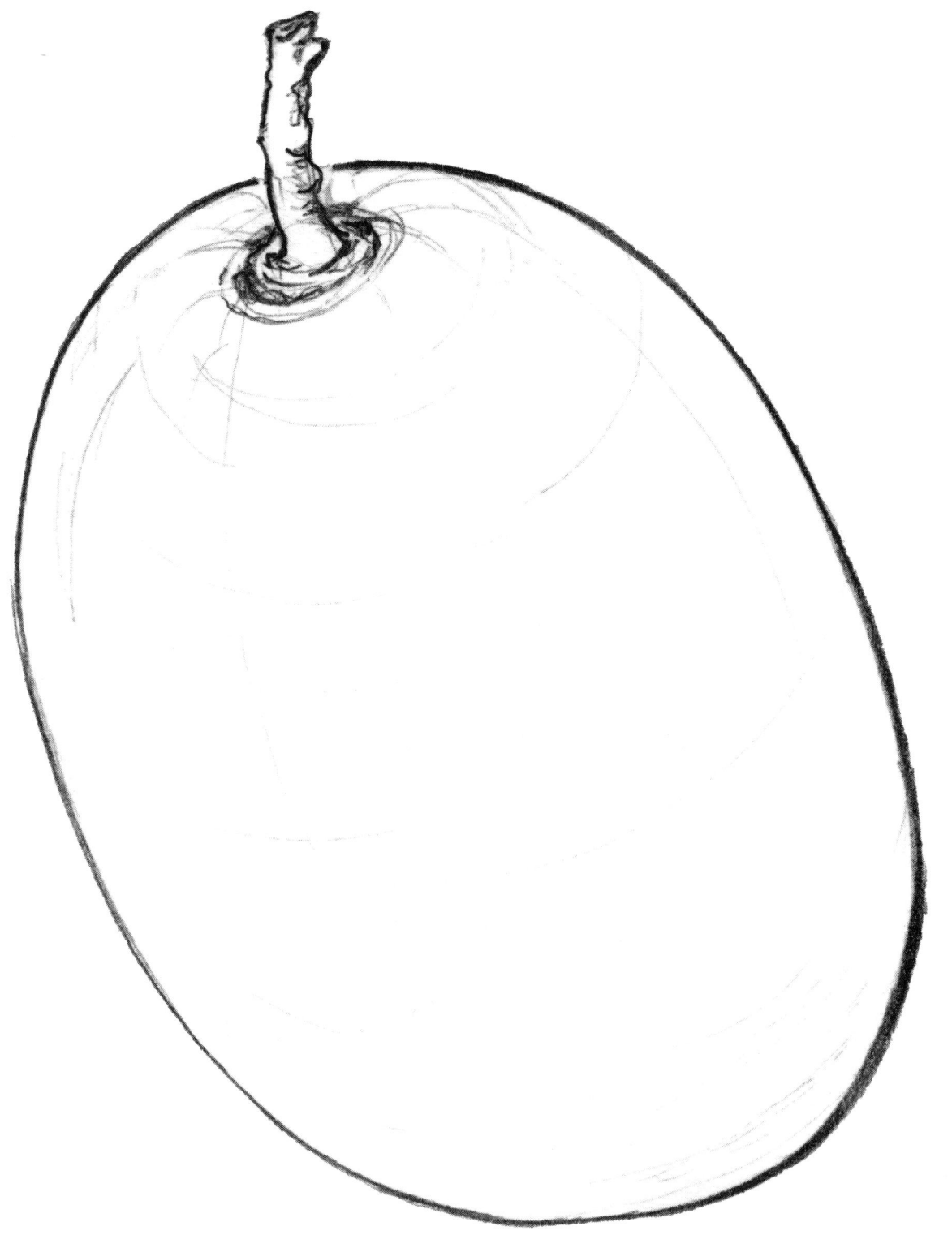

Name	Date

Food-Change Final Data

Look carefully at your food. Describe the food as clearly and with as much detail as you can. Draw as many details as you like on the food outline. Your observations include both your words and the drawing.

Type of Food: _____

Date: _____

Room Temperature: _____

Humidity: _____

  (If your class is not recording the humidity, skip this and go on to the next item.)

Color of the Skin
Describe the color. Use adjectives such as bright, dull, or spotted. Describe any areas where the color changes. Draw any areas where there are color changes on the food outline.

_____

_____

_____

_____

Smell
Describe the smell. Try to use as many adjectives as you can. Some examples are: flowery, musty, moldy, fragrant, sweet.

_____

_____

_____

_____

(continued on next page)

Name   Date

Food-Change Final Data

Texture

Describe how it feels. Use words like firm or soft, smooth, rough. If there are any soft areas, use a ruler to measure the area.

_____

_____

_____

_____

Damage to the Skin

Study the skin carefully. Look for any spots where there are tears or bruises. Use a ruler to measure them. Draw any damaged spots on the food outline.

_____

_____

_____

_____

Conclusion

Look back at your prediction. Compare your prediction to what you observed today. What have you learned? What evidence do you have to support your conclusion?

_____

_____

_____

_____

(continued on next page)

Farm to Table & Beyond
©2008 Teachers College Columbia University

BECOMING FOOD SCIENTISTS : INTERACTING PARTS : FOOD PROCESSING : ENVIRONMENTAL EFFECTS : WASTE : MAKING CHOICES

Name   Date

Food-Change Final Data

Name   Date

Food-Change Final Data

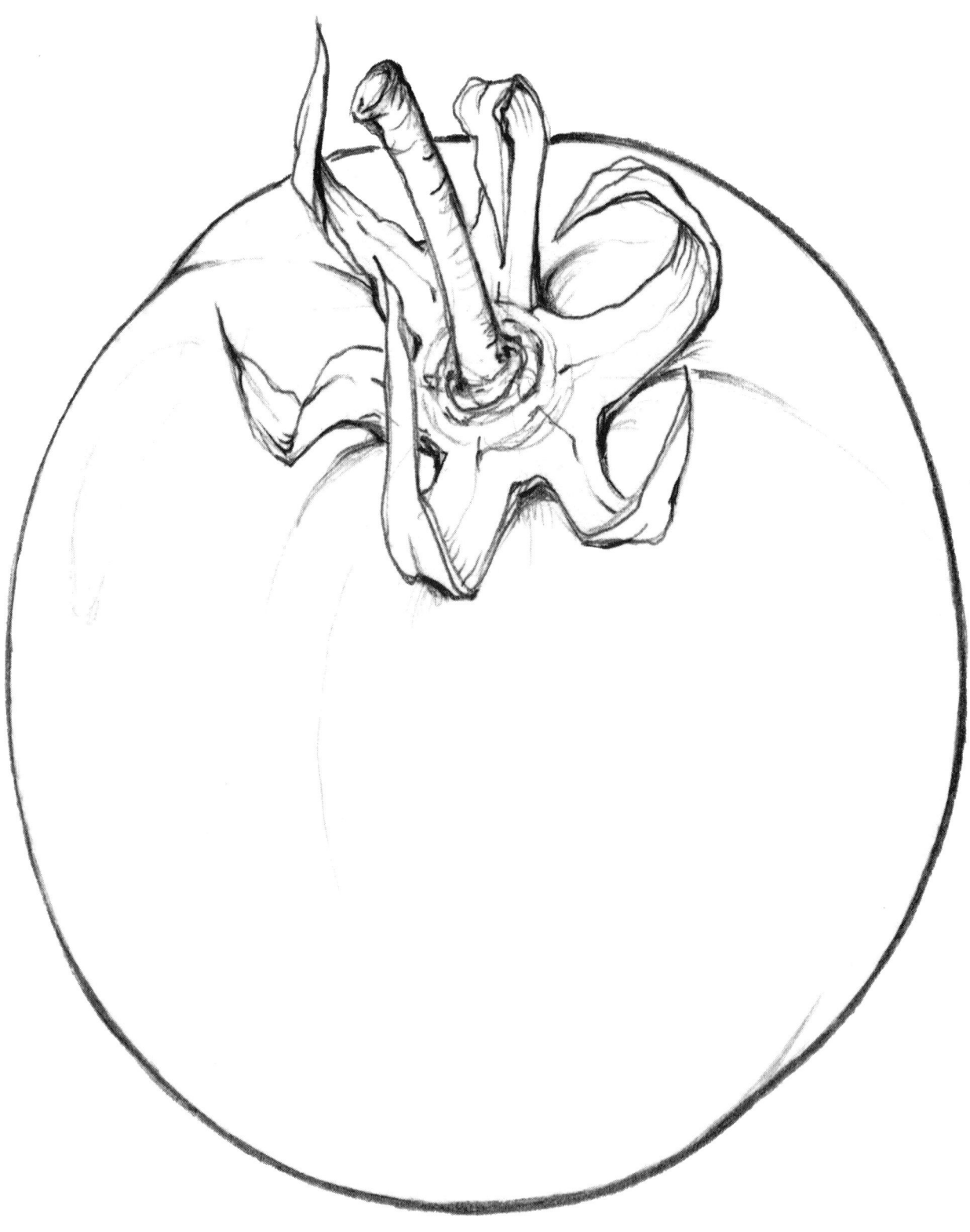

Name

Date

Food-Change Final Data

Materials and Manufacturing

AIM

To learn about food package–manufacturing systems and to understand that raw materials used in the process come from the natural environment.

SCIENTIFIC PROCESSES

- **question, search, theorize, experiment**

OBJECTIVES

Students will be able to:

- **discuss failure in packaging design;**
- **identify natural resources used in food packaging;**
- **describe how different kinds of food packages are made.**

OVERVIEW

In this lesson, students continue to analyze packaging. First, they create a time line of food packaging history. Next, the class tests the strength of different kinds of materials. Then students experiment with these materials and test their performance as a snack wrap. Next, student teams begin the research phase of a long-term design project. In this project, student groups research different kinds of materials used in packaging. They complete the project in Unit 4. In doing their research, students investigate the use of natural resources in packaging. The goal is to increase student awareness of the complexities of the interactions between human-designed and natural systems. Students research a manufacturing process, including the raw materials that are used. They consider the impact on the natural environment during manufacturing, from harvesting or mining to the use of water. The *Food Packaging Research Resources* lesson resource provides links to Web sites to guide their research.

MATERIALS

For the teacher:
- *Product Research Checklist* lesson resource
- *Food Packaging Research Resources* lesson resource
- *Milestones in Packaging History* lesson resource
- *Testing Properties and Performance* experiment sheet
- (Optional) Large index cards or sentence strips
- (Optional) Clothesline
- (Optional) Clothespins or paper clips

For the class:
- Materials from the *Materials and Manufacturing Supply List* lesson resource

- Chart paper
- Markers

For each pair of students:
- *Which Wrap?* activity sheet
- Hand lens
- Ruler
- Scissors

For each student:
- *Trade-offs* student reading (pp. 90–91)
- *Food Packaging Discoveries* student reading
- *Sand to Glass, Plant to Paper, Petroleum to Plastic,* and *Ore to Aluminum Can* activity sheet
- LiFE Log

PROCEDURE

Before You Begin:

- Review the *Testing Properties and Performance* experiment sheet. Gather the materials listed on the *Materials and Manufacturing Supply List* lesson resource.

- Review the Project Planner (p. 25), the *Farm to Table & Beyond Expo* information (p. 27), and the *Product Research Checklist* and *Food Packaging Research Resources* lesson resources. You may wish to have students create portfolios of their work to use at the expo in Unit 6. Consider transferring some of the tips to chart paper and post them at the front of the classroom.

- Copy and review the *Milestones in Packaging History* lesson resource. Cut out the milestones. You may wish to glue them onto index cards so they can be rearranged as students gather more information. Or, copy the information onto sentence strips. Keep the time line to use with the transportation lesson. This will help students make connections between packaging and transportation. Make space on a bulletin board for the time line or stretch a long line, such as a clothesline, along one side of the classroom. If you use a clothes-line, use clothespins or paper clips to attach the events. You will be adding more in future lessons, including dates in the history of transportation, food processing, energy, and garbage.

- Remind students to bring in *Trade-offs,* their homework reading from Lesson 6. If they have not read it, have them read it before beginning this lesson.

- Make copies of the *Food Packaging Discoveries* student reading for each student and the *Which Wrap?* activity sheet for each pair of students.

- Make copies of the *Sand to Glass, Plant to Paper, Petroleum to Plastic,* and *Ore to Aluminum Can* activity sheets for students on each product-research team.

- If you have not already done so, post the Module Question and Unit 2 Question at the front of the classroom.

MODULE QUESTION

What is the system that gets food from farm to table, and how does this system affect the environment?

UNIT QUESTION

What is the system that gets food from farm to table?

 **SEARCHING**

1. Review Module and Unit Questions

Remind students of the Module Question and the Unit Question. Explain that they are going to continue to investigate food packaging. Tell them that they are going to make a time line of a brief history of food packaging and begin to focus on materials. Students will add to the time line in future lessons and will begin to see some more connections between different parts of the food system.

2. Review Student Reading about Trade-offs

If you have not already read *Trade-offs,* read it as a class before going on with this lesson. Invite students to discuss what they have learned about constraints and trade-offs. Brainstorm a quick list of constraints related to food packaging (cost, materials, weight).

3. Introduce Materials Science

Have students help you create the time line of food-packaging and materials history. Use the *Milestones in Packaging History* lesson resource to start the time line. Stretch a piece of chart paper across the bulletin board if you not using a clothesline. Begin about 10,000 B.C. Label this "beginning of agriculture." Have students write the dates and events on the chart paper, or hang the index cards along the time line. Make sure you leave enough room for students to add more events as they come across them in their research. Please note that in addition to any events that students add from their research, there are milestone lesson resources for transportation, food processing, energy, and garbage in future lessons.

Distribute the **Food Packaging Discoveries** student reading. Have students take turns reading the story out loud. Discuss some of the possible trade-offs in using clay urns as food containers. *Do you think it would be possible to store fish in paper containers on a ship like this? What do you think would happen?* (The paper would get wet and disintegrate.) *Do you think clay or metal would make a better food container for a ship?* (Metal would rust, get hot in the sun.) *Can you think of any limitations to using clay?* (Weight; it might break.) *Why do we make new kinds of materials and products?* (New problems to solve; new understanding of materials; new materials.)

4. Testing Properties

Display the waxed paper, plastic wrap, aluminum foil, and paper towels. Refer to the **Testing Properties and Performance** experiment sheet. *Do you think that any of these materials is a perfect design?* (There is no perfect design. There are the best designs for different problems.) *Do you think there were trade-offs when designers created these materials?*

EXPERIMENTING

5. Product Testing

Have students work in pairs. Introduce the experiment. Follow the procedure outlined on the **Which Wrap?** activity sheet. Have students record their observations on the **Which Wrap?** activity sheet. Are any of the materials opaque?

Are any transparent? Is there a difference between transparent and translucent? Are there any translucent materials? Which material is the easiest to squish? Did any of the materials tear when you opened them up again and flattened them out? Were you surprised by any of your observations?

6. Performance Testing

You may wish to do the performance testing as a whole-class activity. Explain to students that they will be considering different conditions that materials need to meet to keep a product from failing — breaking or not performing its specified function. In this case, they are testing a snack wrapper. Discuss the design problem. Students may be familiar with systems failure in the context of computers and phone systems. However, the idea of a material failing may seem odd to them. Use this as an opportunity to check for their understanding of systems. Make sure all students understand the assignment. If you do this as a whole class, divide the class into small groups and have each group design one performance test. Have representatives from each group help you with their performance test. Have the rest of the class record their observations.

THEORIZING

7. Build Theories about Materials

Ask students to think about what they have learned about the four different kinds of materials. Encourage them to think about the questions they were trying to answer about which material does the best job of meeting the snack-wrapper requirements. *Which material meets the most design requirements? What is your evidence?*

APPLYING TO LIFE

8. Product Recommendation

Have students take out their LiFE Logs and write two paragraphs that describe the snack-wrap material that they recommend and why.

Have them explain the tests that they used to determine which material to use. Remind them to discuss any design requirements their recommended material doesn't meet and how they plan to solve that problem.

9. Discuss Research Project

Divide the class into four project teams. Assign each team a product: paper, plastic, glass, and aluminum. Distribute the appropriate activity sheet: *Sand to Glass, Plant to Paper, Petroleum to Plastic,* or *Ore to Aluminum Can*. Review the guiding questions. Elicit the research questions from students. Tell students they will be working on this research project for several weeks. Remind them that communication is an important part of science. In Lesson 22, each project team will present a poster to the class that describes what they have learned about their product. Introduce to students the idea of putting on a *Farm to Table & Beyond* Expo in Unit 6. Begin a planner and checklist for the expo. Plan to include this project as part of the expo. Have students bring in sample products made from their materials.

Tell students that they will be doing Internet searches. Meet with each group and distribute the suggested Web resources from the *Food Packaging Research Resources* or bookmark these Web sites on the computer. Give students time to look over the activity sheet and briefly discuss their project with their team members. Have each team elect a team leader to guide the project and keep track of the details and deadlines. Have teams identify tasks and make assignments. Tell them to begin the project as homework.

Product Research Checklist

Getting Started

1. Begin preliminary research to become acquainted with the subject.

2. Teams divide general topic into subtopics, such as:
    a. Description of the raw material;
    b. Description of the mining, drilling, or harvesting process;
    c. Description of the manufacturing process (sand to glass, etc.);
    d. Description of uses of the finished product.

3. Make a calendar with key dates noted, such as:
    a. Note-taking complete;
    b. Preliminary draft due;
    c. Revised draft due;
    d. Final copy due;
    e. Class presentation.

Information Gathering and Note-Taking

1. Make sure students understand how to appropriately cite reference materials.

2. If the reference is from the Internet, have students cite the complete URL of the particular page, not just the site. Include the date the page was accessed.

3. If students get help from experts, such as contacting a paper manufacturer, remind them to give the expert's full name, title, and the company's name.

Research Tips to Help Guide Students

Here are some tips to help students get started with their research. The example below can be modified for each of the different packaging products: glass, plastic, aluminum, and paper.

What do I want to know about my topic? Example: Sand to Glass
    a. Where does the sand come from that is used to make glass?
    b. Who gets the sand?
    c. How do they get the sand? Is machinery used? What kind of machinery?
    d. Is the environment affected in any way?
    e. Are other resources used in the process of getting the sand? (Water, oil, fuel.)
    f. What happens after the sand is mined?
    g. How does the sand get to the factory to make the glass?
    h. What energy sources are used to make glass?
    i. How is the sand made into glass?
    j. After the sand is heated, is it cooled? IIow is it cooled?
    k. What resources are used in packing the final product to be shipped to a food processor?
    l. How is the finished product shipped?
    m. Can the finished product be recycled? How is it recycled?

Food Packaging Research Resources

These Web sites are a great place for students to begin their project research. Consider discussing this research project with your students' language arts or library media teacher. You may find that this assignment complements work that students are already doing and reinforces skills they are learning. The graphic organizer at the end of this resource will help students compile their notes. Also check the Linking Food and the Environment Web site (*www.tc.edu/life*) for additional online resources that can be viewed, printed to read offline, or emailed as links to students.

Aluminum and Other Cans

• **Beverage cans** — *www.madehow.com/Volume-2/Aluminum-Beverage-Can.html*

This Web site shows technical aspects of aluminum-can manufacturing. The discussion covers the background of aluminum, locations where raw aluminum is found, and the manufacturing process. You and your class will read about cutting the blank, redrawing the cop, trimming the ears, cleaning and decorating, making the lid, and filling and seaming. The final sections cover byproducts, waste, and the future of aluminum cans.

• **A step before cans** — *www.alcoa.com/rigid_packaging/en/about/making_cans.asp*

If your students wonder where the can-makers get the materials to make cans, this site will show them, through a diagram and written explanation, how Alcoa converts primary metals into large rolls of thin aluminum, called ingots, that will eventually become beverage cans.

• **Metal beverage cans** — *www.ball.com/page.jsp?page=15*

Use this site from Ball Corporation to illustrate the step-by-step process of creating beverage cans from large rolls of aluminum. The interactive illustration allows you to click through each stage of the manufacturing as it zeroes in on one task and offers a brief written description. A poster-sized PDF encompassing the whole process is also available here.

Paper

• **The papermaking process** — *www.madehow.com/Volume-2/Paper.html*

Is it the Chinese or the Egyptians who show the earliest evidence of papermaking? Find out at this site, which elaborates not only the history of paper, but also modern paper-manufacturing methods. It details the raw materials used and the several steps of the process, including making pulp, beating, turning pulp into paper, and finishing. The description ends with a discussion of environmental concerns around making paper.

• **Papermaking** — *www.ontariosciencecentre.ca/scizone/e3/paper/default.asp*

This Web site from the Ontario Science Centre shows photographs of a papermaking exhibit, and includes sections on the history of papermaking, how paper is made, and instructions for making paper at home.

- **The papermaking process in color** — *www.individual.utoronto.ca/abdel_rahman/paper/fpmp.html*

A detailed collage will lead you through the papermaking process on this site. Each stage pictured in the illustration has a title, which appears below and clicks through to a description of what takes place. Included in the manufacturing loop is the use of recycled-paper products.

- **Papermaking, illustrated** — *www.wipapercouncil.org/process.htm*

This kid-friendly site from the Wisconsin Paper Council takes viewers on a tour of the papermaking process in five stages. Each illustration shows what the process accomplishes, with a written description alongside.

Plastic

- **Plastic beverage containers** — *www.ball.com/page.jsp?page=21*

The Ball Web site illustrates the step-by-step process of making plastic beverage containers using polyethylene terephthalate (PET). Users can click through each step of the process, viewing an illustration of the machinery used to make the containers and a brief explanation of the purpose of the equipment.

- **Plastic soda bottles** — *www.madehow.com/Volume-1/Soda-Bottle.html*

Plastic soda bottles made from PET are the focus of this page. It gives a brief background, then describes, to the molecular detail, the raw materials used to make PET. It guides readers through the polymerization process during manufacturing, then moves on to bottle-making, and finishes with quality control.

- **How plastic is made** — *www.plasticsresource.com/s_plasticsresource/ sec.asp?TRACKID=&CID=126&DID=228*

This site from the American Chemistry Council defines plastic and explains its raw materials and manufacturing in everyday language. It covers topics such as the basics of manufacturing, the structure of polymers, additives, and the uses of plastics in an understandable manner. At the end is a list of products made from different types of plastic.

Glass

- **Glassmaking** — *www.sgcontainers.com/index.nsf/vwNV4/ B1F3BC5F6D2109C585256D1C00175356?OpenDocument*

The glassmaking process is shown in seven steps on this Web site. Each step includes a diagram or illustration with a written description of the process, the machinery used, and the purpose of the action in everyday language. The steps are raw materials, glass melting, glass conditioning, forming, annealing and coating, inspection, and packaging.

- **Glass-bottle manufacturing** — *www.tynant.com/glass.htm*

This site contains diagrams and photographs of the steps used in glassmaking with brief explanations. It also describes the process of determining the color of a glass bottle.

Research Graphic Organizer

Retrieving information from the complex computer-based information environment can be challenging for students. Have students use this graphic organizer to help guide them through their online research.

SEARCHING   **FOOD PACKAGING RESEARCH**

Web Page Title: _____

Web Address (url): _____

Author's Name (If Known): _____

Date of Last Update, if Available: _____

Date You Last Viewed the Page: _____

Notes (write these in your own words):

_____

_____

_____

_____

_____

Quotes:

_____

_____

_____

_____

_____

BECOMING FOOD SCIENTISTS : INTERACTING PARTS : FOOD PROCESSING : ENVIRONMENTAL EFFECTS : WASTE : MAKING CHOICES

Milestones in Packaging History

3000 B.C.	Glass is used to coat pottery
1500 B.C.	Glass is used for containers in Egypt and Mesopotamia
300 B.C.	Glassblowing begins
1ST CENTURY B.C.	Chinese use paper made from mulberry bark to wrap foods
1200 A.D.	In Bohemia, tin is used to coat metal
1310 A.D.	Papermaking techniques arrive in Great Britain
1690 A.D.	Papermaking arrives in America (Pennsylvania)
1809 A.D.	Napoleon offers a reward to anyone who can feed his army — invention of food preserved in tin cans
1817 A.D.	First commercial cardboard boxes are made, in Great Britain
1844 A.D.	First paper bags made, in Great Britain — paper made from rags
1862 A.D.	In Great Britain, Alexander Parkes invents the first man-made plastic
1866 A.D.	First can opener: before this time, tin cans were opened with a hammer and a chisel
1867 A.D.	Paper is made from wood pulp
1870s A.D.	Glue is used in making paper bags
1872 A.D.	Smith Brothers introduces packaged cough drops
1895 A.D.	Automatic glass blowing machine invented that can make multiple bottles at one time, patented by Michael Joseph Owens, in Ohio
1900 A.D.	Waxed paper is used to wrap sandwiches
1933 A.D.	Cellophane manufactured. Originally used by the military; it is not used in homes until the 1950s.
1947 A.D.	Reynolds Metals introduces Reynolds Wrap (aluminum foil), which changes the way Americans cook and store food
1950s A.D.	Germans make plastic foam that is used for foam cups, trays, and boxes for the food industry
1970s A.D.	Plastics begin to replace paper

Materials and Manufacturing Supply List

TESTING PROPERTIES AND PERFORMANCE (p. 120)

Supplies

For the class
- Dishpan
- Box of saltine crackers
- Box of waxed paper
- Box of plastic wrap
- Box of aluminum foil
- Roll of paper towels
- Water

Testing Properties and Performance

In this experiment, students test different kinds of materials. They compare the properties of waxed paper, plastic, paper towels, and aluminum foil. Next, they experiment with these materials and test their performance as food packaging.

Setup

1. Gather materials.

2. Fill a dishpan about half full of water.

3. Distribute the *Which Wrap?* activity sheet, a ruler, a hand lens, and scissors to each pair of students.

Procedure

1. Have students work in pairs.

2. Give each pair several crackers and a sample of each of the four materials: waxed paper, plastic wrap, aluminum foil, and paper towels.

3. Review the *Which Wrap?* activity sheet with students.

4. Guide students through steps 1 and 2 on the activity sheet. Ask student volunteers to share their observations with the class.

5. Have students complete the tests described in step 3, except the test for Condition #2, on their own. We suggest teacher supervision for the absorbency tests.

6. Remind students to record their observations. You may wish to walk among the student groups to check for understanding.

7. After students have completed their observations, invite them to share their findings with the rest of the class.

Questions

1. *Look closely at each of the materials. What do you observe? Describe the color, weight, and texture of the materials. What else do you observe?*

2. *Which material is easiest to squish?*

3. *Can you stretch any of the materials?*

4. *Do any of the materials tear easily?*

5. *Which material is the most absorbent?*

6. *Based on your observations and your tests, which material do you think makes the best snack wrapper? Explain your answer.*

Name   Date

Food Packaging Discoveries

In 2002, deep beneath the waters of the Black Sea, researchers discovered evidence of food packaging that had survived more than 2,400 years. Scientists in a submersible found a ship filled with ancient clay storage jars called **amphorae.** Ancient Greeks used these clay jars to transport their favorite food from one seaport to another. Scientists believe the ship was traveling from a Black Sea colony to the Greek mainland.

Dr. Robert Ballard, a scientist with the Institute for Exploration at Mystic Aquarium, in Mystic, Connecticut, led the research expedition. The shipwreck is in about 275 feet (84 meters) of water. Scientists recovered one amphora, which is about three feet (one meter) tall by one and a half feet (half a meter) wide. Researchers analyzed the contents of the jar and found that it contained the bones of a freshwater catfish that had been cut up into steaks, dried, and stored in the jar. Catfish was a popular food in Greece. Scientists who studied the jar said that it looks like the style of amphorae made in Sinop, Turkey, which was a Greek settlement in the fourth century B.C.

(continued on next page)

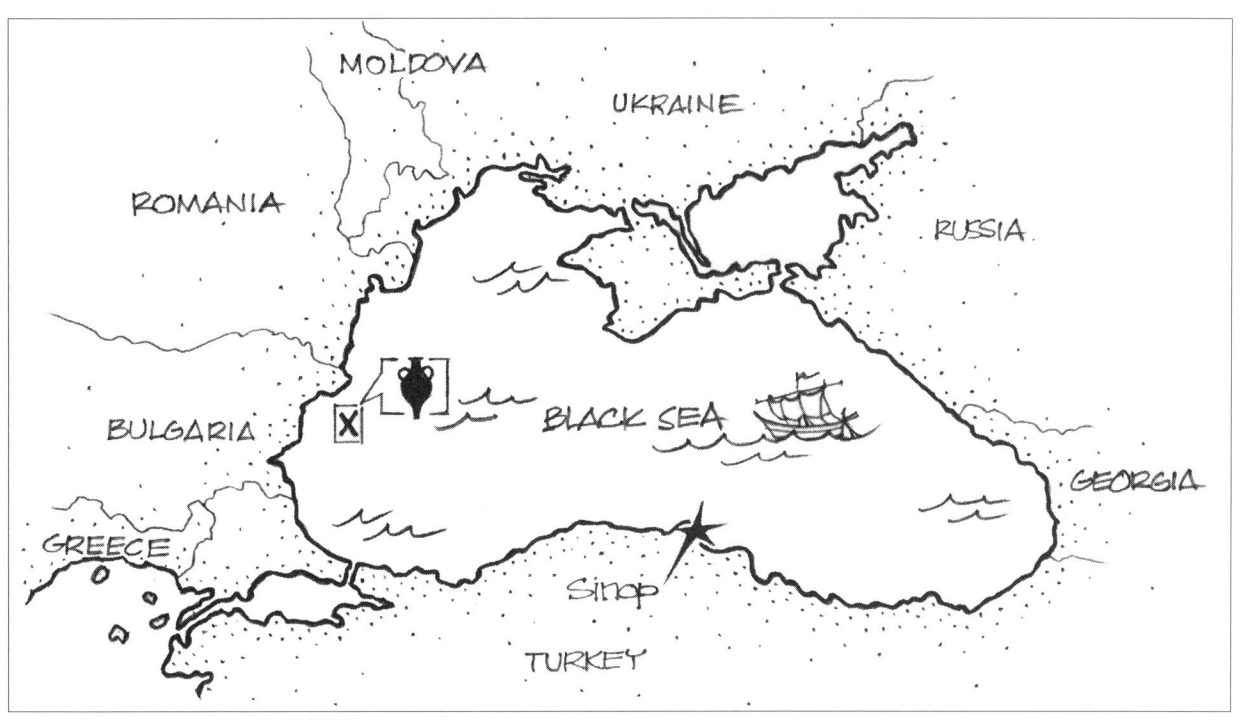

Name   Date

Food Packaging Discoveries

Does it seem odd that the ancient Greeks had food packaging? Not if you think about the purpose of food packaging. The Greeks wanted to protect the food they got in one place, transport it to another, and store it. For as long as there has been a surplus of food, there has been some kind of food packaging. The earliest forms of storage might have been leaves that were wrapped around food.

Eventually, food-storage containers included baskets, animal skins, gourds, pottery, glass, and metal containers. These containers all have something in common. They are made from materials found in nature. You may wonder: If there have been food-storage containers for thousands of years, why do we need new ones, or why change them? Maybe someone discovered a new material. Maybe someone found a way to make a product that is lighter or stronger or will last longer. Think about how lucky you are that you don't have to carry your lunch to school in a huge clay pot.

What makes a product designer choose some materials and not others? Why are there paper bags but not glass bags? Why not gold beverage cans? It's all in the properties of materials and how materials interact with other materi-als. For example, it's easy to bend a sheet of paper, but if you want to store a liquid, would you fold a piece of paper and pour in the liquid? Of course not. However, if you coat the paper with something, like wax, you can make it "liquid-proof."

In this lesson, you are going to learn about some of the ways materials are changed to make something new. You will be thinking like materials scientists, people who study the "stuff" things are made of. You will be learning about the properties of materials used in food packaging, how processing changes them, and how well the materials do their job. You will be learning about materials and manufacturing.

Farm to Table & Beyond
©2008 Teachers College Columbia University

Name	Date

Which Wrap?

1. Carefully examine each of the four pieces of material. Record your observations in your LiFE Log.

    a. Describe the color, the weight, and the texture.

    b. Are any transparent? Translucent? Opaque?

    c. Examine them with your hand lens. Describe what you see.

    d. Squish each one into a tiny ball. Then open up the material and flatten it out again. Describe how easy or hard it was to squish the material. What happened when you opened it up again and tried to flatten it out?

2. With your teacher's help, test each material to see which one is the most absorbent. Record your tests below.

Waxed Paper

_____

_____

_____

_____

Plastic Wrap

_____

_____

_____

_____

Aluminum Foil

_____

_____

_____

_____

Paper Towel

_____

_____

_____

_____

(continued on next page)

Farm to Table & Beyond
©2008 Teachers College Columbia University

BECOMING FOOD SCIENTISTS : INTERACTING PARTS : FOOD PROCESSING : ENVIRONMENTAL EFFECTS : WASTE : MAKING CHOICES

Name	Date

Which Wrap?

3. You have been hired to recommend a material for a company to use to wrap a cracker snack. The company wants a product that can protect the snack from dust and insect pests. It also wants people to be able to identify the snack without opening the wrapper. The company thinks that people might want to carry this snack in a backpack, so they want the wrapper to protect the cracker from getting crushed by other things in the backpack. They also think it would be good to make the wrapper as watertight as possible, just in case the person gets caught in the rain without an umbrella and the backpack gets wet. You take the job. The company gives you sample crackers.

Now you need to design the tests. You need to test different materials under four conditions. The four conditions are listed below. Review each one, then think about how you can test the materials to find out which does the best job of protecting the crackers under the different conditions. Describe your tests. Keep careful records so you can review your results and make your recommendation to the cracker company. Review your experiments with your teacher before you begin.

Condition #1: Keep out dust and insects.

Describe your test:

_____

_____

_____

_____

_____

_____

(continued on next page)

Name	Date

Which Wrap?

Once your test is complete, record your results for each wrapper.

Waxed Paper   **Plastic Wrap**

_____   _____

_____   _____

_____   _____

_____   _____

Aluminum Foil   **Paper Towel**

_____   _____

_____   _____

_____   _____

Condition #2: Keep crackers from getting crushed.

Describe your test:

_____

_____

_____

_____

_____

_____

(continued on next page)

Name   Date

Which Wrap?

Once your test is complete, record your results for each wrapper:

Waxed Paper   **Plastic Wrap**

_____   _____

_____   _____

_____   _____

_____   _____

Aluminum Foil   **Paper Towel**

_____   _____

_____   _____

_____   _____

_____   _____

Condition #3: See the cracker without opening the package.

Describe your test:

_____

_____

_____

_____

_____

_____

(continued on next page)

BECOMING FOOD SCIENTISTS : INTERACTING PARTS : FOOD PROCESSING : ENVIRONMENTAL EFFECTS : WASTE : MAKING CHOICES

Name   Date

Which Wrap?

Once your test is complete, record your results for each wrapper.

Waxed Paper   **Plastic Wrap**

_____   _____

_____   _____

_____   _____

_____   _____

Aluminum Foil   **Paper Towel**

_____   _____

_____   _____

_____   _____

_____   _____

Condition #4: Keep the crackers dry.

Describe your test:

_____

_____

_____

_____

_____

(continued on next page)

Name   Date

Which Wrap?

Once your test is complete, record your results for each wrapper.

Waxed Paper   **Plastic Wrap**

_____   _____

_____   _____

_____   _____

_____   _____

Aluminum Foil   **Paper Towel**

_____   _____

_____   _____

_____   _____

_____   _____

When you complete the experiments, answer the following questions.

1. Which material meets the most design requirements?

_____

2. Are there any conditions that the material you worked with does not meet?

_____

_____

_____

(continued on next page)

LESSON 7: MATERIALS AND MANUFACTURING

Name   Date

Which Wrap?

3. If there are conditions that the material does not meet, can you think of any solutions to the problem so you can recommend a wrapper to the company? *Hint:* It's okay to add something new to the material. For example, you might think of using a photograph or drawing of the cracker snack if the wrapper material you use is not transparent.

_____

_____

_____

_____

4. Write a paragraph in your LiFE Log that describes the snack-wrap material you recommend and why.

Name   Date

Sand to Glass

Look around and you will see glass everywhere. Glass is used to make mirrors, light bulbs, windows, jars and bottles for food and drinks, and many other things. Glass is one of the few materials that can be recycled over and over again while keeping its strength. There are many different kinds of glass and different ways to make them. Your research project is to learn about the raw materials used to make glass. You will investigate the process used to get the raw materials. You will also research how glass bottles are manufactured.

Questions To Guide Your Research

Raw Materials
1. What raw materials are used to make glass?

2. What is sand?

3. What process is used to get sand? Are any other natural resources used in the process of getting sand? *Hint:* If machines are used, do they need fuel? Where does the fuel come from?

4. Are the raw materials used in glass renewable resources?

Manufacturing
1. How is glass made? Are the materials heated? Are other natural resources used to heat the furnaces? Do the materials need to be cooled? Are resources used to cool the materials?

2. Are machines used in the process of making glass bottles? If machines are used, do they need fuel? Where does the fuel come from?

3. Does it take energy to manufacture glass? What kind of energy?

4. Is glass recycled to make new glass?

 Farm to Table & Beyond
©2008 Teachers College Columbia University

Name   Date

Plant to Paper

You probably know that paper is made from trees. But did you know that paper can be made from other kinds of plants, too? Paper is made from cotton, jute, hemp, and bamboo, as well as trees. Thousands of years ago, the Chinese invented the papermaking process. They used mulberry bark as the raw material. The Egyptians made paper from papyrus plants. There are many different kinds of paper and they are designed and manufactured for specific uses. There is writing paper, laser-printer paper, newsprint, art paper, tissue paper, wrapping paper, cardboard, and kraft paper used for paper bags. This research project looks at the manufacturing process that turns trees into paper.

Questions To Guide Your Research

Raw Materials
1. What raw materials are used to make paper from trees?

2. What is wood pulp?

3. What process is used to get trees? Are any other natural resources used in the process of harvesting trees? *Hint:* If machines are used, do they need fuel? Where does the fuel come from? Are trees floated down rivers or lifted out by helicopters? How does the lumber get to the factory?

4. Are the raw materials used in paper renewable resources?

Manufacturing
1. How is paper made? Are the materials heated? Is water used? What happens to any water that is used during the manufacturing process?

2. Are machines used in the process of making paper? If machines are used, do they need fuel? Where does the fuel come from?

3. Does it take energy to manufacture paper? What kind of energy?

4. Is paper recycled to make new paper?

Name   Date

Petroleum to Plastic

While metal, paper, and glass have been used for thousands of years, plastic is a modern material. Plastic wasn't popular in America until after World War I, and the first plastic bottle was made in the 1970s. Now think about this: 30 years later, it is reported that Americans went through 50 billion plastic water bottles in 2006! That's a lot of plastic. Your research project is to investigate the raw materials that are made into plastic. You will also research the plastic-manufacturing process.

Questions To Guide Your Research

Raw Materials

1. What raw materials are used to make plastic?

2. What is plastic? What is petroleum?

3. What process is used to get petroleum? Are any other natural resources used in the process of getting petroleum? *Hint:* If machines are used, do they need fuel? Where does the fuel come from?

4. Are the raw materials used in plastic renewable resources?

Manufacturing

1. How are plastic bottles made? Are the materials heated? Are other natural resources used to heat the furnaces? Do the materials need to be cooled? Are resources used to cool the materials?

2. Are machines used in the process of making plastic bottles? If machines are used, do they need fuel? Where does the fuel come from?

3. Does it take energy to manufacture plastic bottles? What kind of energy?

4. Are plastic bottles recycled? What products are made from the recycled plastic?

Name	Date

Ore to Aluminum Can

How many metal cans do you think are made each year? Would you believe about 131 billion? That's the number, according to the Can Manufacturers Institute, and they come in more than 600 sizes and styles. Food cans are generally made of steel and beverage cans are most often made of aluminum. For your research project, you will investigate the raw materials used to manufacture aluminum beverage cans. You will also research how aluminum cans are manufactured out of the raw materials.

Questions To Guide Your Research

Raw Materials

1. What raw materials are used to make aluminum?

2. What is aluminum?

3. What process is used to get aluminum? Are any other natural resources used in the process of getting aluminum? *Hint:* If machines are used, do they need fuel? Where does the fuel come from?

4. Are the raw materials used in aluminum renewable resources?

Manufacturing

1. How are aluminum beverage cans made? Are the materials heated? Are other natural resources used to heat the furnaces? Do the materials need to be cooled? Are resources used to cool the materials?

2. Are machines used in the process of making aluminum cans? If machines are used, do they need fuel? Where does the fuel come from?

3. Does it take energy to manufacture aluminum beverage cans? What kind of energy?

4. Are aluminum cans recycled to make new aluminum?

Farm to Table & Beyond
©2008 Teachers College Columbia University

BECOMING FOOD SCIENTISTS : INTERACTING PARTS : FOOD PROCESSING : ENVIRONMENTAL EFFECTS : WASTE : MAKING CHOICES

Transporting Food

AIM

To gain an understanding of the role transportation systems play in food systems.

SCIENTIFIC PROCESSES

- question, construct knowledge, apply

OBJECTIVES

Students will be able to:

- discuss transportation as a system;

- describe the role of transportation in our food system;

- discuss the effect of transportation systems on the environment.

OVERVIEW

In this lesson students investigate transportation systems as part of the system that gets food from farm to table. Student readings provide a brief history of transportation and an overview of modern air, land, and water transportation systems. Using a jigsaw learning strategy, students become experts on different modes of transportation and share what they have learned with their group. In Unit 4, students expand on this work by considering transportation's impact on the environment. One goal of these transportation readings is to highlight the complexity of the various systems. Encourage students to think about trade-offs and feedback loops as they consider how food is transported in our global food system. It may be challenging for students to envision the scale of the system. Before students begin their transportation studies, check in with the product-packaging research teams from Lesson 7 and see how their work is progressing. At the beginning of the lesson, students also take a few moments to make detailed observations for their food-change experiment.

MATERIALS

For the teacher:
- *Technology and Science* teacher note
- *Product Research Checklist* lesson resource (p. 114)
- *Transportation-System Jigsaw* lesson resource
- *Milestones in Transportation History* lesson resource
- (Optional) Large index cards or sentence strips
- (Optional) Map of United States

For each group:
- Chart paper
- Markers

For each student:
- *A Brief History of Transportation* student reading
- *Food-Change Monitoring* activity sheet (pp. 100–104)
- *Moving Food by Airplane, Moving Food by Railroad, Moving Food by Inland Waterways, Moving Food by Ocean Freighter,* or *Moving Food by Truck* student reading
- *Transportation System Guiding Questions* activity sheet
- LiFE Log

PROCEDURE

Before You Begin:

- Read the *Technology and Science* teacher note, the *Product Research Checklist,* and the *Transportation-System Jigsaw* lesson resources.

- Review the *Brief History of Transportation, Moving Food by Airplane, Moving Food by Railroad, Moving Food by Inland Waterways, Moving Food by Ocean Freighter,* and *Moving Food by Truck* student readings, and the *Transportation System Guiding Questions* activity sheet. Make enough copies for each student in the assigned "expert" group to have one copy of the appropriate transportation system reading and the activity sheet.

- Copy and review the *Milestones in Transportation History* lesson resource. Cut out the milestones and glue them onto index cards. Or copy the milestones onto sentence strips.

- If you have not already done so, post the Module Question and Unit 2 Question at the front of the classroom.

MODULE QUESTION

What is the system that gets food from farm to table, and how does this system affect the environment?

UNIT QUESTION

What is the system that gets food from farm to table?

 **QUESTIONING**

1. Review the Module and Unit Questions

Tell students that in this lesson they will learn about transportation, one more subsystem in the food system. They will also think through and describe the role of transportation systems and how these systems help move food from the farm to the table.

2. Monitor Projects

Before you begin to investigate transportation, take a few minutes to check in with the product-packaging research teams. Meet briefly with each team to check on their progress. Use the *Product Research Checklist* lesson resource to guide the conversation. While you are meeting with the research teams, have students take turns monitoring their food-change experiment. Remind students to make detailed observations and to note what they observe on their *Food-Change Monitoring* activity sheets. Close this discussion by checking in with students to find out what, if any, food changes they are noticing. Encourage questions, discussion, and debate among the students. Continue the conversation until you feel that most students have developed a new understanding about food change.

3. Discuss Transportation History

Distribute the *A Brief History of Transportation* student reading. Invite students to help you add to the time line of food-packaging and materials history that you started in Lesson 7. Use the *Milestones in Transportation History* lesson resource. If you made cards with the dates, have students add those to the time line. If not, have students record dates and events directly on the food-packaging history time line. Once students have added the transportation milestones to the food-packaging history, invite the class to study the time line. Ask them to think about the changes in food packaging and transportation. *Can you see any developments that caused the food system to change from what it was in the past to what we have today?* (Refrigerated cars; new materials like plastic.) *Did the kinds of food that could be shipped from one place to another change?* (Refrigerated cars made it possible for trains to transport perishable items; airplanes shortened the transportation time for perishable items; food could travel farther.) As a class, read *A Brief History of Transportation.*

Ask students to think about the history of transportation. *People have moved from place to place for a very long time; why do you think there were so many changes in the past 100 years?* (Changes in materials; changes in technology.) Accept all answers.

4. Discuss Food Transportation

Remind students that a system is an interacting group of parts that work together to accomplish a common goal. Ask students to name a few examples of systems so you can check for understanding. Elicit from students their ideas about technology. *Does technology help people?* (It helps extend human abilities; helps people move from place to place and move things from place to place; helps humans measure things.) *What are some examples of technology that you use in your daily life?* (Telephone, computer, car, bus.) *How would our lives be different if we did not have a system for transporting food?* (We would have to have access to locally grown food or we would starve; we would have to be farmers; every family would need a plot of land to grow food.)

Explain to students that different kinds of technology are used in transporting food. As a class, discuss food transportation. *What kinds of vehicles can you think of that transport food?* (Boats, trains, airplanes, trucks.) *Have you ever seen any vehicles deliver food to a market? How do farmers bring their food to the farmers' market? How did explorers transport food in ancient times?* (Ships, animal caravans, pack animals.) *Where did the energy come from to "power" animals?* (Food.) *Where does the energy come from that makes trucks and planes and other modern vehicles move?* (Gasoline, diesel fuel, petroleum, fossil fuels.) Refer to the **Technology and Science** teacher note to guide your discussion.

5. Transportation Jigsaw

Refer to the **Transportation-System Jigsaw** lesson resource for suggestions on ways to use a jigsaw strategy with this part of the lesson. Have students break into groups of five. Appoint one member of each group to become an expert on the following transportation systems: Air, Inland Waterways, Ocean, Railroad, and Truck. Distribute the **Moving Food by Airplane** student reading to the Air experts, the **Moving Food by Inland Waterways** student reading to the Waterways experts, and so forth. Give students enough time to review the student readings and activity sheets. Have them work with their fellow experts to discuss the main points of their assigned transportation system. Give students time to make posters or diagrams that they can share with the members of their "home" group. Encourage students to use the **Transportation System Guiding Questions** activity sheet to organize the information for the poster or diagram. Students may wish to include their transportation system map on their poster or diagram. When "expert" groups have completed their work, have them return to their home group. Have each expert report back to the members of the home group.

6. Homework

Pose the following scenario to students: You grow apples in the Yakima Valley region of Washington State. You have a new customer in Biloxi, Mississippi. *Thinking about all that you have learned about different transportation systems, how will you ship your apples to Biloxi?* (If you have a map of the United States, point out the Yakima Valley and Biloxi.) Have students explain why they have chosen the transportation system or systems that they selected to move the apples from your orchards in Washington to a supermarket in Biloxi. *What are the advantages to using the system or systems you selected? Are there any disadvantages?* Have students write their responses in their LiFE Logs.

Technology and Science

Where would we be without technology? In the eyes of many, we would not be here, for toolmaking is one of the key characteristics that set human beings apart from our early ancestors. From the first stone tools, humans have been using technology to alter the world around them. The book *Science for All Americans* says of technology: "In the broadest sense, technology extends our abilities to change the world: to cut, shape, or put together materials; to move things from one place to another; to reach farther with our hands, voices, and senses. We use technology to try to change the world to suit us better. The changes may be related to survival needs such as food, shelter, or defense, or they may relate to aspirations such as knowledge, art, or control. They can include unexpected benefits, unexpected costs, and unexpected risks — any of which may fall on different social groups at different times. Anticipating the effects of technology is therefore as important as advancing its capabilities." [1]

In exploring and investigating food systems, your students have an opportunity to observe technology, and the effects of technology, in a meaningful way. Research suggests that it is not easy for students in grades 5 through 8 to differentiate between science and technology. As you engage students in discussions of the food system, remind them that science is a way of understanding the world around us. Science explains and predicts. The National Science Education Standards emphasize that "science and technology are reciprocal. Science helps drive technology, as it addresses questions that demand more sophisticated instruments and provides principles for better instrumentation and technique. Technology is essential to science, because it provides instruments and techniques that enable observations of objects and phenomena that are otherwise unobservable due to factors such as quantity, distance, location, size, and speed. Technology also provides tools for investigations, inquiry, and analysis." [2]

As your students explore the technologies involved in food systems, remind them that there are no perfect designs. Technological solutions have trade-offs. Sometimes the solution is costly, or it's not aesthetically pleasing, or it's not efficient, or perhaps it's not safe. Assessing risk is part of technology. In their farm-to-table studies, your students will be challenged to weigh the pros and cons of many aspects of our highly technical food system.

Before you begin this lesson, we encourage you to check for your students' understanding of technology. Research on student learning indicates that students associate technology with environmental crises, such as pollution, or military weapons. If you find this is the case with your students, try brainstorming a list of technologies they use as they get ready for school. Start them off with a toothbrush.

---

[1] American Association for the Advancement of Science, 1990 p. 25
[2] National Science Education Standards, 1996, p. 166

Transportation-System Jigsaw

Consider using a cooperative-learning strategy called a jigsaw strategy for this lesson. After breaking the class into small groups, you assign each member of each group a form of transportation. The "transportation experts" leave their original group and meet with the other experts in their form of transportation — all the air transportation researchers meet together, and so forth — to learn as much as possible about that form of transportation. When the students return to their original, or "home," group, the group members take turns reporting back as experts. With this approach, all the forms of transportation come together in the whole group.

Setting Up a Transportation-System Jigsaw

1. In this lesson, students investigate five modes of transportation (airplane, barge, container ship, railroad, and truck). Divide the class into groups of five students each. These are the "home" groups.

2. Assign one mode of transportation to each student in a home group. If you do not have enough students, you may wish to have some groups work on the Inland Waterways (barges) and others on the Ocean Freighter transportation system.

3. Give students time to read over the student reading and the activity sheet at least twice to become familiar with it.

4. Form temporary "expert" groups by having one student from each home group join other students assigned to the same kind of transportation. You will have five expert groups: airplane, barge, container ship, railroad, and truck. Give students in these expert groups time to discuss the main points of their transportation system and to discuss how they will share this information with their home groups. They may choose to make posters or diagrams. Make sure the expert groups understand that they are working together to learn about their topic. Encourage discussion and debate. The experts will be helping each other develop ways in which to present their understandings to the other members of their home group.

5. When the expert groups have completed their work, each expert presents back to the home group. Encourage others in the group to ask questions for clarification.

Milestones in Transportation History

Add these to the Milestones in Packaging History time line that you started in Lesson 7. Encourage students to think about trade-offs when one form of transportation replaced another. Were there other effects? For example, what happened when food was delivered to a central location for distribution? Was there a need for large-scale markets to handle the food? What other kinds of changes might have occurred? Were new jobs created? Were jobs lost?

4000 B.C.   In South America, Andean peoples domesticate llamas and use them to carry loads. In Mesopotamia and the steppes of southern Russia, ox-drawn sledges are in use.

3500 B.C.   Date of the oldest known wheel, discovered in the Middle East

3000 B.C.   Horses are domesticated and used as a means of transport; the plow and draft animals are being used

2000 B.C.   Wheeled transport is in use in a region stretching from northern Europe to western Persia and Mesopotamia. Egyptian seafaring ships trade with others in the Red Sea region.

1500 B.C.   Camels are domesticated in the Middle East

1000 B.C.   Camel caravans bring goods from India to Mesopotamia and the Mediterranean

300 B.C.   Via Appia is constructed, connecting Rome to the Bay of Naples

1282 A.D.   In China, Marco Polo witnesses kites carrying human beings

1483 A.D.   Leonardo da Vinci begins the first design of flying machines

1815–20 A.D.   Steamboats become important in western trade in the United States

1825 A.D.   Erie Canal is finished; canal-building era (1825–40) begins

1850s A.D.   Steam and clipper ships improve overseas transportation

1855 A.D.   Panama Railroad offers service between Atlantic and Pacific sides of the Isthmus of Panama. The trip from ocean to ocean is reduced from four days to four hours.

1869 A.D.   First North American transcontinental railroad is completed

1870 A.D.   Refrigerated cars are introduced, increasing national markets for fruits and vegetables

1893 A.D.   Five transcontinental rail lines and a web of other railroads link the American West to the rest of the country. By this time, almost any town can receive food and goods from any section of the country within a week or two. Factories can ship their products anywhere. Marketing becomes a nationwide enterprise.

1903 A.D. The Wrights demonstrate the airplane

1910 A.D. Early aviation promoters looking for practical uses for the airplane come up with the idea to use planes to carry freight

1910–25 A.D. Increase in use of automobiles and increase in road-building

1920s A.D. Truckers begin to transport more perishables and dairy products

1930s A.D. Farm-to-market roads are emphasized in Federal road-building programs

1964 A.D. United Airlines becomes the first airline in the United States to offer nonstop transcontinental all-cargo service

1950s A.D. Trucks and barges compete successfully for agricultural products as railroad prices increase

1960s A.D. Agricultural shipments by all-cargo planes increase, especially shipments of strawberries and cut flowers

Name

Date

A Brief History of Transportation

For more than 6,000 years, human beings have depended upon some form of transportation other than their feet. The earliest forms of land transportation were different kinds of domesticated animals. Humans domesticated wildlife that lived in the same environments they did. For example, in the Andes, early civilizations used llamas. In Mesopotamia, the camel was tamed and used to carry **goods** long distances. In Egypt, by 2000 B.C., horses were pulling chariots.

Humans also used waterways to move goods from one place to another. The earliest forms of water transportation included dugout canoes and rafts. By 2000 B.C., archeological evidence tells us, the Egyptians were sailing in the Red Sea, traveling from what is the modern city of Port Safaga to the trading center called Punt. No one knows precisely where Punt was located, but some scientists speculate it was near present-day Somalia, about 1,000 miles away. Thousands of miles away in the Pacific Ocean, evidence from the Fiji Islands suggests that seafarers arrived there, possibly from New Guinea, by about 1500 B.C.

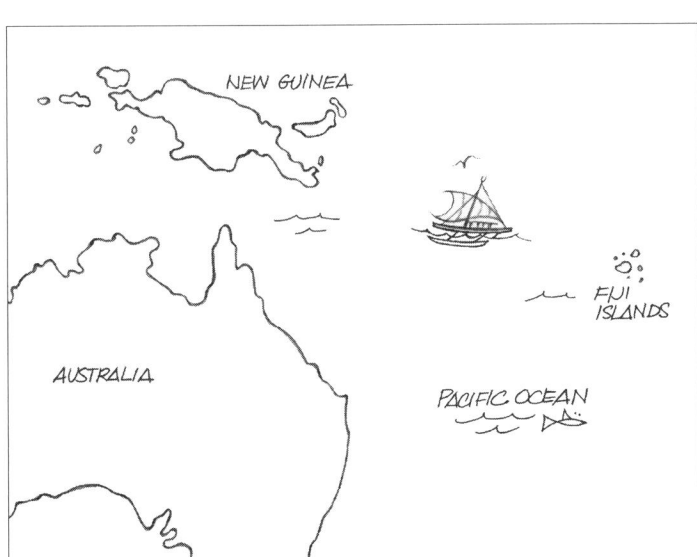

Humans took to the air, although not willingly, 1,500 years ago in China. Historical records indicate that in the sixth century, Emperor Kao Yang had prisoners tied to bamboo mats that served as kites. Once tied to the mats, the prisoners were forced to jump off towers to see if they could fly. Legend has it one prisoner survived this early attempt at flight.

Later, in the thirteenth century, Marco Polo wrote about the practice of Chinese sailors' tying a man to a kite and launching it. If the kite flew straight up, it was a good omen for the sailors' voyage. If the kite crashed, merchants would not load their goods onto the sailors' ship.

(continued on next page)

Name   Date

A Brief History of Transportation

It was about 500 years later before human flight really took off. In 1783 in France, the Montgolfier brothers demonstrated the flight of a hot-air balloon. Two years later, Jean-Pierre Francois Blanchard and John Jeffries, an American, made the first balloon trip across the English Channel. It took 2 1/2 hours. In 1793, George Washington observed the first balloon launch in the United States. The balloon, using hydrogen gas, flew from Philadelphia to Gloucester County, New Jersey — a journey that lasted 45 minutes.

However, in spite of these early attempts to conquer the air, it wasn't until the twentieth century that a flying machine became a reality. It was 1903 when the Wright brothers flew their airplane. Just seven years later, early promoters of air transportation had already started thinking about moving goods by air. The first demonstration of airfreight took place in Ohio in 1910 — it involved flying a bolt of silk from Dayton to Columbus, a distance of about 75 miles. A Columbus newspaper reported that the airfreight had beaten the railroad express between the two cities. Just think about how our transportation system has changed since then.

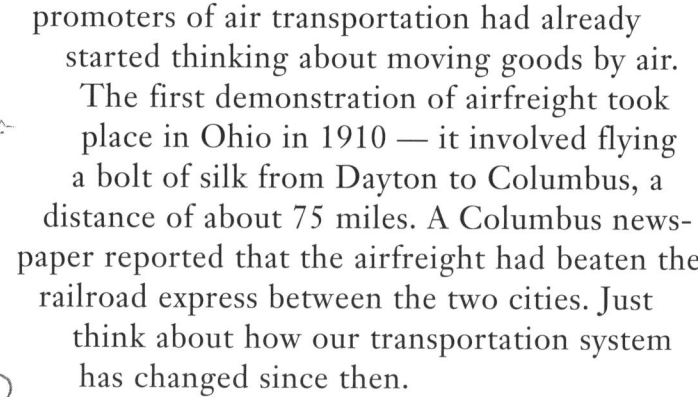
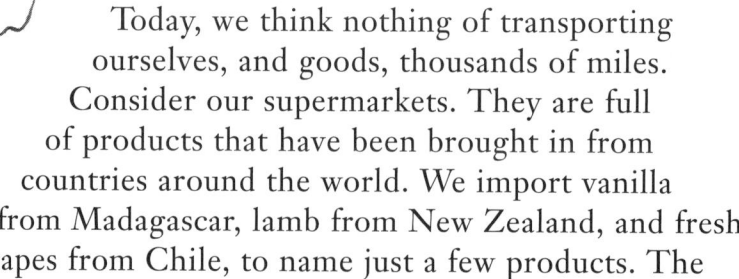

Today, we think nothing of transporting ourselves, and goods, thousands of miles. Consider our supermarkets. They are full of products that have been brought in from countries around the world. We import vanilla from Madagascar, lamb from New Zealand, and fresh grapes from Chile, to name just a few products. The United States not only imports goods, it also exports goods. We ship corn to Mexico and brown rice to Japan. In about 100 years, we have moved from flying silk between two cities less than 100 miles apart to a global food system.

Name   Date

Moving Food by Airplane

What has 6 million parts, 171 miles (247 kilometers) of wiring, 5 miles (8 kilometers) of tubing, 16 main landing gear tires, a tail height of 63 feet 8 inches (19.4 meters) — about as tall as a six-story building — and is made of 147,000 pounds (66,150 kilograms) of high-strength aluminum? A Boeing 747-400, one kind of plane used to transport cargo, including food.

How does an air transportation system work as a whole?
Air transportation goes from airport to airport. Trucks bring the food to the airport where it is loaded onto a plane. Food is shipped in special containers that make it efficient to load and unload the plane quickly. The containers also protect the food so it does not get damaged when it is shipped. Once a plane lands at its destination, the containers are quickly removed from the plane and ready to move again.

What are some of the essential parts of the air transportation system?
Key parts of the air transportation system are the plane, the pilot, other members of the crew, fuel trucks, air-traffic controllers, a communications system, workers to load and unload the containers of food, maintenance workers to take care of the plane, and the airports. You also need computer systems to monitor the flights, trucks and drivers to bring the jet fuel to the plane, and instructors to train the pilots. The plane itself is made of many essential parts. Some of the major parts of the plane system are the weather radar scanner, the landing gear, the engines, the wings, the doors, the tail fin, the flight deck, tires, flaps, toilets, and a galley (kitchen). These are just some of the parts of an air transportation system.

(continued on next page)

BECOMING FOOD SCIENTISTS : INTERACTING PARTS : FOOD PROCESSING : ENVIRONMENTAL EFFECTS : WASTE : MAKING CHOICES

Name   Date

Moving Food by Airplane

How is cargo shipped?

Goods — like packages, mail, food, computer equipment — can be shipped on passenger planes or cargo planes. The freight, or cargo, is put in special containers that are designed to fit inside the cargo hold of the plane. The cargo needs to be in the right kind of packaging and packed properly inside the containers or it might end up damaged. For example, if the flight is a bumpy one, the cargo might bounce around. If there is a rough landing, the cargo might get crushed against other cargo in the container. Imagine what could happen to a container full of strawberries that were not packaged to protect them from being squished. If the strawberries don't make it to the market safely, the shipper loses the sale.

What are the advantages of shipping by air?

Shipping by air is faster than shipping by other modes of transportation. Strawberries can be loaded onto a plane in Los Angeles in the evening and arrive in New York City the next morning. Air transportation also makes it possible to ship perishable fruits and vegetables to international markets. These fruits and vegetables might rot if they had to travel for days in a container on a ship.

What are the disadvantages of shipping by air?

It takes a lot of fuel to fly a plane, which makes it expensive to ship food by air. In fact, it costs more to ship by air than by truck, ship, or train. Another disadvantage to shipping food by air is that planes cannot deliver products directly to supermarkets. This means that when you ship food by air, you also need to use another kind of transportation system to get the food to the airport and to deliver it when it arrives at the destination airport. Trucks move the food to the airport from the farm where it was grown or raised and more trucks deliver it once it is flown to the next airport.

What are some of the challenges of shipping by air?

One big challenge in transportation is scheduling. With air transportation, you have to keep both the planes and the crews moving. Think about those strawberries that you're shipping. The crew and the plane full of fresh strawberries take off from Los Angeles and arrive in New York. The crew goes off and rests. The strawberries are unloaded, and the plane is refueled and ready to go in just a few hours. Another crew that has rested comes to the airport and is ready to fly, but what if there is nothing to ship back to Los Angeles? To be efficient and profitable, when a plane lands at one airport and its cargo is removed, there has to be cargo to load back onto the plane to ship somewhere else.

(continued on next page)

Name

Date

Moving Food by Airplane

Weather is another challenge when shipping by air. What if it was sunny and warm in Los Angeles when the strawberries took off, but there was a blizzard in New York City and the plane had to land in Chicago or Baltimore? What happens to the strawberries? If you were in charge, how would you plan? Do you think that a communication system might be an important part of a transportation system? Do you think you'd use computers and telephones to find a way to help you move the cargo?

Does air transportation have an impact on the environment?

Building an airport, which is an essential part of an air transportation system, affects the environment. Think about it. The land is changed to make the landing strips and the terminals. The natural environment is replaced by a human-designed environment. Native species no longer have their habitat and need to adjust, or move away. In the case of birds, airports sometimes need to design systems to manage and control them. Birds and airplanes most often collide during takeoff or landing. To help reduce bird strikes (collisions), airports try to change the environment to make it less attractive to birds. For example, they will remove trees or use devices to frighten birds away.

There is concern that air transportation contributes to global warming and destruction of the ozone layer. As the demand for air transportation increases, airports need to expand to handle the traffic. This means even more air pollution, and an increase in noise and the amount of traffic on the roads leading to and from the airports. Aircraft engines give off carbon dioxide, nitrogen oxides, carbon monoxide, sulfur oxides, and **particulate matter** (fine particles), which pollute the air.

As air traffic increases, so does the overall noise created by jet engines. One of the major reasons people oppose runway expansions is that they do not want the noise to increase. Airports are trying to cut down on the impact of the noise by soundproofing homes, building noise barriers, and relocating people and businesses.

Removing frost, snow, and ice from planes and runways can affect the environment. Planes have to be deiced to ensure the safety of all onboard. However, the chemicals that are used can contaminate water and kill fish. Airports and airlines have to manage the wastewater so that the chemicals do not enter the water system.

Airports store fuel so they can refuel planes and ground-service equipment. If you have ever looked around an airport, you probably have seen baggage loaders, snowplows,

(continued on next page)

Name   Date

Moving Food by Airplane

tractors, and cargo-moving equipment. All these vehicles need fuel. Storing fuel at the airport means airports need to have plans to prevent oil spills and to control leaks. They also need to have a way to prevent an oil spill from reaching any water that is nearby.

DID YOU KNOW...?

- A 747-400 that flies 3,500 miles (5,630 kilometers) carries 126,000 pounds (56,700 kilograms) of fuel. It uses an average of 5 gallons (19 liters) per mile.

- A 747-400ER flying from Los Angeles, California, to Melbourne, Australia, can carry more than 63,500 gallons of fuel (240,370 liters).

- Airline cargo handlers use the 747-400's cargo-handling system to load and unload more than 65,000 pounds (30,000 kilograms) of cargo in less than 15 minutes.

Farm to Table & Beyond
©2008 Teachers College Columbia University

Name	Date

Moving Food by Railroad

Since the 1800s, trains have played an important role in the development of the United States. It was 1830 when the Baltimore and Ohio Railroad's "Tom Thumb" carried guests 13 miles on its trial run — a trip that took 57 minutes. Just 39 years later, on May 10, 1869, the Central Pacific line and the Union Pacific line joined to create the first transcontinental railroad. Today, two of the largest railroads in the United States are the Union Pacific Railroad and the Burlington Northern Santa Fe Railway.

How does a railroad transportation system work as a whole?

Rail transportation goes from train terminal to train terminal. This means that the food that is being shipped by rail is also moved by at least one other form of land transportation. Think about it. Food is grown or raised on farms. Somehow the food has to get from the farm to the terminal. It may take several steps to get there. For example, it may go by truck from the field to a food-processing plant. From there, a larger truck takes the food to a railroad yard, where it is loaded onto freight cars. When the train reaches its destination, a truck takes the food to a distribution center. From there, other trucks carry the food to markets.

What are some of the essential parts of the railroad transportation system?

Key parts of the railroad transportation system include the trains, the tracks, the railroad terminals, the engineers and crew that operate the trains, and some workers to load and unload the containers of food. There are locomotives that pull freight long distances, and there are locomotives at the train yards that sort incoming railroad cars. In addition to these basic parts, a railroad transportation system includes a communications system (computers, people who operate the computers, information systems), sales and marketing staff, financial systems, and engineers, but not just the engineers on the trains. There are also civil engineers and industrial engineers who help design the railroad cars and maintain and build the network of tracks. Some of the responsibilities of the engineering team include inspecting tracks, operating machinery, maintaining signals,

(continued on next page)

Name   Date

Moving Food by Railroad

and repairing tracks. There are train dispatchers who make the decisions about where to move the trains. There are switchmen who guide the trains on their assigned tracks. And there is the locomotive engineer who operates the train from the railroad station at the beginning of the trip to the final destination.

How is cargo shipped?

There are specialized freight cars to carry different kinds of cargo. For example, a **tank car** carries different kinds of liquids, including molasses. A tank car can hold as much as 31,000 gallons of molasses! A railroad car called a **covered hopper** carries grain, salt, sugar, soybeans, and flour. These products are loaded through hatches on the top of the car and unloaded through chutes on the bottom.

Boxcars have wide doors, which make it easy to load and unload them. They usually are used to carry loads that need to be protected from the weather. Boxcars can be built with insulation to protect the freight from heat and cold. They can be built with cushioned frames to give a smooth ride to fragile freight as the train rumbles down the tracks. If you have ever ridden on a train, you may have noticed how the train shifts from side to side as it moves along. If you have a fragile cargo of eggs, you can imagine how scrambled they might be when you arrive at your destination. Boxcars are designed to carry canned goods, cereals, food that needs to be refrigerated, and bags of flour.

Refrigerated cars, or "rolling refrigerators," keep perishable goods cold. The inside temperature can be kept as cold as 20 degrees below zero. In 1857 a meatpacking firm in Chicago introduced an ice-cooled car to keep shipments from spoiling. By 1872 many strawberries and other fruits were being shipped from southern Illinois to Chicago under refrigeration. In 1885 berries from Virginia were shipped in refrigerated cars to New York.

What are the advantages of shipping by rail?

One big advantage is that a single double-stack train can carry as much freight as 280 trucks. Another advantage is that railroads spend fewer travel hours stalled in traffic than trucks do. The railroad industry reports that if 25 percent of the truck freight were moved by rail instead, by 2025 there would be a savings of 16 billion gallons of fuel and 2.8 billion fewer travel hours wasted in sitting in traffic.

What are the disadvantages of shipping by rail?

One of the major disadvantages is that you can only use a railroad where there are tracks. Of course, it's also true that you can only drive a truck where there is a road, but trucks can take side streets and alternative routes if there is a problem.

(continued on next page)

Name   Date

Moving Food by Railroad

What are some of the challenges of shipping by rail?

When the first transcontinental railroad was completed, on May 10, 1869, it established a West-East system. Today, there is a need for a North-South route system. To build these new routes would require new laws and lots of money. Once a train yard is built and the tracks are laid down, they can't be moved. For people to invest the amount of money that is needed to improve and expand the railroad system, the investors want to be sure that the railroad business will thrive and make a profit. If shippers decide not to use trains, the investors will lose their money.

Does the railroad transportation system have an impact on the environment?

Some historians consider the transcontinental railroad to be one of the United States's greatest technological achievements. It was built in six years almost entirely by hand. Workers blasted through the rock and built bridges so they could ford streams. Thousands of workers, including Irish and German immigrants, former Union and Confederate soldiers, freed slaves, and especially Chinese immigrants, played a part in the construction. However, some people also say that the railroad contributed to the decline of buffalo herds in the 1800s.

Today, noise is one of the most common environmental impacts from railroad transportation systems. It is possible to control train noise by modifying the trains or tracks or by building noise barriers. Train horns also contribute to noise pollution. Train horns are used to warn motorists or pedestrians that a train is approaching a highway rail crossing. The horn has to be loud enough for a car's driver to hear it with the windows rolled up and the air conditioner and radio on.

Trains contribute to air pollution. The engines run even when they are not moving. Trains idle to keep their fuel and water lines from freezing. The exhaust from locomotives pollutes the air with **particulate matter** that includes dust, smoke, soot, pollen, and soil particles. These fine particles can pass through the nose and throat and get stuck in the lungs and cause health problems, including lung damage, asthma, and bronchitis.

However, there is good news. The railroad industry is working hard to become a "green" technology. This means it is researching ways to become an environmentally friendly form of transportation. One example is "the Green Goat." This locomotive is similar to an automobile that relies on both a gasoline engine and a battery-powered electric motor. The Green Goat depends on its large, onboard storage batteries, which are charged by a small diesel engine. This new technology may reduce particulate matter

(continued on next page)

Name   Date

Moving Food by Railroad

by up to 80 percent and reduce the amount of fuel that the locomotive needs to run by about 16 percent. The railroad transportation system is working hard to become the "green" alternative to shipping by trucks.

DID YOU KNOW...?

- In 2001, there were 1,314,136 freight cars in service in the United States.

- Burlington Northern Santa Fe Railway hauls enough fertilizer in one year to fertilize a field the size of the entire state of Kansas.

- Approximately 50 percent of the agricultural commodities that Burlington Northern Santa Fe Railway hauls is transported to export points in the Pacific Northwest, the Gulf of Mexico, Mexico, and the Great Lakes.

- Burlington Northern Santa Fe Railway moves enough sugar to make more than three million batches of cookies a year.

- Burlington Northern Santa Fe Railway transports more than a billion cans of canned goods each year.

- Burlington Northern Santa Fe Railway hauls enough grain to supply 900 million people with a year's supply of bread.

- During 2006, railroads removed more than 12 million trucks from the highways — the equivalent of almost 50 million automobiles.

Farm to Table & Beyond
©2008 Teachers College Columbia University

BECOMING FOOD SCIENTISTS : INTERACTING PARTS : FOOD PROCESSING : ENVIRONMENTAL EFFECTS : WASTE : MAKING CHOICES

Name   Date

Moving Food by Inland Waterways

Imagine that there were no cities or towns in the United States and you were asked to find a location for a city. Where would you put it? What would you think about? If you were like many people throughout history, you would locate the city near a body of water — on the coast or on a river or a lake. Think about large cities like New York; Boston; Washington, D.C.; Los Angeles; Miami; St. Louis; Chicago; or New Orleans. Why do you think cities and towns were built on the banks of rivers? One reason is that early factories used the rivers for power and to move their products to market.

Canals were an important part of the early network of waterways in the United States. With the completion of the Erie Canal, in 1825, there was a route from the Atlantic Ocean to the Great Lakes. This became a crucial route for people moving westward and for products moving from Midwestern farms to Atlantic ports.

How does the inland-waterway transportation system work as a whole?

Farm products, such as grain, are brought to a port, loaded onto barges, and moved to another port. The United States Inland Waterway System includes more than 25,000 miles of navigable waters. It includes the Mississippi River System, which connects the Gulf of Mexico to the Great Lakes, the St. Lawrence Seaway, and Canada. Goods from 27 states leave the United States through ports in Louisiana. Almost all of the rivers and canals that are part of the inland-waterway system are in the eastern half of the United States. Ships can navigate only three rivers on the West Coast: the Columbia River, the Sacramento River, and the San Joaquin River.

What are some of the essential parts of the water transportation system?

Key parts of the inland-waterway system are a towboat; barges; a captain, a pilot, and a crew; and workers to load and unload the barges. A tow is a group of barges hitched together. A single towboat can push a tow of 15 barges — five barges long and

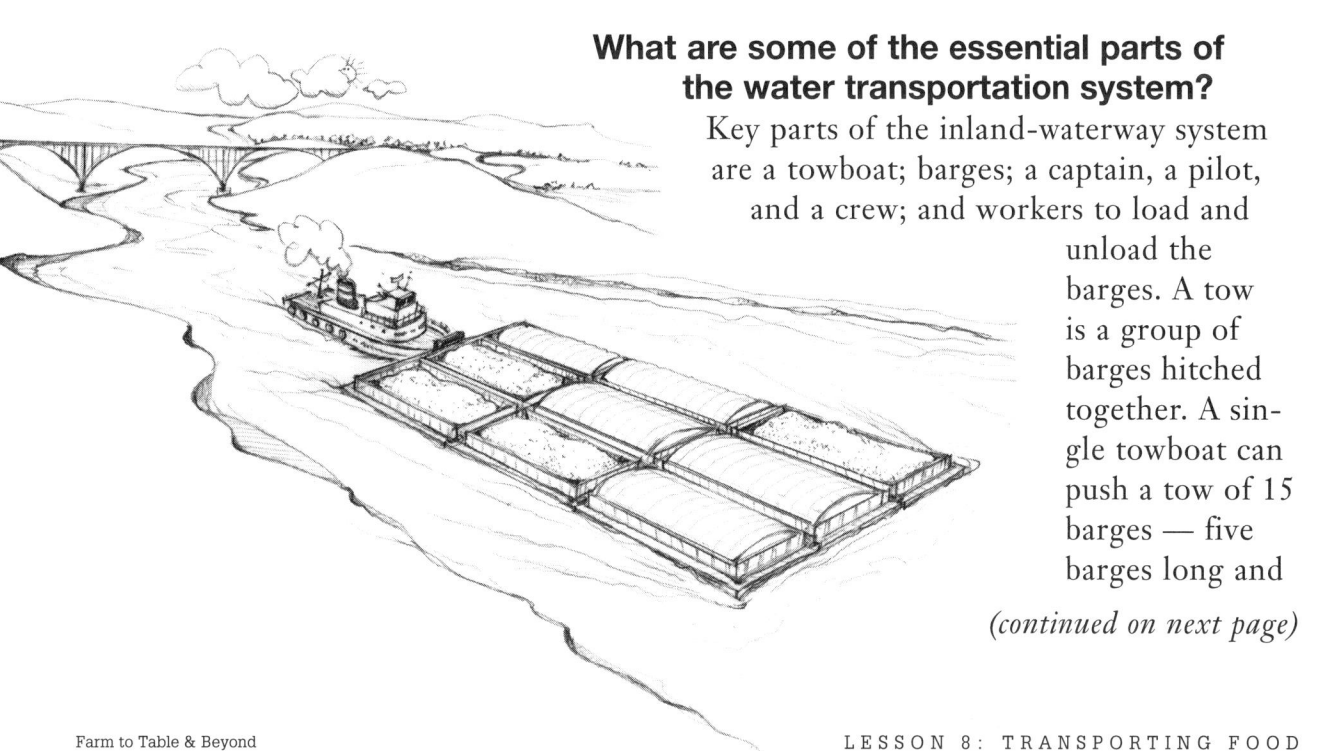

(continued on next page)

BECOMING FOOD SCIENTISTS : INTERACTING PARTS : FOOD PROCESSING : ENVIRONMENTAL EFFECTS : WASTE : MAKING CHOICES

Name   Date

Moving Food by Inland Waterways

three barges wide. The system also includes canal locks, fuel for the towboat, and lock tenders to operate the locks to let barges pass through. Canals are man-made channels for water. In transportation, they are used to connect rivers, lakes, and oceans.

Canals need to be flat, so locks are used to raise and lower boats by adjusting the water levels. A lock makes it possible for a boat to pass from a section of canal at one water level to another section at a different water level. A boat going from a lower level to a higher level enters the lock through the lower gates. The lock tender then closes the lower gates and opens a valve in the upper gates. This lets water into the lock and raises the boat. Once the water level in the lock equals the level in the higher section of the canal, the tender opens the upper gate, and the boat continues its journey. Going in the opposite direction, when a boat enters the lock through the upper gates, the tender closes the gates and the water level is lowered to the level of the canal below. Then the lower gates are opened and the boat goes into the canal. Lock tenders also monitor and inspect canal equipment and make minor repairs. Locks make it possible for vessels to "stair-step" their way through the inland-waterway system to reach inland ports such as Minneapolis, Chicago, and Pittsburgh.

When barges are not in use, they are kept in a holding area, which is like a parking lot. Supplies for the towboats are ferried out in small, fast boats.

There are many different kinds of towboats, both large and small. Some of the parts of a towboat include the pilot house, the deck, the engine, deck lights, navigation lights, rudders, an electronic chart system, radar, life preservers, a gasoline tank for outboard-motor fuel storage, fire stations, a galley (kitchen), bathrooms, a dining room, bedrooms, a laundry room, exhaust vents, a ladder, and an intercom.

How is cargo shipped?

Barges carry unprocessed products like soybeans, corn, wheat, sorghum, and rice. Since barges can't pull up to a field for a farmer to load grain directly onto the barge, how is grain loaded? A farmer delivers the crop to a grain elevator. (The term "elevator" refers to the process that moves the grain up and then lowers it down again.) How does this elevator work? The grain is unloaded from a truck into an area called a pit. A vertical conveyor belt runs from the pit to the top of the grain elevator. Dozens of small buckets attached to the conveyor belt carry the grain from the pit and dump it into a bin at the top of the elevator. When it's time to ship the grain, a slide door is opened and the grain flows out of the bottom of the bin into the barge.

(continued on next page)

Name   Date

Moving Food by Inland Waterways

Another kind of barge, called a tank barge, carries liquid products, like molasses, which is used for cattle feed. About 60 percent of United States farm exports travel through inland waterways.

What are the advantages of shipping by inland waterways?

One of the major advantages of inland waterways is that they allow large amounts of bulk goods to be moved long distances. According to the U.S. Department of Transportation, moving cargo by barge on an inland waterway is both energy-efficient and cost-effective. One barge can carry 1,500 tons. It would take 15 railcars, each carrying 100 tons, or 60 trucks, each carrying 25 tons, to carry the same amount.

The inland-waterway transportation system provides low-cost transportation between the grain-growing part of the United States and markets around the world. The system also provides a low-cost way of moving fertilizer to farms.

What are the disadvantages of shipping by inland waterway?

One of the biggest disadvantages of the lock system in the United States is that it is out of date. It was developed more than 60 years ago and was built to handle 600-foot barge tows. Today, barge tows are twice that length. To get through the lock system, the tow has to be "cut" into sections. This causes delays all along a river, which increases the cost of shipping. The engines keep running while boats wait, which means that additional fuel is burned due to the lock delays.

What are some of the challenges of shipping by inland waterway?

Waterway transportation is much more affected by weather conditions than railroads are. On the upper Mississippi and the Great Lakes and St. Lawrence Seaway, the inland waterways are unusable for three to four months during the winter.

Does the inland-waterway transportation system have an impact on the environment?

Building and maintaining ports and the canal-and-lock system affects the environment. For example, it is necessary to dredge channels to keep them deep enough for navigation. Dredging stirs up pollutants such as pesticides, oil, grease, and heavy metals that may have settled on the bottom of the waterway. These pollutants can harm the ecosystem. Even if there are no pollutants, stirring up the sediment can affect the water quality. Today, all dredging projects have to be designed to protect human and ecological health.

(continued on next page)

BECOMING FOOD SCIENTISTS : INTERACTING PARTS : FOOD PROCESSING : ENVIRONMENTAL EFFECTS : WASTE : MAKING CHOICES

Name   Date

Moving Food by Inland Waterways

Plant and animal species that live in ports can be affected by noise pollution, ship strikes (collisions with ships), oil pollution, and invasive species. Gray water — wastewater from ships' sinks, showers, laundries, and galleys — can carry chemicals. Even though this water is diluted, it still can have an effect on water quality and the health of the ecosystem.

Another unexpected impact is the introduction of invasive species, which compete with native species. Ships take in ballast water to help stabilize the vessel. When the ships reach their destination, they release the ballast water and any organisms that are in the water. The zebra mussel is an example of a European species that stowed away in the ballast water of a ship traveling from Europe to the Great Lakes. First observed in the mid-1980s in Lake St. Clair, zebra mussels have had a huge impact on habitat, water quality, and biodiversity. They are not just a problem in the Great Lakes. They have been carried downstream by the flow of water and commercial and recreational boat traffic. Now zebra mussels inhabit the waters of 20 states and the Canadian provinces of Ontario and Quebec.

The zebra-mussel invasion and its impact on the Great Lakes ecosystem have received a great deal of attention. However, the introduction of exotic species is not a new problem. It is estimated that 130 exotic species have been introduced to the Great Lakes since the St. Lawrence Seaway opened, in 1959.

DID YOU KNOW...?

- A barge can move one ton of cargo 500 miles on a single gallon of fuel. That same gallon of fuel moves a ton of cargo only 200 miles by railcar and only 60 miles by truck.

- Almost 50 percent of the St. Lawrence Seaway traffic travels to and from overseas ports, especially in Europe, the Middle East, and Africa.

- The Erie Canal took seven years to build. It cut through 363 miles of wilderness and included 83 locks. The locks were built to help overcome the 568-foot difference between the level of the Hudson River and that of Lake Erie.

- Two major developments of the nineteenth century increased the economic importance of inland waterways — steamboats and canals.

- In 1842, Joseph Dart, a retail merchant, invented what became known as the grain elevator.

BECOMING FOOD SCIENTISTS : INTERACTING PARTS : FOOD PROCESSING : ENVIRONMENTAL EFFECTS : WASTE : MAKING CHOICES

Name	Date

Moving Food by Ocean Freighter

Imagine that you are the head of a huge transportation company. One day you are standing at a port watching goods being taken off tractor trailers and loaded onto ocean freighters. You wonder if there is a way to move the goods more efficiently. What if you loaded the tractor trailers right onto the ships instead of unloading them? An idea like this changed how cargo is shipped today. This idea led to the development of containers for shipping cargo. The containers could be used on trucks, railroads, and barges. Today, there is a container-cargo handling system for loading and unloading ships. Food is transported in containers that are insulated and have a built-in refrigeration system.

How does the ocean transportation system work as a whole?

Ocean transportation goes from port to port. Food is brought to one port by truck, barge, or railcar, loaded onto a ship, and transported to another port. Food is shipped in containers that make it efficient to load and unload the ship. The port of Los Angeles, as an example, has eight major container terminals and a 20-mile express railway that connects the port to the rail hubs in downtown Los Angeles. It also has 13 acres of warehouses. There are overhead cranes and loaders for lifting the cargo containers. There is a main gate with 16 entrance lanes for container-carrying trucks. All of these parts support the efficient movement of cargo containers between different modes of transportation.

What are some of the essential parts of the ocean transportation system?

What do you need to move apples from a port on the West Coast to a port in Asia? You need an ocean freighter, a port, terminals, containers to hold the cargo, a captain and crew, fuel, and cranes to load and unload the containers of food. You need workers at the port, including workmen to load the ships; exporters and

KANSA CARGO LINES

(continued on next page)

Name   Date

Moving Food by Ocean Freighter

shippers; operators of trucks, railcars, towbarges and tugboats; and government work-ers to inspect the cargo. Ports also need to have temporary storage for products waiting to be shipped.

How is cargo shipped?

Ocean freighters carry edible products like apples, sugar, wheat, corn, soybeans, barley, oats, and flaxseed. In choosing a port, growers need to think about how much it will cost to ship their food to the port plus the cost of shipping overseas. If you are shipping apples, the apples need to be stored in a refrigerated warehouse to help keep them crisp. Apples also have to pass inspection to make sure there are no pests or fungus on them. They can be stored for almost 12 months inside a special warehouse where the temper-ature, humidity, oxygen, and carbon dioxide are monitored and controlled to keep the fruit from ripening too quickly.

Containers may be ventilated, insulated, or refrigerated. A container may be 20, 40, 45, 48, or 53 feet long; either 8 feet or 8.5 feet wide; and 8.5 or 9.5 feet high.

What are the advantages of shipping by ocean?

The marine transportation system opens the door to a global marketplace. This means that growers can sell their food not only in the United States, but also overseas. Shipping by ocean is less expensive than shipping by air.

What are the disadvantages of shipping by ocean?

It takes longer to ship food by ocean freighter than it does by air. It also takes more than one mode of transportation for goods to make the entire journey from the field to the market at the final destination. For this reason, ports are one part of a rail, highway, and inland-waterway transportation system. Ports have to have access to other forms of transportation if they are to function.

What are some of the challenges of shipping by ocean?

One of the greatest challenges is the amount of time it takes to ship by ocean. If you want to ship perishable products, like fruit and vegetables, the food must be packed correctly and shipped in refrigerated containers so that it is kept in a chilled environment during the voyage. Suppose your job was to ship apples from Washington State to Taiwan. What would you need to know and do? You'd want to know how long it might take for the apples to arrive. You'd find out that some ships take longer than others and that you need

(continued on next page)

Name   Date

Moving Food by Ocean Freighter

to know the schedules for the different ships. One ship might take 15 days and another might take as long as 21 days because it stops at other ports on the way to Taiwan. You might find out that apples will stay fresh for 22 days in a refrigerated container. While the apples might still be fresh on the ship that will take 21 days, you decide to ship them on the ship that is scheduled to arrive in Taiwan in 15 days, just to be sure that the apples arrive in good condition.

Does the ocean transportation system have an impact on the environment?

Building and maintaining ports and terminals affects the environment. Ports are built at a point where the land and the water meet. Developing and expanding a port often requires major changes to the environment. For example, it may require dredging to make the channels deeper so that ships can enter. The ocean sediment can be contaminated with pesticides, oil, grease, and heavy metals like lead and mercury. Dredging can affect water quality and the organisms that live in the ecosystem. The dredged material has to be disposed of properly. If the material is not contaminated, it can be used for different purposes such as habitat restoration or used in the construction of airports, parking lots, and roads.

Other environmental concerns include air quality, oil pollution, and water pollution. Gray water — wastewater from ships' sinks, showers, laundries, and galleys — can carry chemicals. Even though this water is diluted, it still can have an effect on water quality and the health of the ecosystem.

Plants and animals live in or use the port environment. For example, the endangered Florida manatee is found along Florida's coastline. These mammals use the waters of Port Canaveral for playing, resting, and feeding and as a travel corridor between the Atlantic Ocean and the Indian River Lagoon. To protect the manatee, Port Canaveral started a manatee-protection program that includes having manatee observers during dredging operations and designating all harbor waters as a slow speed zone. The North Atlantic right whale travels to the coastal waters off Port Canaveral during the winter months. The whales are often close to the surface of the water and can collide with ships. Volunteers watch for and report whale sightings so approaching ships can be notified.

Another unexpected effect of ocean transportation is the introduction of invasive species, which compete with native species. A ship takes in ballast water to help stabilize the vessel. When the ship reaches its destination, it releases the ballast water and any organisms that are in the water. The zebra mussel, the spiny water flea, and the muffin

(continued on next page)

BECOMING FOOD SCIENTISTS : INTERACTING PARTS : FOOD PROCESSING : ENVIRONMENTAL EFFECTS : WASTE : MAKING CHOICES

Name   Date

Moving Food by Ocean Freighter

crab are some of the animals that have been brought in through the ballast water of transoceanic freighters.

Oil pollution is one of the most serious environmental problems in the marine environment. While some oil pollution is caused by tanker spills, most of it takes place in ports during loading, tank-washing, and wastewater discharging. Oily discharges come from grease and oils used to maintain engines and shipboard machinery and devices used to clean oil-carrying cargo tanks. Oil can smother tide pools and kill marsh grass, shellfish, and invertebrates. Sea and shore animals can be killed when their feathers or fur are covered with oil.

DID YOU KNOW...?

- Ships classified as Panamax are the maximum dimensions that will fit through the locks of the Panama Canal.

- Apples have to be shipped with a core temperature of 34 to 35 degrees to maintain their crispness.

- Ships have an expected life span of 20 to 30 years and can cost more than $100 million to build.

Name	Date

Moving Food by Truck

Before there were trucks, goods moved from place to place by horse-drawn wagons traveling on unpaved roads. Deliveries that now take days could take weeks. However, the internal-combustion engine and trucks changed all this.

In the 1890s, the Mack brothers built engines that burned gasoline fuel, and in the early 1900s, they built what they called the Manhattan Truck. The trucks later became known as Mack Trucks. Other companies followed and began to make trucks, too. By the 1930s, the trucking industry grew as modern highways began to replace dirt roads. In the 1940s, Congress created a National System of Interstate Highways. Trucks were able to move along at a steady pace on these four-lane highways. As the roads improved, the trucks improved. New lighter, stronger materials were developed and used in manufacturing trucks and truck cabs were designed to be more comfortable for drivers.

Today, there are more than 4 million miles of public roads in the United States and more than 200 million vehicles being driven on them. Highways are used to move freight both locally and long-distance. Almost all freight shipments make at least part of their farm-to-table journey on highways.

How does a highway transportation system work as a whole?

A truck loaded with freight moves from a supplier at one location to a receiver at another location. The two places may be across the country, thousands of miles apart, or they may be just a few miles away in the same state or county. The trucks used to move food come in many different sizes. Smaller trucks may move the food from the field to a temporary warehouse. At the warehouse, the food is packed and prepared to load onto larger trucks that may haul the food across the country. A long-haul trucker can drive as far as 500 miles a day. Trucks used for a long haul, like a **big rig,** have a compartment called a sleeper behind the cab where the driver can sleep. The highway transportation system is the most interconnected of all the different ways of transporting goods.

(continued on next page)

Name   Date

Moving Food by Truck

What are some of the essential parts of the highway transportation system?

Key parts of the highway transportation system are trucks, drivers, fuel stations, a system of highways, oil refineries, a communications system, tires, a traffic-control system, and workers to load and unload the food. Each of these key parts of this system also has subparts. For example, some of the main parts of a truck are the engine, gears, the drive shaft, the brakes, the suspension, hookups, the body, a computer, and radios. School is part of the highway transportation system, too. Truck drivers, also called truckers, go to school to learn how to drive a truck. When the trucker is on the road, the system includes truck stops, gas stations, radio programs, and restaurants.

How is cargo shipped?

A long-haul trucker may drive a big rig, which is sometimes called an 18-wheeler, a semi, or a tractor trailer. These trucks can be as big as a house. They travel the highways delivering goods, including food. A big rig is a truck that has two parts — a tractor and a trailer. The driver sits in the tractor or cab and the trailer is hooked to the tractor. The freight is carried in the trailer.

There are different kinds of trailers, depending on the goods that the truck is carrying. A **reefer van,** which is like a refrigerator on wheels, carries food that can spoil. Inside the reefer, a gas-powered motor keeps the air cold. A stainless-steel **tanker** carries liquid loads, including milk. The load inside a tanker can slosh around. When this happens, it can throw off the truck's balance and make it hard to drive the truck. Walls inside the tanker, called **baffles,** slow down the sloshing. The next time you see a tanker, look for numbers printed on the truck. These numbers are a code that tells firefighters what's inside the tanker.

What are the advantages of shipping by truck?

The highway transportation system provides the most flexibility of all the transportation systems. Highways and roads are everywhere, which makes it possible for trucks to take routes that planes, barges, freighters, and trains can't take. If drivers discover there is a problem ahead, they can take alternative routes.

What are the disadvantages of shipping by truck?

One of the biggest disadvantages is traffic congestion. It is estimated that about 60 percent of truck delays are the result of events like construction work zones, crashes, breakdowns, and extreme weather conditions. When trucks are stuck in traffic, they sit with the engines idling, which wastes fuel.

(continued on next page)

Name   Date

Moving Food by Truck

Most big trucks use diesel engines, which burn diesel fuel. Diesel fuel is not like gasoline. It is heavy and sooty. Because of this, diesel engines can be hard to start. When it is cold, the fuel becomes jelly-like. Some drivers keep their engines running all night to keep the diesel fuel warm and fluid.

What are some of the challenges of shipping by truck?

Driving safely is one challenge. Sometimes truck drivers cause accidents by falling asleep at the wheel. Today there are laws that say how many hours per week drivers can work and how often they have to stop and rest. Sometimes one driver will rest in the sleeper while someone else drives. Drivers have to keep logs that show when they drove and when they rested. Truckers work on weekends because the traffic is lighter and they can travel farther. Truckers need to pay attention to the weather — rain, fog, heat, and blizzards.

Drivers also have to take care of their trucks. They need to know what to do if the truck breaks down far away from a city or town. Truckers keep an eye on their truck's tires. They look for cuts and bubbles that might show there is a problem with a tire. They also make sure there is enough air in the tires and enough tread to grip the road. Trucks need regular checkups by a mechanic.

Does the highway transportation system have an impact on the environment?

Transportation is a form of human activity, and human activity has an impact on the natural environment. For example, the process of building a road destroys habitat. To prepare the landscape to put in a road, trees are cut down and removed. Any wildlife that interacted with the trees will be affected when the trees are chopped down. Birds or small animals that nested in the trees need to find a new habitat. Sometimes wetlands have to be drained to put in highways. Wetlands are a critical habitat for many kinds of plants and animals. They are also important in terms of filtering and cleaning water. Environmental laws now require that wetlands be created somewhere else to replace wetlands that are destroyed. However, the plants and animals that depended on the destroyed wetland will still be affected, even if a new wetland is created elsewhere.

Once the highways are built, there are other effects, including air pollution, noise pollution, and water pollution. Have you ever noticed the solid walls that are sometimes built between a highway and the homes along it? These walls are noise barriers, which can help reduce traffic noise by about half. To work, they have to be tall and long with no openings.

(continued on next page)

Name   Date

Moving Food by Truck

Exhaust fumes contain nitrogen oxides and tiny particles called **particulate matter** that pollute the air and cause health problems. Particulate matter can pass through the nose and throat and get stuck in the lungs and cause lung damage. The fine particles can also make respiratory conditions like asthma and bronchitis worse.

When rainwater or melting snow washes off roads and bridges, it picks up dirt, dust, rubber, and metal deposits from tire wear, antifreeze, engine oil that has dripped onto the pavement, fuel that has spilled at fueling stations, pesticides and fertilizers used on roadside vegetation, and litter and carries them into lakes, streams, rivers, and oceans. In regions where it snows and salt is used on the roads, runoff containing salt can create high sodium and chloride concentrations in ponds, lakes, and bays. This changes the water chemistry and can kill fish.

DID YOU KNOW...?

- The longest Interstate highway is I-90, which runs from Seattle, Washington, to Boston, Massachusetts, a distance of about 3,020 miles.

- The shortest Interstate route segment is I-95 in the District of Columbia, which is 0.11 mile long.

- Alaska is the only state without any Interstate routes.

- East-west Interstate route numbers end in an even number. North-south routes end in an odd number.

- Interstates carry about 60,000 people per route mile a day, 26 times the number of all other roads, and 22 times the number of rail passenger services. Over the past 40 years, that's the equivalent of a trip to the moon for every person in California, New York, Texas, and New Jersey combined.

- About 1.5 billion pounds of dynamite and other explosives were used to make tunnels, cut through hills, and for other purposes for the Interstate system.

- Almost 2.5 billion tons of aggregate (sand, gravel, and crushed stone of various sizes) have been used in building the Interstate system. This would make a stockpile two miles in diameter and one mile high.

Name   Date

Transportation System Guiding Questions

Your teacher has asked you to become an expert on one kind of transportation. Your assignment is to create a poster or a diagram that helps you teach the others in your "home" group what you have learned. Use the questions below to guide your work.

1. What kind of transportation are you learning about? _____

2. What kind of vehicles does your transportation system use?

_____

_____

_____

3. What two points does your transportation system connect?

_____

4. Name three jobs that humans perform to support this kind of transportation.

     a. _____

     b. _____

     c. _____

(continued on next page)

Name   Date

Transportation System Guiding Questions

5. Name five parts of your transportation system.

a. _____

b. _____

c. _____

d. _____

e. _____

6. Describe how these parts interact.

_____

_____

_____

_____

_____

7. How does your transportation system interact with parts of the environment?

_____

_____

_____

_____

_____

(continued on next page)

Name	Date

Transportation System Guiding Questions

8. How does your transportation system impact the environment?

_____

_____

_____

_____

_____

9. What are the advantages of using your transportation system?

_____

_____

_____

_____

10. What are the disadvantages of using your transportation system?

_____

_____

_____

_____

Food-System Synthesis

AIM

To synthesize what we have learned about the food system.

SCIENTIFIC PROCESSES

- construct knowledge, reflect

OBJECTIVES

Students will be able to:

- respond to the Unit 2 Question;

- identify parts of the food system;

- demonstrate understanding of the interacting parts of the food system.

OVERVIEW

Throughout this unit, students have been learning about parts of the food system and the functions they serve in getting food from farm to table. The emphasis has been on interacting parts. In this culminating lesson, students synthesize what they have learned thus far. They create a brace map that lets them demonstrate what they have learned about the food system, parts of the food system, and the relationships between the parts and the whole. As a class, students brainstorm a list of food-system parts they will include in their brace map. The ***Mapping Relationships*** lesson resource provides some sample maps. Next, students work in groups to create a flowchart that illustrates what happens to one food on its journey between the farm and the table. The goal is for students to show as many interactions as possible. The lesson concludes with students reflecting on what they have learned and writing in their LiFE Logs. In Unit 3, students will conclude their food-change observations and investigate different kinds of processes used to extend the shelf life of food.

MATERIALS

For the teacher:
- *Mapping Relationships* lesson resource
- *Food-System Flowchart* lesson resource

For the class:
- Materials from the *Food-System Synthesis Supply List* lesson resource

For each group:
- Chart paper
- Markers
- *Brace-Map Organizer* activity sheet

- *Flowchart Organizer* activity sheet
- (Optional) Index cards
- (Optional) Tape

For each student:
- LiFE Log

PROCEDURE

Before You Begin:

- Review the *Mapping Relationships* and *Food-System Flowchart* lesson resources.

- Gather the materials listed on the *Food-System Synthesis Supply List* lesson resource. Students will refer to these products when they work on their flowcharts.

- Make copies of the *Brace-Map Organizer* and *Flowchart Organizer* activity sheets for each group. If students will be working in pairs, make copies for each pair. Review the activity sheets.

- If you have not already done so, post the Module Question and the Unit 2 Question at the front of the class.

MODULE QUESTION

What is the system that gets food from farm to table, and how does this system affect the environment?

UNIT QUESTION

What is the system that gets food from farm to table?

 THEORIZING

1. Review Module and Unit Questions

Explain to students that this is the final lesson in the unit, and it is an opportunity to reflect on and synthesize what they have learned about the Unit 2 Question.

2. Introduce Brace Maps

Tell students they will be creating diagrams that show the food system and the parts of the food system that they have been studying. Explain to students that creating this diagram is how they will begin to answer the Unit 2 Question. They will be thinking about all that they have learned about the parts of the food system and how these parts interact. You may wish to allow students to review their LiFE Log notes from earlier lessons, as well as their student readings. Remind students that they have been learning about not only the food system as a whole, but also about the parts and how they interact.

Explain that brace maps are graphic tools that students can use to organize and represent what they know. If students are not familiar with this type of visual tool, sketch an example on the board as a model. Refer to the *Mapping Relationships* lesson resource for sample maps.

3. Brainstorm Parts

Ask a student volunteer to read the Unit 2 Question out loud. Tell students that this is the question their brace maps will answer. Explain that the map starts with the whole. *What do we call the system that gets food from farm to table?* (The food system.) *What are some parts of the food system?* (Transportation, airplane, trains, packaging, farming.) Record student responses on chart paper or the board.

Distribute the *Brace-Map Organizer* activity sheet to students. Have students work in pairs or in small groups. Tell students that the brace map on the activity sheet is to help them get started. They may find they want to add lines to it. Students should customize the map to fit what they know about the food system and its parts.

When students have completed their map, invite volunteers to share their maps with the class. Encourage students to discuss the maps and ask questions. *Does anyone disagree with what is shown on this map? Does anyone have a map that shows different parts?* Students may show different parts or may think of other interactions. If students disagree or map the parts and the whole in a different way, remind them to cite the evidence they have to support their view. Display student maps.

4. Model Flowchart

Have students work in groups of four to create a flowchart that illustrates what happens to one food between the farm and the store. Explain that their goal in making this flowchart is to show as many interactions as possible. Tell them to think about the kind of feedback that there might be at each step of the process and who would benefit from this feedback. Tell students that they will be comparing these diagrams to the ones they made in Lesson 4. By noting the differences in the two diagrams, students can assess what new knowledge they have gained about the food system.

Walk students through an example. *If your flowchart is showing what happens to an apple between the farm and the store, and the apples get bruised during shipping, who finds out the apples are bruised?* (Store owner, consumer, possibly a quality-control person somewhere in the system.) *Who needs to get the feedback about the bruised apples so it doesn't happen again?* (Person who packages the apples, or the person who packs them in boxes — it depends on where the apples were bruised.) *What happens to bruised apples if they reach the store?* (Store owner can't sell them or consumer can't eat them.) *What happens if the apples can't be sold or people can't eat them?* (Store owner loses money; maybe the store owner won't buy apples from the farmer again; consumer may not buy apples from the store again; apples become waste.) The *Food-System Flowchart* lesson resource models one way to show feedback on the flowchart diagram. Demonstrate this for students.

5. Make Flowchart

Have students work in small groups. Distribute the *Flowchart Organizer* activity sheet or have students record their descriptions on index cards and arrange them on chart paper with arrows showing the flow. You may find it works best to use a roll of paper or to give student groups several sheets of chart paper to tape together to have enough room for their flowcharts. Hold up the materials from the *Food-System Synthesis Supply List* lesson resource. Help each group choose a food, or assign one food to each group — some ideas are: dried apples, canned vegetable soup, corn-flakes cereal, frozen corn, tomato sauce, and packaged animal cookies. Foods that have one or only a few main ingredients work best for this activity. If a group chooses a food with many ingredients, have them focus on one or very few main ingredients. Explain that students should draw or describe in words what the food — or the main ingredient — looks like when it is harvested and ready to leave the farm. This information goes in the first box. Next, ask students to fill in the steps that happen to the food between the farm and the store. Have students briefly describe what happens at each step. Encourage students to indicate each place along the farm-to-store system where there would be feedback — both positive and negative.

Consider having students use the cards for the written description. Then have students place the cards on the chart paper and indicate the flow with arrows. Above each of the cards, have students use pictures to show what is happening at this particular point in the system. For example, if the card indicates that workers are harvesting apples, students would draw an orchard with workers and write why this step is an important part of the whole system. Below the arrow and descriptive card, students describe how this part of the system could break down and the affect this breakdown would have on the other parts of the system. Review the *Food-System Flowchart* lesson resource for a sample of how the flowchart could look.

As student groups work on their diagrams, help them by posing questions to prompt their thinking: *What is the main ingredient in this food? Is it in the same form as when it grows? How has it been physically changed since it left the farm? Where did this change take place? Who or what changed the food? Is it packaged? If so, what is the package made from? Where was the food packaged? How did the food get inside the package? How did the food travel from place to place? Where is energy being used? How many stops did the food have to make?*

6. Presenting Food-System Flowcharts

Ask a representative from each group to present the group's diagram to the class. Have each group's presenter describe what is happening, starting with the farm and ending with the store. Encourage students to ask questions. You may wish to consider hanging these diagrams in the classroom so students can modify their drawings as they learn more about food systems. Students can look back at these diagrams and track how their thoughts change as they complete the lessons in this module.

7. Homework

Ask students to write in their LiFE Logs and reflect on what they have learned. Have them write several paragraphs that respond to the statement "The most interesting thing that I have learned about the food system and its parts is …"

Mapping Relationships

Graphic organizers are visual tools that help students organize information and communicate what they have learned to others. The brace-map graphic organizer helps students understand relationships between a whole and its parts. The examples below illustrate some of the possible brace maps that students can create to show different aspects of interacting parts of the food system.

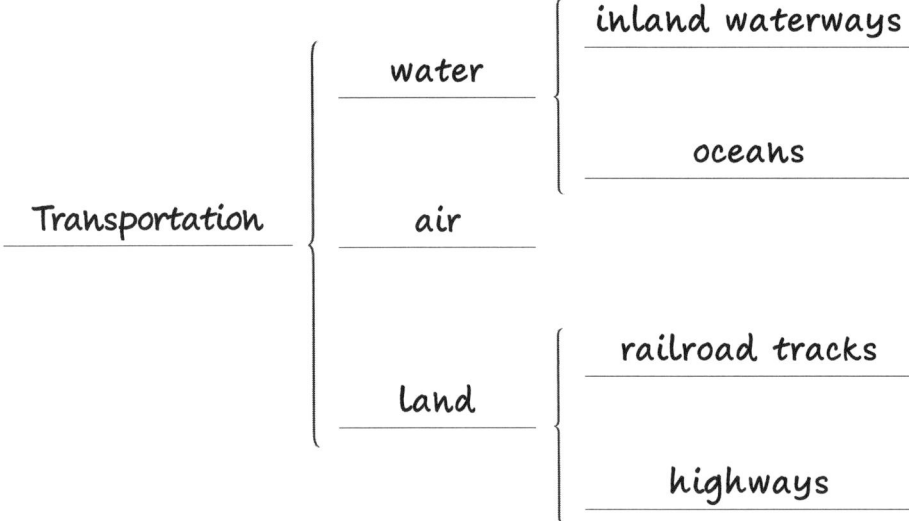

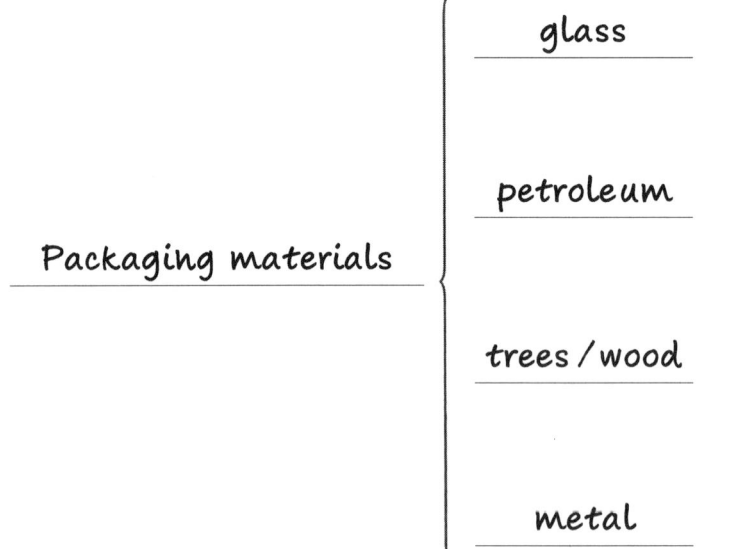

Food-System Flowchart

Use this sample flowchart of an apple moving through the farm-to-table system. On the sample, we show one way to incorporate student drawings into the flowchart by placing the illustrations above the flowchart. Feel free to incorporate them in any way that works for your classroom. We've found that students enjoy illustrating the steps; some have illustrated the entire flowchart. Do what works for your students. Don't forget to have students include feedback loops.

FOOD: APPLESAUCE

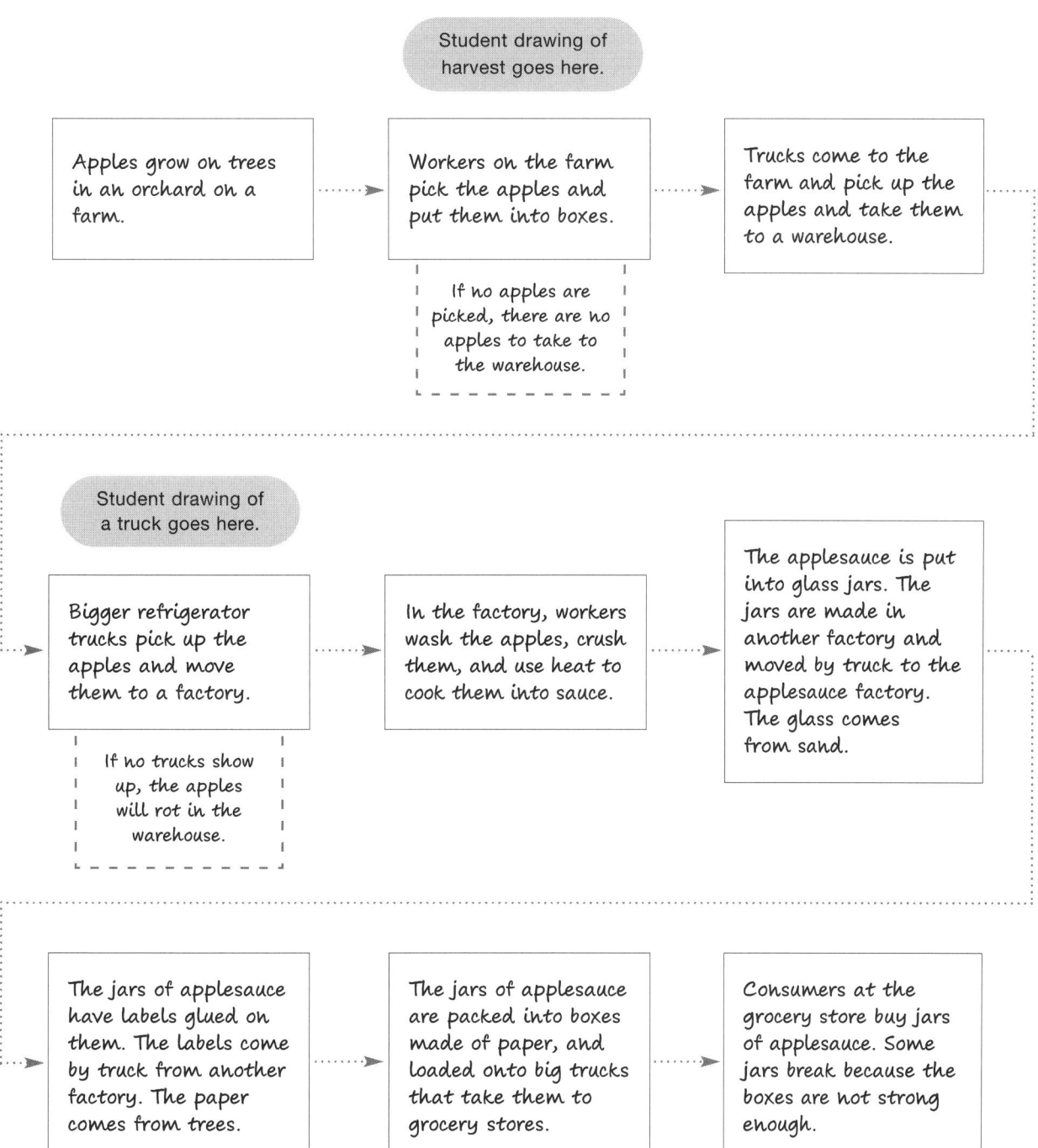

Food-System Synthesis Supply List

MAKE FLOWCHART (p. 174)

Supplies

For the class
- 1 package dried apples (or other dried fruit)
- 1 can vegetable soup
- 1 box corn-flakes cereal
- 1 package frozen corn
- 1 can tomato sauce
- 1 package animal cookies

Name   Date

Brace-Map Organizer

Use this map to help you organize your thinking about the system that brings food from the farm to the table.

Name   Date

Flowchart Organizer

Use this flowchart organizer to organize your thoughts about what happens to food as it moves from the farm to the table.

Food Processing

Analyzing Food-Change Data

AIM

To develop theories about why food changes.

SCIENTIFIC PROCESSES

- **gather data, compare, build theories, apply**

OBJECTIVES

Students will be able to:

- **analyze their food-change data;**

- **describe how the food changed;**

- **articulate theories about the food changes;**

- **discuss and debate the theories of others.**

OVERVIEW

This lesson returns to the food-change observations started in Lesson 6. Students make final observations of their tomatoes, strawberries, and grapes and then summarize their data. After analyzing their data, students build their own theories about what caused the food spoilage, and what the microorganisms on the food need to grow. Challenge your students to think hard about what they have observed. Remind them to compare their observations over time to their baseline data. Student groups share their theories with the class. Encourage questions, discussion, and debate among students. This lesson gives students the foundation they will need as they study how and why we dry, can, and freeze food later in this unit. For homework, students look in their homes for examples of foods that have been preserved and record them in their LiFE Logs.

MATERIALS

For the teacher:
- *Developing Theories about Food Changes* sample conversation
- *Preserving Food* teacher note
- *Students' Food-Change Theories* lesson resource

For each group:
- Strawberry, tomato, and grape from food-observation experiment (Lesson 6)
- Chart paper
- Markers

For each student:
- Completed *Food-Change Monitoring* activity sheets
- *Food-Change Final Data* activity sheet (pp. 105–109)
- *Food-Change Theories* activity sheet
- LiFE Log

PROCEDURE

Before You Begin:

- Gather materials.

- Review the ***Developing Theories about Food Changes*** sample conversation, the ***Preserving Food*** teacher note, and the ***Students' Food-Change Theories*** lesson resource.

- Make copies of the ***Food-Change Final Data*** activity sheet for each student if you did not already do so in Lesson 6. Make copies of the ***Food-Change Theories*** activity sheet for each student.

- Gather the completed ***Food-Change Monitoring*** activity sheets from earlier lessons to have available for students to use to analyze their data and develop their theories.

- If you have not already done so, post the Module Question and Unit 3 Question at the front of the classroom.

MODULE QUESTION

What is the system that gets food from farm to table, and how does this system affect the environment?

UNIT QUESTION

What happens to food as it moves from farm to table?

 QUESTIONING

1. Review Module and Unit Questions

Introduce the Unit 3 Question. Tell students that in Unit 3 they will be studying how we change food from the time it is grown to the time we eat it. Explain that in this lesson students are going to analyze the food-change data they have been collecting.

 EXPERIMENTING

2. Final Observations

Explain to students that during this activity, they will make the final observations of the food they have been studying. Remind students to make detailed observations and to handle the foods carefully to minimize damage. *What signs of change in the food should you be looking for?* (Changes in color, smell, texture, and damage to the skin.) If your class has worked with a compost box, discuss the connection between that kind of decomposition and the tomato, strawberry, and grape spoilage.

Have students work in their food-change observation groups and complete the ***Food-Change Final Data*** activity sheet. Give each group a sheet of chart paper and markers. When students have finished collecting their data, have each group record its observations of what happened to the foods on the chart paper.

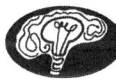 THEORIZING

3. Developing Theories

Distribute the ***Food-Change Theories*** activity sheet. After students have finished summarizing their data, invite them to develop theories to explain why the observed changes occurred. Have them record their observations and theories on the activity sheet. This part of the activity may be challenging. Remind students that theories are their ideas, based on evidence they have collected, about why they think micro-organisms grew on the food. The illustrations of the tomato, strawberry, and grape on students' monitoring sheets should help them develop theories to explain their results. The illustrations highlight parts of the foods that could specifically contribute to the growth of microorganisms, such as damage near the stem or a bruised area.

Explain that although developing theories may be difficult, it is an important part of being a food scientist. Encourage students to discuss ideas with their group members and to think openly. Remind them that a theory is an idea about what happened. It is not a fact that is correct or incorrect. It is an explanation based on evidence. Be sure to stress that the groups' theories must be supported by observation. Refer to the ***Students' Food-Change Theories*** lesson resource when leading this discussion.

4. Share Results

When students have recorded their theories, invite each group to share its observations and theories with the class. Remind students that they are food scientists presenting data from their study of food change. Encourage them to challenge the explanations proposed by others if they disagree. Tell them to ask presenters for more information if they don't understand another group's theory. This type of work helps students develop skills in giving evidence to support their arguments and in participating in scientific debates. See the ***Developing Theories about Food Changes*** sample conversation for an idea of how this discussion might go.

Explain that learning how fresh fruits and vegetables change when they are left out at room temperature will help students understand why we preserve food and build a foundation to understand how and why different methods of preserving food work.

APPLYING TO LIFE

5. LiFE Logs

Ask students to take out their LiFE Logs and respond to this focus question: "The most interesting thing I have learned about food change is..."

6. Homework

Explain to students that in this unit they will learn about how microorganisms grow, and what we can do to prevent them from growing on food, thus making food last longer. This is called preserving food. Ask students to find five foods in their homes that have been preserved. Give them a hint that if the food is not fresh, like fresh fruit or vegetables, it has been preserved. Have students list the foods and how they think the foods are preserved in their LiFE Logs.

Developing Theories about Food Changes

This sample conversation in the **Theorizing** phase of the QuESTA cycle will help you guide your students through a conversation that will enable them to construct new knowledge. As you engage students in your own conversation, encourage them to debate the interpretation of the results of their experiment, to explain their own theories about what they learned, and to recognize how they might use this new knowledge in their daily lives. This is a guide. Feel free to adjust your questioning to the needs of your class and to continue the conversation beyond what is presented below.

MR. B: Let's think about the results of our food-change experiments. *Hilary, why do you think that your group's strawberry got smaller and wrinkly over the course of the experiment?*

HILARY: It's like it was getting old. It looked like my grandmother! She's wrinkly like a raisin.

MR. B: Okay, that's an interesting idea. *Can you say more?*

HILARY: Well, my grandmother complains to me she is shrinking and getting wrinkled.

MR. B: *I like how you're thinking about this, but how is the strawberry like your grandmother? And what's your idea about a raisin?*

JORDAN: I think it's like a balloon that after the first day gets smaller and smaller 'cause the air inside it escapes.

MR. B: Good thinking. *So what's escaping from the strawberry?*

HILARY: But the strawberry isn't full of air like a balloon. I think the juice is escaping and it's drying out, like raisins. Grapes are juicy but raisins are small and wrinkly.

MR. B: Great connection! So your theory is that the juices inside the strawberry started drying out and the inside got smaller so the skin started to wrinkle. *Now, what about that fuzzy stuff on your tomato, Andrew?*

ANDREW: It looks like fur. It started out small and now it's bigger. Maybe something furry is growing on it, like the stuff that makes peaches fuzzy on the outside.

MARA: But peaches start out fuzzy and taste good. The stuff on the tomato looks gross, like mold or something.

MR. B: *What do you mean by mold, Mara?*

MARA: I don't know but once I found an old piece of cheese in the refrigerator and it had all this green stuff on it and my mom said it was mold and we threw it away.

MR. B: Okay, so mold is not something you want to eat. *Any ideas about why it grew on the tomato?*

ANDREW: Well, that spot where the mold is was kind of cut open and oozing a little bit the last time we observed the food. Maybe it liked the juice inside of the tomato.

MICHELLE: That happened to our grape, too. At the top, where it came off the stem, it was oozing a little bit and then today it has all this fuzzy mold on it.

MR. B: So it sounds like we have some evidence for your theory, Andrew. *Does anyone disagree? Were any group's results different than Andrew's and Michelle's?* Remember that as scientists when you disagree with someone else's theory, you need to support your ideas with evidence from your experiment.

Preserving Food

This is the first lesson in the Food Processing unit. In this unit, students learn how and why food is preserved and processed, another part of the farm-to-table food system. In Food Processing, students add to their knowledge of food systems by investigating what happens to food after it is harvested and before it arrives at the store. In this first lesson of the unit, students analyze the data they have collected in the *Food Change* experiment and begin to develop their own understanding of why food is processed and preserved.

As soon as food is harvested, it begins to deteriorate. Once food spoils, it is no longer edible. If any of the fruits that students studied were damaged or bruised, they were particularly susceptible to microorganisms that enter through the damaged spot and begin to break down the tissue. One of the main reasons that spoilage occurs is the growth of fungus and bacteria on food.

In this unit, we introduce food change and spoilage through the lens of food processing and preservation. Our emphasis is on the interactions of chemicals and the use of food-processing technology to slow down the growth of microorganisms. Because this module was developed for students in grades 5 or 6, we introduce microorganisms rather than investigate microbial activity. Research suggests that studying spoilage as the result of microorganisms should wait until middle school. Studies of student learning have indicated that students were not aware that microbes initiate the process of decay.[1] Students tended to think that insects break up the material after it begins to rot of its own accord. Even after instruction, 15- and 16-year-olds used the terms "bacteria," "fungi," and "decomposers," but were not able to describe what they do.[2]

Preserving food serves multiple purposes. Preserved foods taste different and add variety to our diets. Sometimes they are more convenient. Opening a can of beans is less time-consuming than soaking dry beans and cooking them. Preserved foods also last longer on store and kitchen shelves than fresh foods do. At one time, people depended on food from sources that were relatively local and were available during their growing season. People "put food by" to have it outside the growing season. Throughout this unit, students will learn about different ways that foods are processed and preserved to extend their life span. Salting, drying, smoking, heating, and cooling are some of the ways to slow down the growth of microorganisms.

---

[1] American Association for the Advancement of Science, 1993, p. 184
[2] Driver, 1994, pp. 65-66.

Students' Food-Change Theories

Below are some examples of results and explanatory theories your students might record after their final observations of the food-change experiment. Remind students to look at their baseline data and to compare that data to the observations they made as they monitored the food. Encourage your students to think hard about the changes they have observed. They should be as thorough as possible in articulating their results and coming up with theories. Explain that the more descriptive they are in summarizing their results the easier it will be to come up with theories that explain those results.

TOMATO	
Observations	**Theories to Explain Observations**
1. Where there was a cut in the skin there was a long line of white mold growing.	**1.** The cut allowed some of the juice in the tomato to get out. The mold grew in the juicy area.
2. A soft spot that started out really small ($1/8$ inch wide) grew really big ($7/8$ inch wide).	**2.** The soft spot was caused by some bacteria growing on the tomato and the bacteria multiplied during the experiment.
3. The smell changed from a sweet, fresh, crispy tomato smell to a strong, unpleasant smell that reminded us of garbage.	**3.** At the beginning we were smelling the tomato, but by the end we were smelling the mold and bacteria that were growing on the tomato.
4. The color changed from a uniform bright red to darker red with several really dark areas.	**4.** The bacteria and mold that grew on the tomato caused changes inside the tomato that caused the skin to look darker.
5. The skin went from firm and smooth to wrinkled.	**5.** Some water evaporated from the tomato, shrinking the inside and causing the skin to wrinkle.

Name   Date

Food-Change Theories

Look at the summaries of the food-change data you have been collecting. Based on the evidence you have, why do you think the food changed? Look back at the drawings of the food and think about what might have caused the changes. Be sure to look back at your description of how the food looked on the first day you observed it.

Observations

Use these questions to guide your observations. Record your answers on the table on the next page.

1. Did you observe any damage or bruises on the food? Where were they?

2. How does the food look now? Has anything about it changed? Does the food look different? What looks different? Is it a different color? Does the food look dry instead of juicy?

3. Does the food look different? What looks different?

4. Is it a different color?

5. Does the food look dry instead of juicy?

Theories

Use these questions to help develop your theories.

1. Why do you think the food changed the way it did?

2. What is the evidence to support your theory?

(continued on next page)

Farm to Table & Beyond
©2008 Teachers College Columbia University

BECOMING FOOD SCIENTISTS : INTERACTING PARTS : FOOD PROCESSING : ENVIRONMENTAL EFFECTS : WASTE : MAKING CHOICES

Name	Date

Food-Change Theories

NAME OF FOOD_____	
Observations	**Theories to Explain Observations**

BECOMING FOOD SCIENTISTS : INTERACTING PARTS : **FOOD PROCESSING** : ENVIRONMENTAL EFFECTS : WASTE : MAKING CHOICES

Becoming Food Processors

AIM

To begin to develop an understanding of food processing.

SCIENTIFIC PROCESSES

- **question, experiment, analyze, apply**

OBJECTIVES

Students will be able to:

- **explain what it means to process food;**

- **describe a process for grinding grains;**

- **describe how heavy cream is turned into butter;**

- **discuss how our lives would be different without food-processing technology.**

OVERVIEW

In this lesson, students begin their studies of different ways to change food. Students move through a series of food-processing stations that provide them with hands-on experience processing grains and turning heavy cream into butter, which they use in Lesson 12 when they make pancakes. The purpose of this lesson is to give students an opportunity to experience grinding as one form of food processing and to begin to extend their thinking of technology to include tools, including kitchen tools. By gaining experience with the labor-intensive activity of grinding flour by hand, students begin to appreciate what motivated people to look for more efficient ways to grind grains. For homework, students read about food-processing technology and consider how they process food in their own homes. Throughout this unit, students explore humankind's use of food-processing technology, such as pounding, salting, cooking, and fermenting.

MATERIALS

For the teacher:
- *Learning-Center Tips* lesson resource
- *Food-Processing Learning Centers* experiment sheet
- *Flour and Butter Demonstration* experiment sheet

For the class:
- Materials from the *Becoming Food Processors Supply List* lesson resource
- Chart paper
- Markers
- 2–3 hand lenses for each station

For each student:
- *Food-Processing Technology* student reading
- *Grains* student reading
- *Grain Observations* activity sheet
- *Milling Observations* activity sheet
- LiFE Log

PROCEDURE

Before You Begin

- Review the *Food-Processing Learning Centers* and *Flour and Butter Demonstration* experiment sheets and gather all the listed materials on the *Becoming Food Processors Supply List* lesson resource. Review the *Learning Center Tips* lesson resource.

- Soak the dried dent corn or hominy for the learning centers before class begins. Let it soak for about an hour. Set up the learning centers.

- Test the coffee grinder to make sure it works and set up the hand grinder, if you are using it.

- Review and make copies of the *Food-Processing Technology* and *Grains* student readings and the *Grain Observations* and *Milling Observations* activity sheets for students.

- If you have not already done so, post the Module Question and Unit 3 Question at the front of the classroom.

MODULE QUESTION

What is the system that gets food from farm to table, and how does this system affect the environment?

UNIT QUESTION

What happens to food as it moves from farm to table?

 **QUESTIONING**

1. Review Module and Unit Questions

Remind students of the Module and Unit questions. Tell students that in this lesson they process, or change, three different kinds of grain: wheat, corn, and oats. They also process heavy cream. Explain that in this lesson, they

are processing ingredients in preparation for cooking pancakes in the next lesson. Invite students to speculate on what products are made from wheat berries, corn, and heavy cream. *What happens to grains before we eat them? How do you think the heavy cream will change?* Accept all answers and record them on the board.

 **EXPERIMENTING**

2. Introduce Learning Centers

Refer to the *Food-Processing Learning Centers* experiment sheet. To do the experiment, divide the class into groups of four. One group begins at the grains station and a second group works at the milling station. Tell students that in the *Food Processing* experiment, they are going to be working in learning centers to observe grains, and then grind them. Make certain that students understand that they will all rotate through each center and have a chance to do each activity. Explain the team jobs and how they will rotate through the stations. Give each team leader a copy of the experiment sheet. If you would like to have a third station, create a reading station where students read the *Grains* student reading and quietly discuss what they have read. Use an egg timer so teams can keep track of their time at each station.

Before you begin, engage students in a brief discussion of technology. Make sure they understand that they will be using kitchen technology in this lesson. Students may be surprised to learn that kitchen tools are examples of technology. They may think of technology in terms of computers, cell phones, and MP3 players. As you begin this lesson and throughout the unit, check for student understanding of technology. Remind students that one of the distinguishing characteristics of human culture is our ability to shape tools. Tools — technology — have made it possible for humans to change the world to suit our needs, including our food needs.

3. Review Observations

Once students have completed their observations, have group reporters share their team's observations. Next, have students take out the *Grains* student reading. Remind students that people have different ways of thinking about the world, including natural resources, like plants. Explain that farmers and botanists look at wheat, corn, and oats differently. Tell students botanists study plants. When a botanist looks at wheat, corn, and oats, she sees three members of the grass family. Look at the plant illustrations. *Imagine you are a botanist, what can you learn about these plants? How are they alike or different?* Point out that the seed head is one feature that helps botanists identify grasses.

Remind students that farmers grow food. Where a botanist sees grasses, a farmer sees food crops. Farmers use the term "grain" to describe any plants that produce grains we eat. Explain that farmers use different terms than botanists to describe grains. *Look at the kernels; what parts do you see?* Explain that the following terms refer to parts used in foods: bran used in cereal and flour, oil from the germ, and flour from the endosperm. If students have not read *Grains,* assign it as homework.

4. Grinding Flour

Move on to the *Flour and Butter Demonstration* to show students how to grind the wheat and corn with the hand grinder or coffee grinder. Discuss how grinding changed the wheat and the corn and the differences in grinding these two grains. Invite students to share their observations. Encourage them to talk about the time and energy it took to grind the flour, both by hand and with the grinder.

5. Making Butter

Show students the heavy cream and record their observations of it on the board.

Tell students that heavy cream and milk both come from cows. Heavy cream is at least 36 percent milk fat while whole milk is at least 3.25 percent milk fat. Follow the steps on the *Flour and Butter Demonstration* experiment sheet. *Do you recognize the whipped cream? What happens to it as the jar gets shaken more? Do you know what the liquid is inside the jar with the butter? Any ideas?* Explain that the heavy cream first changed to whipped cream and then into butter. You will know when you have butter when there is a solid clump in the jar. While students are making butter, you or a few students may want to grind more wheat and corn. You will need a total of one cup of each for making the pancakes in the next lesson.

6. LiFE Log

Have team members review their activity sheets from this lesson and record the step-by-step process for grinding flour from corn, wheat, or oats in their LiFE Logs. Invite volunteers to share their descriptions.

6. Homework

Have students read the *Food-Processing Technology* student reading. Ask students to identify a tool or a machine in their home that was invented to help people process food. Have students list ways in which this technology is helpful and list any problems that this tool has caused. Ask them to think about whether or not the tool requires some form of energy to run. Remind students to record this information in their LiFE Logs.

Learning-Center Tips

Collaborating with others by working in learning centers is fun and engaging, especially if you follow some simple tips. We urge you to establish rules for working in learning centers. Use the rules below, or you may wish to brainstorm a list with the class. Post the rules on chart paper and review them with students before they begin working in the centers.

If your students are not familiar with working in centers, take a few minutes to walk them through the process. Make sure all students understand both the rules and the process.

Rules

1. Wait until the teacher is finished working with another group to ask questions.

2. Stay at your station until you are instructed to move to the next station.

3. If you have a question:

   a. Reread the information sheet.

   b. Ask someone in your group for help.

   c. Write your question on a piece of paper and continue to work until you see that your teacher is no longer working with another group and is free to help you.

Rotating through the Stations

Design a rotation strategy that can be displayed in front of the classroom for easy reference.

Job Assignments

If possible, have students work in teams that are no larger than four. Assign each student in the group a job. Having a job will help students stay focused. Here are some suggestions:

1. Project leader

2. Materials gatherer

3. Science researcher

4. Reporter

1. Project Leader

• Make sure everyone is on task.

• Watch the time and voice level of group members.

(continued on next page)

- Make sure that your team completes its work.

- Take care of the cleanup before you move on to the next station.

2. Materials Gatherer

- Gather the materials for your group.

- Return all materials when finished.

- Report any spills or accidents to the teacher.

3. Science Researcher

- Gather all the team's activity sheets and student readings.

- Report to the teacher if your team needs help understanding the activity.

- Turn in your team's work.

4. Reporter

- Complete all the activity sheets.

- Summarize your group's findings.

- Report your group's findings to the class.

Becoming Food Processors Supply List

We suggest using corn, wheat, and oats in the *Food-Processing Learning Centers* experiment. You only need corn and wheat for the pancake recipe in the next lesson. We added the oat kernels because they are relatively soft and students will be able to grind them into rolled oats using the rolling pin. You should be able to find different kinds of grains in a health-food store. If you have difficulty, you can substitute nuts and have students grind the nuts.

GRAIN-OBSERVATION STATION (p. 191)

Supplies

For each group of 4 students
- 3 small paper plates
- Several kernels of dent corn (hominy), oats, and wheat berries (soft wheat kernels)
- (Optional) 1 ear of dent corn
- (Optional) Wheat stalk
- (Optional) Oat stalk
- Small bowl of water

MILLING STATION (p. 192)

Supplies

For each group of 4 students
- 1 one-quart mixing bowl
- 2 cups water
- 1 wooden rolling pin
- 1 mortar and pestle
- 1 cutting board (or other hard surface to crush the kernels)
- 1/4 cup soft wheat berries (soft wheat kernels)*
- 1/2 cup dent corn (hominy) or dried sweet corn; (not soaked)**
- 1 cup dent corn (soaked)
- 1/4 cup oat kernels
- 4 small bowls or cups to hold grain

* You should be able to find wheat berries at health food stores and some supermarkets. If possible buy soft wheat as it will be easier for the students to grind.

** We have used both hominy and dried sweet corn. Most supermarkets carry hominy. For dried sweet corn, we have used "Just Corn." It is sold at health food stores or online at: *www.justtomatoes.com*, *www.amazon.com* or *www.wildernessdining.com*. If you buy dried sweet corn, save any that is left over to use in Lesson 15, when students explore different ways to preserve food.

(continued on next page)

FLOUR AND BUTTER DEMONSTRATION (p. 193)

Supplies

For the class
- Coffee grinder
- 3–4 cups wheat berries
- 3–4 cups dried corn ("Just Corn" works well)
- Measuring cups
- 2 containers with lids
- 1 cup heavy cream (If you use ultra-pasteurized heavy cream, it will take longer to "churn" into butter than unpasteurized heavy cream. We have been successful at finding unpasteurized heavy cream at local dairies.)
- 1 32-ounce jar with lid
- (Optional) Grain hand grinder

Food-Processing Learning Centers

In this experiment, groups of four students rotate through stations to observe three different grains and investigate grinding as a method of food processing. These learning centers provide students with hands-on experience in observing the edible part of cereal plants, or grains, and processing food.

GRAIN-OBSERVATION STATION

Students closely examine different kinds of grains and record their observations. They use hand lenses as tools to extend their vision.

Setup

Label three paper plates: "Wheat," "Corn," and "Oats." Put out the plates, the bowl of water, and the hand lenses.

Procedure

1. Review the ***Learning-Center Tips*** lesson resource with students. Brainstorm a list of rules with the class and post them on chart paper.

2. Have the materials gatherer for the group pick up the wheat, corn, and oat kernels.

3. Have the science researcher pick up the activity sheets.

4. Have the project leader review the activity and discuss tasks with the rest of the team members.

5. All team members use hand lenses to make observations about the three different grains. The reporter records the team's observations on the ***Grain Observations*** activity sheet.

6. After completing the activity sheet, the team cleans up the station and moves on to the next one.

Questions

1. *How are the different grains alike?*

2. *How are they different?*

MILLING STATION

Students experiment with technology by using tools to grind grains by hand.

Setup

1. If you are using dried dent corn, place 1 cup dent corn and 2 cups water in the mixing bowl 1 hour before class begins. If using dried sweet corn, do not soak it.

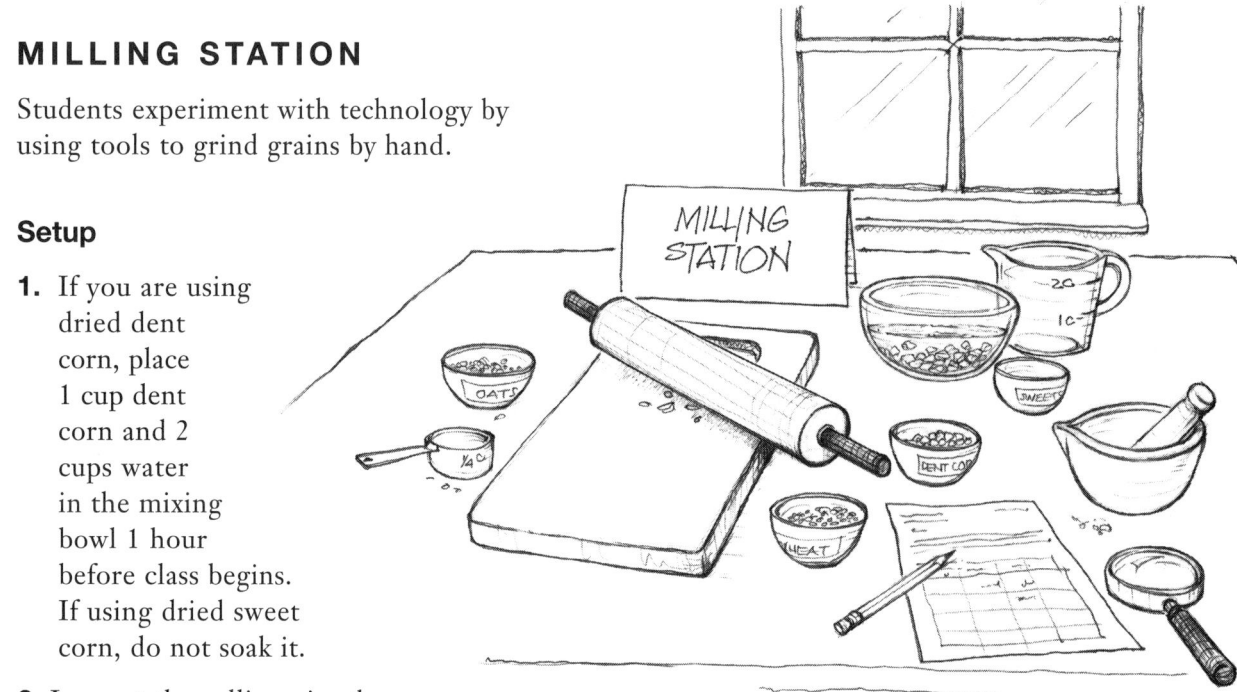

2. Lay out the rolling pin, the mortar and pestle, and the cutting board.

Procedure

1. Have the materials gatherer for the group pick up the wheat, oat, and corn kernels. You can either measure out the materials for each group, or have the materials gatherer do this task. Be sure to keep the soaked corn separate and label it so students will know which corn they are trying to grind.

2. Have the science researcher pick up the activity sheets.

3. Have the project leader review the activity and discuss tasks with the rest of the team members.

4. Team members grind the unsoaked hominy in the mortar and pestle. The reporter records what the team does and their observations. Next have team members grind half of the soaked hominy. If you use only dried sweet corn, students will grind only these unsoaked kernels.

5. Team members use the rolling pin and cutting board to grind the wheat, oats, and the remaining half of the soaked corn kernels. The reporter records what the team does and their observations.

6. Team members use the hand lenses to examine the kernels. The reporter records their observations on the *Milling Observations* activity sheet.

7. After completing the activity sheet, the team cleans up the station and moves on to the other station. If students have rotated through all the stations, have them read the *Grains* student reading.

Questions

1. *How would you describe the process of grinding the grain? Was it difficult? Did it require much energy?*

2. *What was similar and different about grinding the corn, oats, and wheat?*

3. *What tools or technology did you use to help you grind the grain?*

Flour and Butter Demonstration

In this demonstration, students observe another tool that can be used to grind grains into flour. Students also learn how to change cream into butter.

Setup

If you have a hand grinder, use it to grind some wheat and corn. If you do not have a hand grinder use a coffee grinder.

Procedure

1. Grind the wheat berries and the corn. If time permits, have student volunteers assist you. You will need one cup of cornmeal and one cup of wheat flour for the pancakes. Briefly discuss what it was like to grind the wheat and corn with a grinder and compare it to students' experiences of grinding by hand. Store the flour for later in the lidded containers.

2. Show students the heavy cream. Have them describe what they see in their LiFE Logs.

3. Pour the cup of heavy cream into the 32-ounce jar. Screw the lid tightly on the jar. Explain that to process the cream into butter, students will take turns shaking the jar. Pass the jar around and have each student shake the jar about five times and quickly pass to the next person. Stop and check the cream every so often. Have students record what they see in their LiFE Logs. Eventually the heavy cream will stick to the sides of the jar. This is the butter taking form.

Tip: The more vigorously students shake the jar, the faster the cream will turn into butter. This is a great way for students with some extra energy to work it off!

4. Keep shaking until butter starts covering the sides of the jar. The liquid you see in the jar is buttermilk.

5. Pour the buttermilk out of the jar. You can use it to bake something or you can drink it as it is. After you pour out the buttermilk, you will have a jar of butter.

Questions

1. *Did it take more energy or less energy to make the flour using a grinder?*

2. *Are there any trade-offs when you use an electric grinder instead of grinding flour by hand?*

3. *Did you notice any difference between the amount of flour from the wheat kernels and the amount of flour from the corn kernels? What ideas do you have to explain any differences?*

4. *What are the trade-offs if you buy butter in the store instead of making it at home and by hand?*

Name   Date

Grains

Guiding Questions

- *What are three grains that we eat?*

- *What are some foods that we get from grains?*

- *Where does wheat grow?*

- *Where does most of our corn grow?*

- *Where do oats grow?*

- *Why do we have to process grains before we can eat them?*

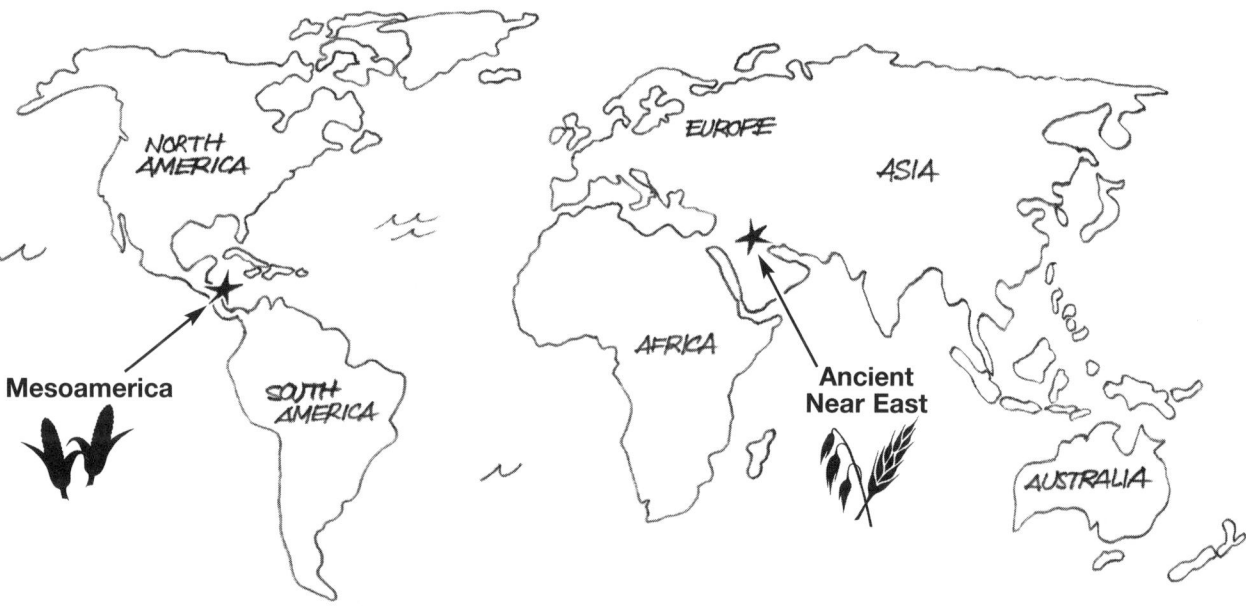

If you ate **cereal** for breakfast, you were eating grass. Seriously. Cereal, or grain, is a type of grass. Wheat, barley, rye, oats, and corn are some of the most popular cereal grains. The name "cereal" comes from the Roman goddess Ceres, who was the protector of grain. Wheat and barley were the grains that farmers first planted more than 10,000 years ago. Corn is the one grain that was originally from Mesoamerica; wheat, barley, rye, and oats came from the Ancient Near East.

In this lesson, you will be investigating three grains that are popular in the United States: wheat, corn, and oats. Read on to learn more about them.

(continued on next page)

Farm to Table & Beyond
©2008 Teachers College Columbia University

Name   Date

Grains

Top 10 Producing States for Each Grain*

KEY

Wheat   Corn   Oats

2007 USDA Crop Production Data

Wheat

Wheat is grown all over the United States. It is grown mostly in the Great Plains
states, including North Dakota. In fact, farmers in North Dakota grow more wheat
than any other crop. There are many different kinds of wheat. They are named by
color, texture, and growing season. For example, there is winter and spring wheat, red
and white wheat, and hard and soft wheat. Hard red spring wheat is used as flour to
make yeast breads, dinner rolls, bagels, and pizza crust. Winter wheat is planted in the
fall. During the winter, its seedlings are **dormant,** which means they are not growing.
Winter wheat needs this cold period before the plants start growing again in the
spring. Winter wheat is used for flour, too.

Wheat is an **annual** grass that grows between two and four feet high. When the plant
is ready to harvest, it turns a golden color. The roots hold the plant firmly in the soil.
Leaves capture energy from the sun to make food for the plant through the process of

(continued on next page)

Name   Date

Grains

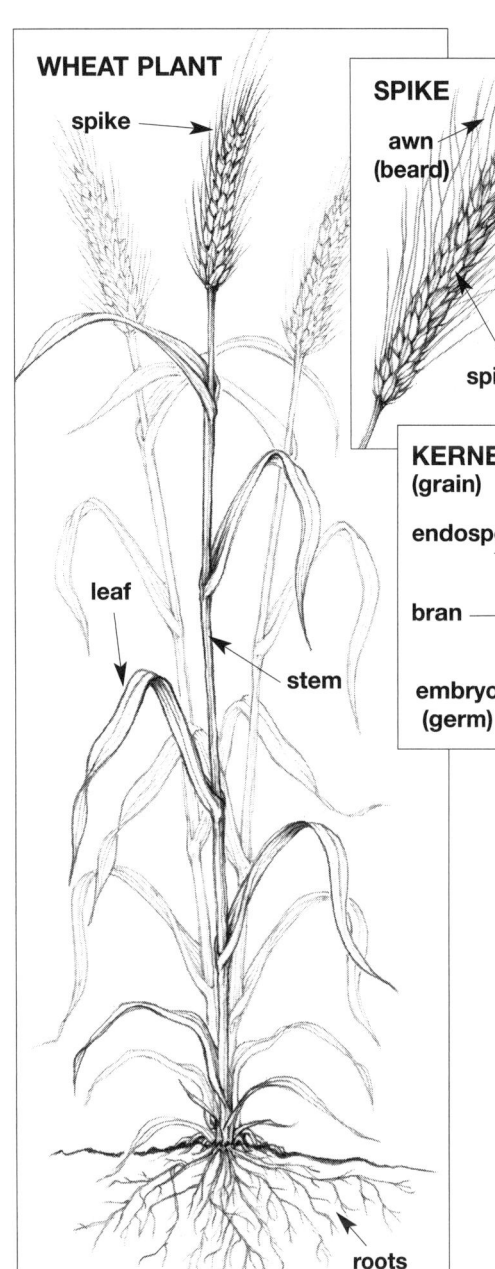

WHEAT PLANT
spike
leaf
stem
roots

SPIKE
awn
(beard)
spikelet

KERNEL
(grain)
endosperm
bran
embryo
(germ)

photosynthesis. Grasses have flowers. On the wheat plant, the stalk of flowers is called the **spike.** The flowers grow in small clusters called **spikelets.** The fruit that wheat flowers produce is the "seed" we eat as a grain. Each grain, or kernel, has three parts: the **bran,** the **germ,** and the **endosperm.** The bran is the protective outer layer that contains most of the fiber, along with vitamins and minerals. The endosperm is the starch inside that makes up most of the berry. It is the food that the seed uses to germinate. The germ is the part of the seed that sprouts. It is rich in vitamins, minerals, and oil.

The process of grinding wheat into flour is called **milling.** At the mill, the wheat is cleaned. First, it passes through magnets to remove any metal. Next, machines remove weed seeds. Then water is used to wash away dirt and stones. Next, heavy rollers crush the grain into flour. Finally, the flour is sifted many times before it is packaged and transported to markets. To make white flour, the bran and the germ are separated from the endosperm through sifting. Whole-wheat flour contains all three parts of the seed. It is more nutritious than white flour since it includes the parts of the kernel that contain the vitamins and minerals.

Corn

In the United States, we produce more bushels of corn than any other grain crop. In 2005, Iowa corn farmers grew almost 2.2 billion bushels of corn on 12.5 million acres of land. A bushel of corn, without husks and cobs, weighs about 56 pounds. Most of the corn grown in the United States is grown in what we call the Corn Belt in the Midwestern states.

(continued on next page)

LESSON 11: BECOMING FOOD PROCESSORS

Name   Date

Grains

There are three basic kinds of corn grown in the United States: dent corn, sweet corn, and popcorn. Most of the corn grown here is **dent corn.** Yellow dent corn is used as animal feed or made into cornstarch, corn syrup, and other products. White dent corn is used to make corn flour, hominy, and grits. We eat sweet corn as corn on the cob, frozen, or in cans, and we pop popcorn.

Corn plants can grow to between seven and ten feet tall. Each plant has a **tassel** at the top with hundreds of small flowers that produce pollen. Long swordlike leaves grow outward from the stalk. An ear of corn grows where the leaves join the stalk and each plant usually has one or two ears of corn. The kernels, or seeds, are found in rows on the ears.

Like the wheat kernel, corn kernels have parts. Some of the parts are the same: the bran, the endosperm, and the germ. The bran, or **pericarp,** is the hard outer coating of the seed. The endosperm is the starchy inside of the kernel. The germ is the part of the seed that sprouts. Corn kernels also have a **tip cap,** which is the pointy part where the kernel was attached to the cob.

Corn has to be processed before it is sold in the market, too. When corn is made into flour or corn oil or grits, first it is cleaned. Stones and metal are removed by passing the corn through mesh screens and magnets. Next, water is used to clean the corn to remove dirt and dust. Then corn millers, or processors, use hot water and steam to soften the bran and germ so they can be separated from the endosperm. Next, the corn is dried, cooled, and sifted. Finally, rollers and sifters create smaller bits of corn that are made into grits, cornmeal, and flour.

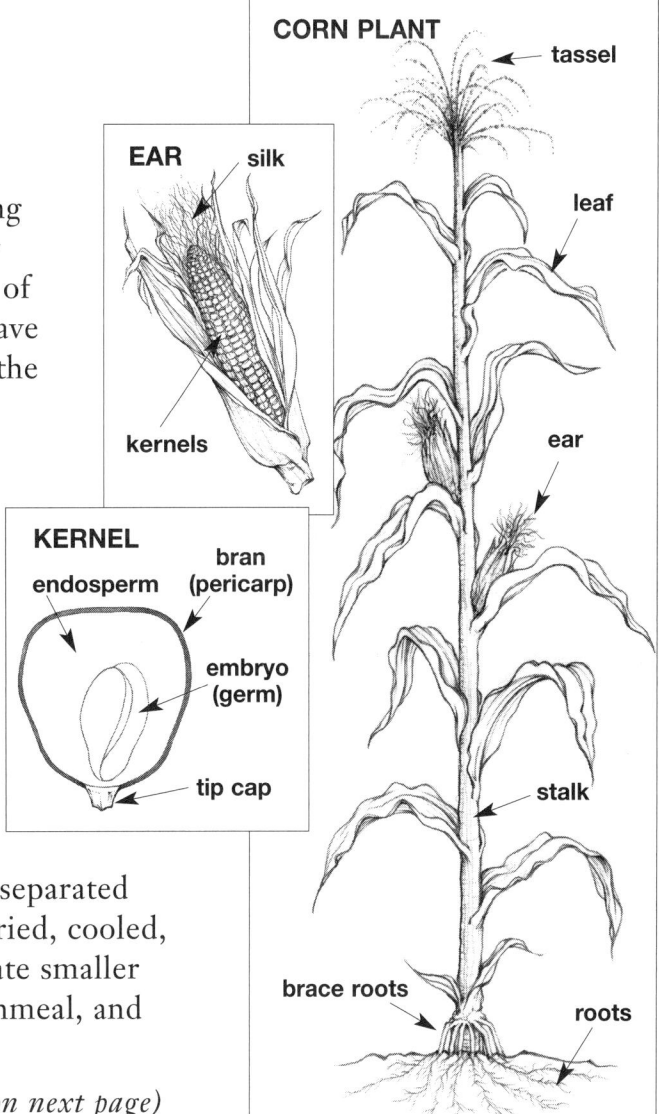

(continued on next page)

Name	Date

Grains

Oats

In the United States, most oats are grown in the north-central states. Oats need lots of moisture and relatively cool weather to grow. Like wheat, oats have either a spring or winter growing season. Red and gray oats are planted in the fall, like winter wheat. White oats are planted in the spring, like spring wheat. More than half of the oats grown are used for grain. Oats and oat bran are most often eaten at breakfast and in snack foods. Oats are a major ingredient in granola cereal and bars.

OAT PLANT
panicle
leaf
main stem
roots

PANICLE
pedicel (stalk)
spikelet

KERNEL
endosperm
bran (seed coat)
embryo (germ)

On an oat plant, the stalk of flowers is called a **panicle.** The oat spikelets are attached to the branches of the panicle by **pedicels,** or stalks. Most oat spikelets are made up of two or more flowers. Like the wheat flowers, the oat flowers produce fruit, which is the "seed" or kernel we eat. One oat plant can produce more than 100 kernels.

Oats are cleaned to remove unwanted materials. The oats pass through machines that take out any material that isn't an oat. Next, machines screen the oats and separate them according to length and width. A large oat kernel is about half an inch long. Then air is used to remove loose hulls and other remaining unwanted material.

The next step is the hulling process. The outer shell, or hull, is removed from the inner kernel and the oats are cleaned. Next the oat kernels are cut and put in packets. Kernels that are used in rolled oats are partially cooked and then passed through machines that roll them into flakes. Next the flakes are cooled, weighed, and packaged.

BECOMING FOOD SCIENTISTS : INTERACTING PARTS : **FOOD PROCESSING** : ENVIRONMENTAL EFFECTS : WASTE : MAKING CHOICES

Name   Date

Food-Processing Technology

Guiding Questions

- *How does technology help us process food?*
- *Why do we process food?*
- *List different kinds of food-processing technology.*

We use technology every day to process our food. Sometimes the tools we use are quite simple, like a knife to chop vegetables. Sometimes the technology is very complex, like the milling process for grinding wheat kernels into flour. But whether the technology is simple or complex, the end result is the same. It changes the food. The chopped vegetables are in bite-sized pieces for us to enjoy. The hard wheat kernels are ground and come out as a soft, silky flour.

People have been processing food since the first time humans made tools. Early cultures used stone tools to process foods. Hunters and gatherers would grind nuts and seeds to make flour. Some of the kinds of tools that **archaeologists** have found include **mortars, pestles,** and grinding stones. After the wheel was invented, people used the wheel to help them grind grains.

Heating food also changes it. Think about the development of cooking technology since early humans first cooked with fire. Long ago, people cooked in large fireplaces. They had pots that hung over the fire on a hook. Today, we have electric stoves, gas stoves, convection ovens, microwave ovens, and solar cookers. All of these are different ways of heating and cooking food. We also use heat to preserve foods. Have you ever canned fruits or vegetables or made jams or jellies? If you have, you used heat to process food.

We also use cold to preserve food. Think about the freezer in your refrigerator at home. What do you keep in there? You may have frozen juice or vegetables or fruit. Someone froze those foods when they were fresh. Now you can eat them or drink the juice whenever

(continued on next page)

Name   Date

Food-Processing Technology

you want. Frozen foods and canned foods make it possible for us to have food when it's not in season. For example, you can thaw frozen berries and eat them when it's snowing outside. Or, you can have orange juice when there are no oranges on the trees.

Think of all the pasta sauce, canned tomato sauce, jam, and jelly that you have eaten. All of these are processed foods. If you think about it, it's amazing how much processed food you eat. Sometimes several ingredients are mixed together. If you can still recognize the ingredients, then this food is processed a medium amount. Think about a bowl of vegetable soup. When you look at it, you can still recognize the different vegetables. They have been cut up, mixed together, and cooked, but you can still recognize them.

Sometimes food is processed so much that it is very hard to tell what the original ingredients were. This food is processed a large amount. Think about a cake or pancakes. When you look at them, you can't recognize the flour, sugar, milk, or baking powder that were used.

When you go to a supermarket, many of the foods that are sold there are called "processed food products." This often means that the food was made in a factory and is made up of several different ingredients. As a food scientist, you are going to explore different kinds of food-processing technology and learn more about different ways that people change food.

Name   Date

Grain Observations

Complete the table below. If you have a hand lens, use it as a tool to extend your vision. The hand lens magnifies objects, so the grains will look larger and you will see more detail. For other observations, you might consider the weight of the grain, or if it floats in a cup of water.

OBSERVATIONS				
Kind of Grain	Color	Shape	Texture (hard, soft, smooth)	Other Observations
Wheat				
Corn				
Oats				

Sketch each grain and label the sketches.

BECOMING FOOD SCIENTISTS : INTERACTING PARTS : FOOD PROCESSING : ENVIRONMENTAL EFFECTS : WASTE : MAKING CHOICES

Name   Date

Milling Observations

Follow the procedures below. Take turns grinding the grains. Make sure that each team member has a chance to try grinding.

MORTAR AND PESTLE

Wheat

1. Take about 5 grains and place them in the mortar.

2. Use the pestle to grind the grains. Pound them very carefully. Try to keep all the grain inside the mortar.

3. After the grains have broken, use the hand lens to examine them. Describe what you see.

_____

_____

4. Continue pounding the grain until it is like flour. Look at the flour with the hand lens and describe what you see.

_____

_____

5. Clean out the mortar.

Dried Corn

1. Take about 5 kernels of dried corn and place them in the mortar.

2. Use the pestle to grind the kernels. Pound them very carefully. Try to keep all the kernels inside the mortar.

(continued on next page)

LESSON 11: BECOMING FOOD PROCESSORS

Farm to Table & Beyond
©2008 Teachers College Columbia University

BECOMING FOOD SCIENTISTS : INTERACTING PARTS : **FOOD PROCESSING** : ENVIRONMENTAL EFFECTS : WASTE : MAKING CHOICES

Name   Date

Milling Observations

3. After the kernels have broken, use the hand lens to examine them. Describe what you see.

_____

_____

4. Continue pounding the kernels until they are like flour. Look at the flour with the hand lens and describe what you see.

_____

_____

5. Clean out the mortar.

Soaked Corn

1. Take about 5 kernels of soaked corn and place them in the mortar. Be sure to leave some of the soaked corn kernels for use later.

2. Use the pestle to grind the kernels. Pound them very carefully. Try to keep all the kernels inside the mortar.

3. After the kernels have broken, use the hand lens to examine them. Describe what you see.

_____

_____

(continued on next page)

Name   Date

Milling Observations

4. Continue pounding the kernels until they are like flour. Look at the flour with the hand lens and describe what you see.

_____

_____

5. Clean out the mortar.

ROLLING PIN AND CUTTING BOARD

Oats

1. Take about 5 kernels and place them on the cutting board.

2. Use the rolling pin to grind the oats. Use pressure to push down on the rolling pin as you roll over the oats. Try to keep all the oats on the cutting board.

3. After you have crushed the oats, use the hand lens to examine them. Describe what you see.

_____

_____

4. Continue rolling the oats until they are like flour. Look at the flour with the hand lens and describe what you see.

_____

_____

5. Clean off the cutting board.

(continued on next page)

         Farm to Table & Beyond
©2008 Teachers College Columbia University

BECOMING FOOD SCIENTISTS :: INTERACTING PARTS : **FOOD PROCESSING** : ENVIRONMENTAL EFFECTS :: WASTE :: MAKING CHOICES

Name	Date

Milling Observations

Soaked Corn

1. Take about 5 kernels of soaked corn and place them on the cutting board.

2. Use the rolling pin to grind the corn. Use pressure to push down on the rolling pin as you roll over the corn. Try to keep all the corn on the cutting board.

3. After you have crushed the corn, use the hand lens to examine the pieces. Describe what you see.

_____

_____

4. Continue rolling the corn kernels until they are like flour. Look at the flour with the hand lens and describe what you see.

_____

_____

5. Clean off the cutting board.

Questions

Read the questions below. Write your answers in your LiFE Log.

1. Which grain was the easiest to grind?

2. Which tool did you think was the easiest to use to grind the grains?

3. How would your life be different if there were no tools to grind grains?

4. How would your life be different if you and your family had to grind all the grains that you ate?

Pancake Science

AIM

To investigate some of the physical and chemical changes that occur when cooking pancakes.

SCIENTIFIC PROCESSES

- question, experiment, apply

OBJECTIVES

Students will be able to:

- list the source of all the ingredients used to make pancakes;

- describe how the wet and dry ingredients are combined to make a thick liquid pancake batter;

- discuss how the cooking process transforms the liquid batter to a solid pancake;

- describe a chemical reaction that occurs when making pancakes.

OVERVIEW

In the previous lesson, students were food processors and prepared some ingredients to make pancakes. In this lesson, students continue to explore food processing by investigating physical and chemical changes. First students observe a kitchen chemistry demonstration. They observe what happens when two common baking ingredients — baking soda and baking powder — are placed in vinegar and in water. Students then participate in a cooking activity and observe baking powder in action. At the end of the lesson, students think about and discuss how much the wheat and corn kernels changed as they were processed from grains to pancakes and what processes were involved in those changes. For homework, students investigate different pancake products that are available in stores as a way of learning that the same food can be processed differently. Students will apply what they have learned in the next lesson as they uncover the degree of food processing in different forms of store-bought pancakes.

Note: Always check for food allergies before cooking.

MATERIALS

For the teacher:
- *Pancake Ingredient Chart* lesson resource
- *Cooking Tips* lesson resource
- *Kitchen Chemistry* experiment sheet
- *Wheat-and-Corn Pancakes* teacher recipe

For the class:
- Materials from the *Pancake Science Supply List* lesson resource
- (Optional) One or more copies of *Pancakes, Pancakes!* by Eric Carle

- (Optional) Overhead projector and transparency

For each group:
- Hand lenses
- Markers

For each student:
- *What Causes the Fizz?* student reading
- *Kitchen-Chemistry Observations* activity sheet
- *Pancakes* student recipe
- LiFE Log

PROCEDURE

Before You Begin:

- If possible, have another adult assist you with this lesson. Students cook over heat and another adult supervisor is most helpful.

- Read the *Kitchen Chemistry* experiment sheet and the *Wheat-and-Corn Pancakes* teacher recipe. Gather the materials listed on the *Pancake Science Supply List* lesson resource.

- Review the *Pancake Ingredient Chart* and the *Cooking Tips* lesson resources.

- (Optional) Check with your school or community librarian and get one or more copies of Eric Carle's *Pancakes, Pancakes!* (Simon & Schuster, 1990). Although written for younger students, Carle's book is an engaging way to review the process of making pancakes — from growing the wheat to the finished pancake product.

- Make an overhead transparency of the *Wheat-and-Corn Pancakes* teacher recipe or make enough copies of it to distribute to student groups.

- Have students wash their hands just before you begin this lesson.

- Make cooking setups for student groups.

- Review and make copies of the *What Causes the Fizz?* student reading, the *Kitchen-Chemistry Observations* activity sheet, and the *Pancakes* student recipe for each student.

- If you have not already done so, post the Module Question and Unit 3 Question at the front of the classroom.

MODULE QUESTION

What is the system that gets food from farm to table, and how does this system affect the environment?

UNIT QUESTION

What happens to food as it moves from farm to table?

 **QUESTIONING**

1. Review Module and Unit Questions

Remind students of the Module Question and the Unit Question. Ask student volunteers to list a few ways that food changes as it moves from the farm to their tables. Invite students to share some of the different kinds of food-processing technology they found at home. *What are some of the tools you found at home that your family uses to process food?* Have students list ways in which each tool is helpful and any problems that it has caused. Ask them to think about whether or not the tool requires some form of energy to run. Remind students to record this information in their LiFE Logs.

 **EXPERIMENTING**

2. Kitchen Chemistry

Explain that to learn about chemical reactions, the class is going to observe a demonstration to see what happens when two kinds of baking ingredients are mixed with liquids. Use the *Kitchen Chemistry* experiment sheet to guide your demonstration. Depending upon the size of your class and the amount of time you have, either do this demonstration twice so all students can gather around you to watch or do the demonstration and have student volunteers walk around the class to show the other students the acid-base reactions. Have students complete the ***Kitchen-Chemistry Observations*** activity sheet as you conduct the demonstration.

After completing the activity sheet, read the ***What Causes the Fizz?*** student reading as a class. Have students compare their observations and ideas to what they learn in the reading.

APPLYING TO LIFE

3. Pancake Ingredient Chart

Follow the directions on the ***Pancake Ingredient Chart*** lesson resource for making the chart on the board. Invite student volunteers to help you fill in the sources in the left-hand column. Help students understand that you are looking for the ingredients in their original form, such as a wheat plant. Post the chart at the front of the classroom.

4. Review Recipe

Divide the class into four groups. Display the ***Wheat-and-Corn Pancakes*** teacher recipe on the overhead or give each group a copy. Explain the instructions and point out that students will need to look at the recipe to determine how much of each ingredient they will need. They also need to carefully measure the ingredients. If you think there are students in your class who are not familiar with measuring spoons and measuring cups, this is a good time to review them. You want to avoid having a student put 1/4 cup of salt into the batter when it should have been 1/4 teaspoon of salt.

Pose the following questions: *Have you ever made pancakes before? Did you make the pancakes the same way as we are going to make them today? Did you use all the same ingredients in your pancakes? Where did the ingredients come from? Did you process them yourself?*

5. Establish Safe Cooking Practices

No doubt your students will enjoy cooking and eating the pancakes. However, cooking can be messy and challenging. Based on what we have learned from cooking with students, we have added tips to make cooking a success for you and your class. Review the ***Cooking Tips*** lesson resource as a class and ask students to make additional suggestions. Post the tips at the front of the classroom.

6. Make Pancake Batter

Tell students that they are going to make the pancake batter. Explain that the term "batter" refers to the ingredients mixed together before they are cooked. Give each of your four groups a large bowl. Make sure that each group has the appropriate set of cooking tools and ingredients (two groups will have the dry ingredients and tools and two groups will have the wet ingredients). Explain how to share utensils and ingredients if more than one group needs a tool or ingredient that you only have only one of, such as a bottle of vanilla extract. Once the groups are finished mixing their ingredients, each dry group can combine its ingredients with one of the wet groups into one bowl. When you finish, there will be two bowls of batter ready to cook. Turn back to the ***Pancake Ingredient Chart*** and invite student volunteers to fill in the right-hand columns.

7. Cook Pancakes

After the batter is ready, tell students that they are ready to cook the pancakes. Ask them to think about the batter and watch it very carefully as they pour it onto the griddle. *What kind of ingredients did you mix together to get the batter?* (They mixed dry and wet ingredients.) *What state of matter is the pancake batter?* (Fluid or liquid). *How can you tell that the batter is a liquid?* (It takes on the shape of the container.) *What would happen if we did not heat the griddle before we poured the batter into the pan?* (It would keep flowing until it reached the edges of the pan.) *Why does heat change what happens?* (The batter begins to get solid as the heat starts to cook it.)

Look closely at the batter as it cooks. The pancake batter becomes a mixture of solid and gas. Baking powder, the leavening agent, produces carbon dioxide when it is mixed with the wet ingredients. When heat is added, more carbon dioxide bubbles are formed and the "trapped" carbon dioxide expands. Point out that baking powder is called "double-acting." This means that some of the gas is given off when the baking powder comes in contact with moisture. More gas is given off when the batter is exposed to heat. The gas trapped in the batter causes the bubbles and they expand when they are heated.

Encourage students to examine the pancakes as they cook. Make sure they do not get too close to the heat source. Engage students in a discussion as they watch the pancakes cook: *Why does one side look smoother than the other? Which side cooked first? Do you think that makes a difference? Can you see where the bubbles were on the second side?*

(Optional) While you continue making pancakes, invite students to read Eric Carle's book *Pancakes, Pancakes!* Explain that the book will help students review what they have learned about pancakes and food processing.

8. Eat and Enjoy

Pass out clean plates and forks. Be sure to allow enough time for students to sit down, relax, and eat. As students get their pancakes, offer the fresh butter that they made in Lesson 11 and maple syrup to top the pancakes. You may also want to offer strawberry jelly for students who would like to eat the way the characters do in *Pancakes, Pancakes!* Remember to eat with your students and encourage any other adults who are in the room to join you. Part of the cooking process should include relaxing and enjoying what you have made.

9. Clean Up

After everyone finishes eating, have students help collect the plates and throw them away or save them for use in your compost bin. Collect the forks and knives, wash them, and put them away for future use. Recruit students to wipe off all tables, pick up anything on the floor, and sweep if necessary.

10. LiFE Logs

Ask students to take out their LiFE Logs and write at least a two-paragraph summary, in their own words, of all the steps it took to make the pancakes. Remind students to include grinding the wheat and corn flour and "churning" the butter.

11. Homework

Distribute the *Pancakes* student recipe to each student. Remind students that the pancakes they made in class were made from scratch. People can also buy pancakes at the supermarket. With these pancakes, some of the processing has already been done. Ask students to list as many different ways to buy pancakes as they can in their LiFE Logs. If possible, have them research different kinds of pancake mixes, batters, and prepared pancakes available in a local market. If it isn't possible to visit a store, have students look for ads in magazines or interview a family member who does the food shopping. You may need to give them hints: mixes (those that you add only water to or add your own milk, eggs, and oil to), premade refrigerated batters, frozen pancakes, and so on.

Pancake Ingredient Chart

On the board, draw a table as shown below. Label the three columns across the top: "Source," "Ingredient," and "How They Are Mixed." List the ingredients in the middle column. Leave the other two columns on the board blank.

PANCAKES		
Source	**Ingredient**	**How They Are Mixed**
Plant (Wheat)	Whole-wheat flour	Dry ingredients are mixed together, making a blended powder
Plant (Corn)	Cornmeal	
Plant (Wheat)	White flour	
Mineral	Baking powder	
Mineral	Salt	
Animal (Chicken)	Eggs	Wet ingredients are mixed together and then combined with dry ingredients to make a thick, liquid batter. This is cooked to make solid pancakes.
Animal (Cow)	Milk	
Plant (Corn)	Corn oil	
Animal (Bees)	Honey	
Plant (Vanilla bean)	Vanilla	
Plant (Maple tree)	Maple syrup	Put on top of the pancakes after they are cooked.
Animal (Cow)	Butter	

Pancake Science Supply List

KITCHEN CHEMISTRY (p. 214)

Supplies

For the class demonstration
- 4 eight-ounce clear plastic cups
- 4 labels for the plastic cups
- 2 index cards
- 2 small plates
- 2 labels for the plates
- 4 stirring sticks or spoons
- 3 tablespoons baking soda (sodium bicarbonate, $NaHCO_3$)
- 3 tablespoons baking powder (sodium bicarbonate; potassium bitartrate, $KHC_4H_5O_6$; and corn starch)
- Measuring spoons
- $1/4$ cup measuring cup
- Small bottle of vinegar
- Warm water
- Small plastic dishpan or plastic tub

WHEAT-AND-CORN PANCAKES (p. 216)

Ingredients

Yields 16–32 small pancakes
- $1/2$ cup white flour
- $1/2$ cup whole-wheat flour, ground in Lesson 11
- $1/2$ cup cornmeal, ground in Lesson 11
- 1 tablespoon baking powder
- $1/2$ teaspoon salt
- 1 egg
- $1 1/2$ cups milk
- 3 tablespoons honey
- 1 tablespoon corn oil
- $1/2$ teaspoon vanilla extract
- Cooking spray or butter to grease the pan
- "Churned" butter (from Lesson 11) for the top of the pancakes
- (Optional) Strawberry jelly if you want to eat the way the characters do in Eric Carle's book *Pancakes, Pancakes!*
- Real (not artificial) maple syrup, preferably from a local source

(continued on next page)

Supplies

For the class
- 2 sets of measuring cups
- 2 sets of measuring spoons
- 2 large bowls
- 2 (or more) large mixing spoons
- 1 hot plate and frying pan or 1 electric griddle
- 1 spatula

For each student
- 1 small paper plate
- 1 plastic fork
- 1 plastic knife
- 1 napkin

Notes:

- Use basic white paper plates, with no wax or dyes, if you want to save them to add to your compost box, if you have one. Wash the plastic forks and knives and save them to use again.

- If you have access to a kitchen or sink, use washable plates and flatware.

- If you work with two wet-ingredient and two dry-ingredient groups, you will need twice the ingredients listed here.

Cooking Tips

Cooking with others is fun, especially if you follow these simple tips. They are not hard and will help you keep your cooking adventures safe, healthy, and enjoyable for everyone. Bon appétit!

Sanitation

1. Wash your hands before you begin cooking. After you have washed your hands, be sure you do not touch anything except the cooking materials. If you do, wash your hands again.

2. Try to keep food from falling on the floor. If food does fall on the floor, be sure to throw it away.

3. If you feel a cough or a sneeze coming, turn away from the food and cover your mouth. Be sure to wash your hands after coughing or sneezing.

4. Wash your hands again if you:

– scratch your head;

– wipe your nose;

– touch the floor;

– touch anything that might make your hands dirty.

Safety

1. Be careful with the hot plates and hot skillets. Do not touch hot plates or skillets if they are hot. Make sure an adult is always nearby when hot plates or skillets are in use.

2. Walk — do not run, jump, or skip in the classroom.

3. When you pass materials to others, do it with dignity and respect.

Making Cooking Enjoyable for All

1. Check for any allergies to any of the ingredients.

2. Treat everyone with respect.

3. When it comes time to eat, if you don't like something, politely say, "No thank you." Please don't say it is "gross" or "nasty." Remember, you worked together as a class to prepare this food. Others want to enjoy it.

4. It's okay to talk quietly with your cooking partners. But when an adult calls for your attention, please stop talking and listen.

5. Be sure to compliment each other on a job well done.

Kitchen Chemistry

Students observe what happens when baking soda and baking powder, two leavening agents, are mixed in different liquids. Depending upon the size of your class and the amount of time you have, either do this demonstration twice so all students can gather around you to watch or do the demonstration and have student volunteers walk around the class to show the other students the acid-base reactions.

Setup

1. Gather the materials.

2. Set this up as a demonstration at the front of the room.

3. Make four labels, one for each plastic cup. Label two of the cups "Water" and the other two cups "Vinegar." Tape the labels to the cups and arrange in pairs, with one "Water" cup and one "Vinegar" cup in each pair.

4. Write "Baking Soda" on one index card and place it on the desk or table next to one pair of plastic cups. Write "Baking Powder" on the second index card and place it next to the other pair of cups.

5. Label the plates "Baking Soda" and "Baking Powder."

6. Place a small amount of baking soda on one plate and a small amount of baking powder on the other. About 1/2 teaspoon of each is enough.

Procedure

1. Place 1/4 cup vinegar in each of the plastic cups labeled "Vinegar." Place 1/4 cup warm water in each of the "Water" cups.

2. Have the first group of students gather at the front of the room. Ask students to record what they observe and what they smell.

3. **Baking Soda Demonstration.** Measure 1/4 teaspoon baking soda. Add it to the first plastic cup of vinegar and lightly stir with a stirring stick. Have students record what they observe.

4. Measure 1/4 teaspoon baking soda. Add it to the first plastic cup of water and lightly stir. Have students record what they observe.

5. Have students use the hand lens to examine the baking soda that is on the plate. Tell them to record what they observe.

6. Baking Powder Demonstration. Measure 1/4 teaspoon baking powder. Add it to the remaining plastic cup of vinegar and lightly stir. Have students record what they observe.

7. Measure 1/4 teaspoon baking powder. Add it to the remaining plastic cup of water and lightly stir. Have students record what they observe.

8. Have students use the hand lens to examine the baking powder that is on the plate. Tell them to record what they observe.

9. Pour the liquids into the dishpan. If you are working with two groups of students, set up the demonstration for the second group.

Questions

1. *What do you think will happen when the baking soda is placed in the vinegar?*

2. *What will happen when it is placed in the water?*

3. *What do you think will happen when the baking powder is placed in the vinegar?*

4. *What will happen when it is placed in the water?*

5. *Use the hand lens to look at the baking soda and the baking powder. Do they look alike or do they look different? Describe what you see.*

Wheat-and-Corn Pancakes

Yields 16–32 small pancakes

DRY INGREDIENTS

1/2 cup white flour

1/2 cup whole-wheat flour, ground in Lesson 11

1/2 cup cornmeal, ground in Lesson 11

1 tablespoon baking powder

1/2 teaspoon salt

WET INGREDIENTS

1 egg

1 1/2 cups milk

3 tablespoons honey

1 tablespoon corn oil

1/2 teaspoon vanilla extract

Cooking spray or butter to grease the pan

Maple syrup and "churned" butter (from Lesson 11) for the top of the pancakes

(Optional) Strawberry jelly

DIRECTIONS

1. Stir together all dry ingredients in a large bowl.

2. Combine all wet ingredients; mix well.

3. Add wet ingredients to dry all at once, and stir until just blended and still a little lumpy.

4. Pour about 1/8 cup* batter onto a hot, lightly greased frying pan or griddle.

5. Turn when pancake has a bubbly surface and slightly dry edges.

6. When cooked through, remove pancake from pan, let cool.

7. Add churned butter and maple syrup or strawberry jelly as desired, and enjoy!

*This recipe yields about 32 two-inch pancakes if you use 1/8 cup batter for each pancake, and about 16 four-inch pancakes if you use 1/4 cup for each pancake. Remember, if you have two dry-ingredient groups and two wet-ingredient groups, you will need twice the ingredients listed here.

Name	Date

Pancakes

Serves 4–6

In class, we have been learning about food processing. To celebrate what we have learned, we made pancakes. Take this recipe home and make it with your family. Tell them what you learned about food processing.

DRY INGREDIENTS

1/2 cup white flour

1/2 cup whole-wheat flour

1/2 cup cornmeal

1 tablespoon baking powder

1/2 teaspoon salt

WET INGREDIENTS

1 egg

1 1/2 cups milk

3 tablespoons honey

1 tablespoon corn oil

1/2 teaspoon vanilla extract

Cooking spray or butter
  to grease the pan

Maple syrup and butter
  for the top of the pancakes

DIRECTIONS

1. Stir together all the dry ingredients in a large bowl.

2. In a second bowl, combine all the wet ingredients and mix well.

3. Add the wet to the dry ingredients all at once. Stir until just blended and still a little lumpy. Try not to over-stir the batter.

4. Pour about 1/8 cup* batter onto a hot, lightly greased frying pan or griddle.

5. Turn when pancake has a bubbly surface and slightly dry edges.

6. When cooked through, remove pancake from pan, let cool.

7. Add butter and maple syrup as desired, and enjoy!

*This recipe yields about 32 two-inch pancakes if you use 1/8 cup batter for each pancake, and about 16 four-inch pancakes if you use 1/4 cup for each pancake.

Name	Date

What Causes the Fizz?

Did you notice the fizz when the baking soda and baking powder were mixed with the vinegar? What was that fizz? It was carbon dioxide (CO_2) bubbles. Where did the CO_2 come from?

That fizz is the result of a chemical reaction. Another name for baking soda is sodium bicarbonate. Baking soda is a **base.** Vinegar is an **acid.** The fizzing happens when the base, sodium bicarbonate, is poured into the acid, vinegar. The acid and the base interact and produce the fizz.

What about the baking powder? Baking powder generally has three parts: sodium bicarbonate, cream of tartar, and cornstarch. Since the sodium bicarbonate is baking soda and you already know that baking soda and vinegar make a fizz, you probably figured out why the vinegar and baking powder fizzed.

But what about the water and the baking powder? The water and baking soda didn't fizz, so what caused the baking powder and water to fizz? The cream of tartar is an acid, but as part of baking powder, it is a dry acid. Add some warm water and the acid is no longer dry. The sodium bicarbonate and the cream of tartar cause an acid-base reaction. The bubbles form from the acid-base reaction, producing fizz.

You might think that this is an interesting hypothesis, but you would like to test it. How could you design a test? What if you had the ingredients to make your own baking powder? Then you could make a **control** and **variables.** Here's what you could do:

Control: Baking powder with all three parts: 1/4 teaspoon baking soda, 1/2 teaspoon cream of tartar, and 1/4 teaspoon cornstarch.

Powder #1: 1/4 teaspoon baking soda and 1/2 teaspoon cream of tartar.

Powder #2: 1/4 teaspoon baking soda and 1/4 teaspoon cornstarch.

Powder #3: 1/2 teaspoon cream of tartar and 1/4 teaspoon cornstarch.

Mix the control with warm water and record what happens. Then mix Powder #1 with warm water. What happens? Repeat with each of the remaining powders. Analyze your results and see what you find out.

BECOMING FOOD SCIENTISTS :: INTERACTING PARTS :: **FOOD PROCESSING** :: ENVIRONMENTAL EFFECTS :: WASTE :: MAKING CHOICES

Name	Date

Kitchen-Chemistry Observations

Vinegar   Water

Color _____   Color _____

Smell _____   Smell _____

Baking Soda

1. What happens when the baking soda is mixed into the vinegar?

_____

_____

2. What happens when the baking soda is mixed into the water?

_____

_____

3. Use the hand lens to observe the baking soda on the plate. Describe what you see.

_____

_____

_____

(continued on next page)

Name	Date

Kitchen-Chemistry Observations

Color _____   Color _____

Smell _____   Smell _____

Baking Powder

1. What happens when the baking powder is mixed into the vinegar?

_____

_____

2. What happens when the baking powder is mixed into the water?

_____

_____

3. Use the hand lens to observe the baking powder on the plate. Describe what you see.

_____

_____

_____

(continued on next page)

BECOMING FOOD SCIENTISTS : INTERACTING PARTS : **FOOD PROCESSING** : ENVIRONMENTAL EFFECTS : WASTE : MAKING CHOICES

Name	Date

Kitchen-Chemistry Observations

Conclusions

1. Did the baking powder and the baking soda react the same way when they were mixed into the liquids? Describe any differences that you noticed.

_____

_____

_____

Application

1. You will be using baking powder in your pancake batter. Do you think the baking powder will change the pancake batter? What evidence do you have?

_____

_____

_____

2. Do you think there will be a reaction when the pancake batter is cooking? What evidence do you have?

_____

_____

_____

3. What do you think would happen if the pancake batter did not have any baking powder in it? What is your evidence?

_____

_____

_____

Degrees of Food Processing

AIM

To explore different degrees of food processing in various store-bought foods.

SCIENTIFIC PROCESSES

- question, speculate, build theories, apply

OBJECTIVES

Students will be able to:

- discuss the different degrees of processing used for pancake mixes and pancakes;

- describe the characteristics that determine how much a food is processed;

- discuss convenience food;

- discuss pros and cons of buying processed foods.

OVERVIEW

In this lesson, students explore pancakes at different degrees of processing. The lesson begins with students sharing the different forms of pancakes that they found. Next, students compare store-bought pancake mixes that require adding water, the kind that require eggs and milk, refrigerated pancake batters, and frozen pancakes to contrast the degree of processing of each product. With this activity, students learn about the different degrees to which foods are processed and gain an understanding that in today's society, most of the food for sale in larger markets is processed. Students then discuss convenience foods and food processing, and think about the reasons people purchase convenience foods. They weigh the pros and cons of having control over what is in our food by processing food ourselves versus having much of our food processing done in factories.

MATERIALS

For the teacher:
- *Processing Food* lesson resource
- *Categorizing Processed Pancakes* lesson resource
- *Milestones in Food-Processing History* lesson resource

For each group:
- *Pancake Cards* lesson resource
- Chart paper
- Markers
- Tape

For each student:
- Homework from Lesson 12
- LiFE Log

PROCEDURE

Before You Begin:

- Review the *Processing Food, Pancake Cards* and *Categorizing Processed Pancakes* lesson resources. On chart paper, record the descriptions of highly processed, moderately processed, and minimally processed foods found on the *Processing Food* lesson resource.

- Make a copy of the *Pancake Cards* lesson resource for each group of students.

- Remind students to bring in homework from Lesson 12.

- Copy and review the *Milestones in Food-Processing History* lesson resource. Cut out the milestones and glue them onto index cards.

- If you have not already done so, post the Module Question and Unit 3 Question at the front of the classroom.

MODULE QUESTION

What is the system that gets food from farm to table, and how does this system affect the environment?

UNIT QUESTION

What happens to food as it moves from farm to table?

 **QUESTIONING**

1. Review Module and Unit Questions

Remind students of the Module and Unit questions. Explain that in this lesson, students focus on one kind of food, pancakes, and explore different ways that the ingredients used to make pancakes are processed.

2. Discuss Commercial Pancakes

Invite students to share their findings from the Lesson 12 homework assignment. *How many different kinds of pancakes did you find in the supermarket?* (Mixes, frozen, refrigerated batter.) Record student findings on the blackboard. Explain to students that throughout this lesson we are going to build on the knowledge we already have about food processing and learn even more. Explain that we can classify pancake mixes based on the degree of processing done to the product. This can range from being processed a small amount to being processed a large amount. *What do you already know about food processing? What does it mean for a food to be processed?* (Food that is changed is processed; ingredients mixed together and no longer recognizable in their original form are highly processed; when you still recognize ingredients, the food is moderately processed; a single food can be minimally processed — like applesauce.)

Remind students that the term "processed" refers to a product that has been changed in some way to make it different than it was before. Post the sheet of chart paper with the descriptions of the different degrees of food processing taken from the *Processing Food* lesson resource. Review these descriptions with students. Invite students to describe in their own words the difference between being minimally processed, moderately processed, and highly processed. Encourage students to think about the kinds of foods they eat at home — foods that are familiar to them — and compare them. This may help students categorize the different degrees of processing in food products. Ask students which foods they think are the most convenient for consumers.

3. Classify Food Processing

Explain to students that they are going to use the pancake cards to explore their understanding of food processing. Give each group of four students a sheet of chart paper; several markers; tape; and the eight cards copied from the **Pancake Cards** lesson resource with the names of pancake mixes, preparation instructions, and lists of ingredients. Ask students to make three columns on their chart paper: "Processed a Small Amount," "Processed a Medium Amount," and "Processed a Large Amount." Invite them to carefully study the list of ingredients on each card and the preparation instructions. Explain that, in general, we need to add fewer ingredients and do less work with a food that has been processed a large amount. Invite students to classify the different pancake brands based on how much they think the product has been processed. Have students tape the cards onto the appropriate columns, or use markers to write the name of the pancake mix in the appropriate column. The name of the mix is found on the card. Refer to the **Categorizing Processed Pancakes** lesson resource for guidance in determining the amount of processing in each product. However, on many of the products opinions may vary.

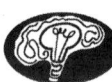

4. Discuss Results

Have each group hang its chart paper at the front of the classroom. Have group representatives briefly share with the class how they classified the eight pancake mixes into the three different categories. Have the group representatives share their reasons for placing the pancake mixes in the categories that they did. After all groups have presented their decisions, ask for volunteers from the class to provide reasons for why they agree or disagree with the categories that their classmates chose for the different pancake mixes.

Encourage debate, but explain to students that it is acceptable to have pancake mixtures classified in different categories as long as a group has a reasonable explanation for the categories it chose. Since processing is really a continuum, students' decisions about which categories to place the different pancakes in may differ from each other and your class's final decision about where to place the different pancakes may differ from what is suggested in the lesson plan. Explain that the goal of this activity is for students to demonstrate what they have learned about how foods can be processed to various degrees.

5. Discuss Convenience Foods

Explain to students that a convenience food is one that requires very little preparation or no preparation at all. Pose the following questions for discussion. *How does the amount of processing differ for each of the pancake mixes? Which pancake mixes were made in factories? Think about how grain is made into flour — do you think the natural environment is affected by the processing done in factories? How? What kinds of processed foods would you purchase? Why?*

6. LiFE Logs

Invite students to add events from the **Milestones in Food Processing History** lesson resource to the time line. Briefly discuss the effects these innovations may have had on how people lived their lives and the foods they ate. Ask students to answer these questions in their LiFE Logs: *What are the benefits of food processing technology for me? What are some of the trade-offs?*

Engage students in a discussion that motivates them to think about the consequences of purchasing and eating convenience foods. As a class, brainstorm a list of some of the costs and benefits for individuals, communities, and the environment.

Pancake Cards

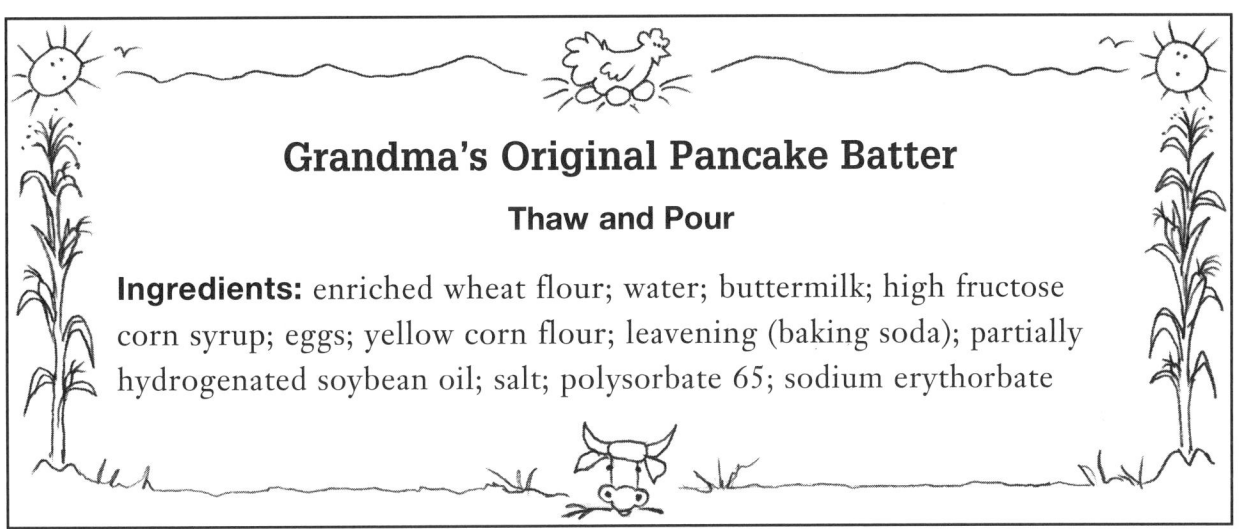

Grandma's Original Pancake Batter

Thaw and Pour

Ingredients: enriched wheat flour; water; buttermilk; high fructose corn syrup; eggs; yellow corn flour; leavening (baking soda); partially hydrogenated soybean oil; salt; polysorbate 65; sodium erythorbate

Uncle Rob's Buttermilk Pancakes

Add Water

Ingredients: enriched bleached flour; sugar; leavening (baking soda); partially hydrogenated soybean oil with mono- and diglycerides; dextrose; dried buttermilk; dry whole eggs; salt; corn syrup solids; defatted soy flour; calcium carbonate; nonfat dry milk; soy lecithin; sodium caseinate; soybean oil

Grandma's Buttermilk Pancake Mix

Add Milk, Eggs, and Oil

Ingredients: enriched bleached flour; sugar; leavening (baking soda); dried buttermilk; salt; calcium carbonate

Early Riser Frozen Buttermilk Pancakes

Heat and Eat

Ingredients: enriched wheat flour; water; high fructose corn syrup; soybean oil; buttermilk; eggs; leavening (baking soda); salt; soy lecithin; added vitamins and minerals

Busy Bee Shake & Pour Pancake Batter

Add One Cup of Cool Water to Plastic Container

Ingredients: enriched flour; degermed yellow corn flour; sugar; leavening (baking soda); partially hydrogenated soybean and/or cottonseed oil; salt; egg whites; dextrose; defatted soy flour; soy lecithin; nonfat milk; maltodextrin

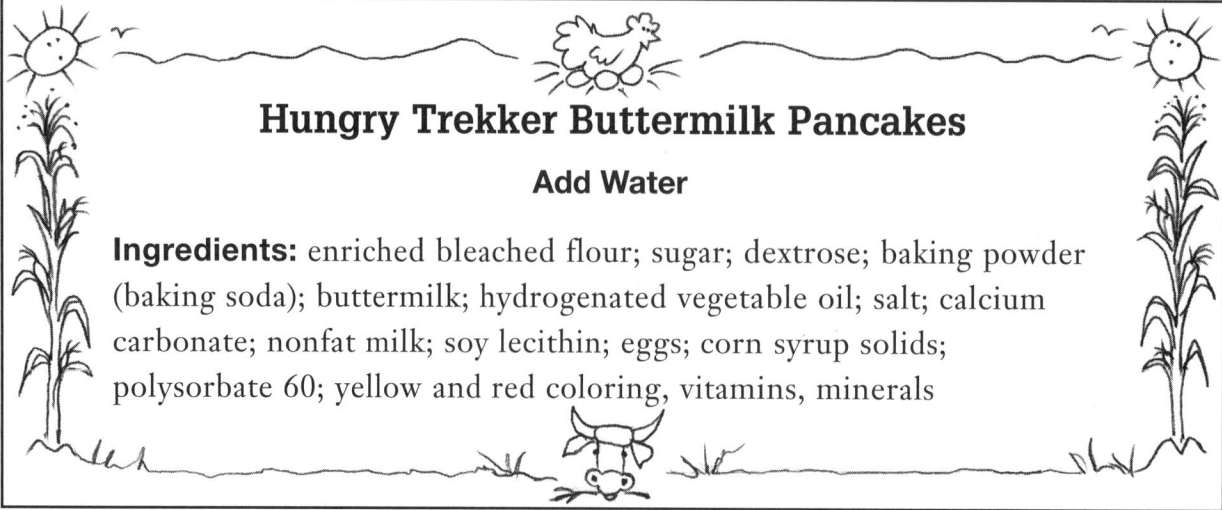

Hungry Trekker Buttermilk Pancakes

Add Water

Ingredients: enriched bleached flour; sugar; dextrose; baking powder (baking soda); buttermilk; hydrogenated vegetable oil; salt; calcium carbonate; nonfat milk; soy lecithin; eggs; corn syrup solids; polysorbate 60; yellow and red coloring, vitamins, minerals

Busy Bee Original Pancake & Waffle Mix

Add Milk & Eggs

Ingredients: enriched bleached flour; partially hydrogenated soybean and/or cottonseed oil; leavening (baking soda); dextrose; salt

Hungry Trekker Original Pancake Mix

Add Milk, Eggs, & Oil

Ingredients: enriched bleached flour; sugar; rice flour; baking powder (baking soda); partially hydrogenated soybean and/or cottonseed oil; salt; calcium carbonate; nonfat milk; soy lecithin; added vitamins and minerals

Processing Food

In our current food system a huge number of processed foods are available. In fact, the vast majority of the 50,000 different foods for sale in large supermarkets are processed. We generally take these processed foods for granted and seldom think about how a food system with such an array of processed foods affects us as individuals, not to mention how food-processing technology affects the health of our communities and the natural environment.

The amount of processing varies from food to food. "Processed" means that the food has been changed in some way, either physically or chemically. Foods can be highly processed — changed a large amount from the way the original ingredients were grown or raised. Foods can be moderately processed — changed somewhat since the ingredients left the farm. Or foods can be minimally processed — changed a small amount since the ingredients left the farm. To determine the level of processing of a food product, think about the following questions: *How far removed from the original source are the ingredients in the food?* For example, *how many steps away from fresh milk is dry milk?* Also ask, *how many ingredients are on the ingredient list?* The more processed a food is the more ingredients it probably contains. *How many ingredients do you recognize? How many can you pronounce the names of?* The more ingredients you do not recognize or cannot pronounce the more processed it probably is.

Highly Processed: Many highly processed foods are mixed foods, often with many ingredients that are mostly or fully prepared in the factory. This means they do not need lots of preparation at home. Examples include hot dogs, chips, and breakfast cereals. Most of the time the original ingredients are not even recognizable in the final product. The term "convenience foods" is often used to describe highly processed foods. A convenience food is a food for which most of the preparation has been done at the factory and little or no preparation needs to be done before eating. For example, highly processed pancake mix has been almost completely prepared in the factory. In fact, some pancake batters are sold completely prepared and refrigerated, and just need to be poured into a skillet. One step further, some varieties of pancakes are completely cooked and frozen, and only need to be heated and eaten.

Moderately Processed: In this category, food has been changed a medium amount from the form in which it was grown or raised. If products have several ingredients, most of them are still recognizable and some preparation must still be done before eating. Examples include a jar of spaghetti sauce and a can of soup. For the pancakes in this category most, not all, of the preparation work has been done in the factory, since the mix contains all the ingredients and at home the water is added, the batter is made, and the pancakes are cooked.

Minimally Processed: In this category, food has been changed a small amount after it was grown or raised on the farm. Examples include applesauce, peanut butter, pre-washed and torn lettuce, and dry milk powder. For the pancakes in this category some of the work has been done in the factory — the mix contains only some of the necessary ingredients, typically only the dry ingredients, such as flour, baking powder, and salt, which are already in a powdery form. Preparation at home includes adding other ingredients such as milk, eggs, and oil, mixing, and cooking the pancakes.

Categorizing Processed Pancakes

In this lesson students work with eight different kinds of pancake products, of varying degrees of processing. The way these products are grouped below is just one suggestion. Your students may decide to group them in another way. This is fine, as long as students provide evidence to support their decisions.

Highly Processed

1. Early Riser Frozen Buttermilk Pancakes (Heat and eat)

2. Grandma's Original Pancake Batter (Thaw and pour)

Sometimes the pancakes are even made in the factory — frozen varieties — or the batter is complete and just needs to be poured onto a skillet — refrigerated batters — and cooked at home. Therefore, we categorized these products as highly processed. The pancakes and batter also contain ingredients that we would not use if making pancakes at home from scratch. Specifically, they have added vitamins and minerals, high fructose corn syrup, and partially hydrogenated oils. The products also contain ingredients — polysorbate 65 and sodium erythorbate, for example — that the home cook would not find in the kitchen.

Moderately Processed

1. Hungry Trekker Buttermilk Pancakes (Add water)

2. Busy Bee Shake & Pour Pancake Batter (Add water to plastic container)

3. Uncle Rob's Buttermilk Pancakes (Add water)

These products contain all the necessary ingredients. You just add water to make the batter. They are dry mixes because the liquid ingredients, like eggs and milk, have been dehydrated. Since the pancakes are not completely made in a factory — mixing and cooking must be done at home — they are moderately processed. However, the ingredient list shows that the products contain ingredients not found in a home kitchen, including defatted soy flour, soy lecithin, corn syrup solids, mono- and diglycerides, polysorbate 60, and artificial colors.

Minimally Processed

1. Hungry Trekker Original Pancake Mix (Add milk, eggs, and oil)

2. Busy Bee Original Pancake & Waffle Mix (Add milk and eggs)

3. Grandma's Buttermilk Pancake Mix (Add milk, eggs, and oil)

These products contain the usual dry ingredients for pancakes. Almost all the ingredients are similar to the ones that would be used at home to make pancakes from scratch. There are very few ingredients that would be foreign to a home kitchen. These mixes are minimally processed because they contain only some of the necessary ingredients, the dry ingredients, and most of the work is done at home. The home cook adds the liquid ingredients, mixes the batter, and cooks the pancakes.

Milestones in Food-Processing History

Add these to the *Milestones* time line that you started in Lesson 7. Encourage students to think about the kind of technology that was used with each kind of processing. What was the source of energy that was used to change the food? What effect do you think these milestones had on the food system at that time?

4000 B.C.	Egyptians use yeast to make bread
3000 B.C.	Fermentation of dough, grain, and fruit juices is in practice
2700 B.C.	Chinese use salt to preserve food
1000 B.C.	Chinese use ice cellars to refrigerate food
500 B.C.	First written evidence of sugar crystal production in northern India
AFTER 1497 A.D.	European fishermen begin processing fish in America to trade for sugar, molasses, and rum
1797 A.D.	First marmalade factory is built in Scotland
1806 A.D.	Frederic Tudor, the "Ice King," ships ice from Boston to Martinique, beginning a global trade in ice
1830s A.D.	Frederic Tudor ships ice to all the West Indies as well as India and Australia
1858 A.D.	Home canning jars are invented
1862 A.D.	Louis Pasteur invents pasteurization
1877 A.D.	Refrigerated meat is first shipped from Argentina to Europe
1880 A.D.	First automatic can-making machinery is introduced into Britain
1894 A.D.	Kellogg brothers make first flaked cereal
1906 A.D.	Battle Creek Toasted Corn Flakes Company produces Kellogg's Toasted Corn Flakes
1911 A.D.	General Electric sells a wooden refrigerator invented by a French monk
1920 A.D.	Small-scale, hand-baked bread is replaced by mass producers like Wonder Bread
1921 A.D.	First drive-in restaurant in the United States opens
1922 A.D.	Canned baby food is made in New York
1928 A.D.	Bread-slicing machine is invented

1930s A.D.	Clarence Birdseye launches retail sales of frozen food
1941–45 A.D.	Frozen food becomes popular
1945 A.D.	Microwave oven is patented in Massachusetts
1948 A.D.	Pace Foods begins bottling and selling salsa in San Antonio
1950s A.D.	The frozen TV dinner becomes popular
1980s A.D.	The typical American now eats fast food nine times a month
2004 A.D.	One-third of American children eat fast food daily
2007 A.D.	17,000 new foods, mostly processed, are introduced in supermarkets

Growing Microbes

AIM

To investigate conditions necessary for microorganisms to grow.

SCIENTIFIC PROCESSES

- **experiment, gather data, hypothesize, identify variables**

OBJECTIVES

Students will be able to:

- **discuss the conditions microorganisms need in order to grow on food;**

- **outline the steps of a procedure to test the conditions microbes need to grow;**

- **analyze data they have collected;**

- **determine the results of their experiments.**

OVERVIEW

In this lesson, students explore microorganisms and identify what they need to grow. Students investigate different growing conditions and gather data that they use to develop theories about microbial growth. They apply this scientific evidence to food preservation by thinking about what microorganisms need to grow and what people can do to limit the growth of food-spoiling microorganisms to extend the shelf life of food. Students test whether or not the addition of preservatives helps prevent food from spoiling. They also investigate the necessary growing conditions for yeast. The emphasis on the yeast experiment is the experimental procedure and data collection. Throughout this lesson, prompt students to think about interacting parts. For homework, students read about methods of preserving food and write their responses to review questions. They also look for five preserved foods in their homes, and describe them in their LiFE Logs. This homework is used in the next lesson.

Alert: Keep the bread samples inside the sealed plastic bags. Some people are allergic to mold and should not smell it or breathe it.

MATERIALS

For the teacher:
- *The World of Microorganisms* teacher note
- *Do Preservatives Affect Mold Growth?* experiment sheet
- *Investigating Microbe Growth* experiment sheet

For each group:
- Materials from the *Growing Microbes Supply List* lesson resource
- Hand lenses
- (Optional) Microscope

For each student:
- *Putting Food By* student reading
- *Mold-Growth Predictions* activity sheet
- *What Makes Microbes Grow?* activity sheet
- *Putting Food By Review Questions* activity sheet
- LiFE Log

PROCEDURE

Before You Begin:

- Review *The World of Microorganisms* teacher note, *Do Preservatives Affect Mold Growth?* and *Investigating Microbe Growth* experiment sheets, and the *Mold-Growth Predictions* and *What Makes Microbes Grow?* activity sheets. Make copies of the activity sheets for each student.

- Read the label on the commercial bread that you are using and record the name of the preservative or preservatives that were used. Look for chemical preservatives such as calcium propionate and sodium benzoate.

- Gather materials listed on the *Growing Microbes Supply List* lesson resource. Make sure you use active dry yeast, not rapid-rise yeast. The rapid-rise yeast does not require the use of warm water to activate the dried yeast cells, which will affect the results of this experiment.

- Review the *Putting Food By* student reading and the *Putting Food By Review Questions* activity sheet that students will read and complete for homework. Make copies for students and set aside to distribute at the end of the lesson.

- If you have not already done so, post the Module Question and the Unit 3 Question at the front of the classroom.

MODULE QUESTION

What is the system that gets food from farm to table, and how does this system affect the environment?

UNIT QUESTION

What happens to food as it moves from farm to table?

QUESTIONING

1. Review Module and Unit Questions

Review the Module and Unit questions with students. Briefly recap what students have been learning in this unit thus far. *What did we learn from our food-change investigation?* (Food rots, spoils.) *What are some of the ways that people change food?* (Process it, grind it, cook it.) Explain that in this lesson the class is going to learn about other ways that food changes as it moves from the farm to the table.

2. Think about Microorganisms

Write the word "microorganism" on the board. Tell students that they are going to investigate what microorganisms need to grow and reproduce. Engage students in a discussion of microorganisms to assess their current understanding. Make certain that students understand that "microorganism" is a term that includes many different kinds of organisms that are of microscopic size, including bacteria, fungi, protozoa, algae, and viruses. Tell students that they will be investigating just a few different types of organisms and ones that are particularly important in terms of food spoilage.

Have student think back to their investigations of food changes. *What did you observe during the food-change experiment? Did you observe microorganisms growing or an increase in the number of microorganisms on the food?* If necessary, remind students that microorganisms are living things. *Based on what you know about living things and what they need to grow, what do you think microorganisms need to grow and reproduce? What do living things need to grow and reproduce?* (Food, water, air, space.) Record students' ideas on the board or on chart paper.

EXPERIMENTING

3. Investigate Growing Conditions

Tell students that they are going to begin to identify what makes some kinds of microorganisms grow. Have the class work in small groups. Distribute one copy of the **Mold-Growth Predictions** activity sheet to each student. In this experiment, students inoculate two slices of bread with spores — one slice made with preservatives and one made without. Then they add moisture, and observe what happens over a period of several days.

Review the **Do Preservatives Affect Mold Growth?** experiment sheet with students. You may wish to have a student volunteer read the description of the investigation at the top of the experiment sheet. Ask students what the word "preservative" means to them, specifically as it relates to food. Record the name of the preservative or preservatives that were used in the commercial bread sample on the board or on chart paper at the front of the class. These preservatives are added to help prevent the growth of mold. Encourage students to use a hand lens and look at both samples of bread. Ask if they can see any evidence that one slice of bread was made with preservatives and the other was not. If you have access to a microscope, have students examine the bread samples at high magnification so they can see the nooks and crannies that form the environment where microorganisms can live and grow.

4. Bread Investigation

Have students collect samples of dust from around the room. The dust contains spores from microorganisms. Engage students in a discussion about why they are inoculating the slices of bread. *Why do we want spores from microorganisms on the bread?* (To try to grow microorganisms.) *Why do you think we sprinkled water on the bread? What do living things need to grow and reproduce?* (Water, air, space, food — the bread is the food.)

5. Make Predictions

What do you predict will happen? Encourage students to discuss this among their group members before making a prediction in front of the class. *Do you think you will be able to grow microorganisms on the bread? Do you think they will grow on both slices of bread? What evidence do you have to support your thinking?*

Put the bread-mold growth experiment aside for at least 48 hours, or until you begin to see microorganisms growing on the bread. If possible, have students make daily observations in their LiFE Logs. When there is evidence of microorganisms growing on the bread, have students draw what they observe in their LiFE Logs. Encourage them to use colored pencils or markers for their drawings, especially if the microbial growth displays color. Color is one way to help identify the microbes. For example, greenish-blue molds on bread may be from the genus *Penicillium*, and a blackish color may indicate a colony in the genus *Rhizopus*. Tell students to use as much description as possible when they record what they observe. Have them analyze their data and compare their results to their predictions.

6. Investigate Growth

Distribute the **What Makes Microbes Grow?** activity sheet to each student. Review the **Investigating Microbe Growth** experiment procedure with students. Tell them they are going to investigate the conditions that yeast, a type of fungus, needs to grow. *Is yeast a food that we eat? How do we use yeast?* (In breads, pastries, fermented drinks.) *What does yeast need to grow?*

Use the following questions to help guide your students' thinking as they experiment with the conditions that yeast needs to grow. *How can we set up an experiment to find out what yeast needs to grow? What are the steps in the experiment? What do you predict will happen to the yeast?*

7. Homework

Distribute the *Putting Food By* student reading and the *Putting Food By Review Questions* activity sheet to each student. Briefly go over the activity sheet with students to help them see what to look for in the reading. Remind students to bring in the completed activity sheet and the reading to use in the next lesson.

Ask students to find five foods in their homes that have been preserved. Have students list the foods in their LiFE Logs and describe how they think the food was preserved.

The World of Microorganisms

Exploring food preservation opens the door to the world of microorganisms. If we know that microorganisms such as mold and bacteria cause food to spoil, is it possible to alter the environment so these microorganisms can't survive? Our overarching question is, What do microorganisms need to grow and reproduce, and can we alter conditions to prevent them from thriving or to use them to our advantage? This part of Unit 3 is a particularly vivid demonstration of the Applying to LiFE phase of QuESTA. Students learn about the conditions that microorganisms need to grow and reproduce and then use this new knowledge in the real world.

In this lesson, students continue to investigate why food spoils and some of the methods humans have designed to prolong the shelf life of food. The lesson's emphasis is on the environmental conditions that microorganisms need to grow and reproduce. We use scientific information to develop technologies that will extend the life of food and increase the variety in our diet. These food-processing and -preserving methods are examples of ways in which human beings shape and control the natural world to suit their needs.

In this lesson we focus on molds that grow on breads and yeast that is used in baking breads. Molds grow from tiny spores that float around in the air. When spores fall onto a damp surface, such as the damaged spot on a piece of fruit, the spores grow into mold. What do molds need to grow? They need a source of food, suitable temperature, moisture, air (but not in all cases), and a suitable pH.

When you observed fruit in the ***Food-Change Observations*** experiment in Lesson 6, you may have noticed fuzz on the fruit, especially if there was a bruised spot or other damage. That fuzz was mold. There are thousands of different kinds of mold. In this lesson, we grow bread mold. This common mold starts out looking white and fuzzy and then begins to turn black. The black spots are spores, which will eventually grow and reproduce to make even more bread mold.

We also explore the growing conditions for yeast, a microorganism that is a type of fungus. The active dry yeast used in the experiment is a dried form of live yeast. Hieroglyphics from Egypt suggest that more than 5,000 years ago, humans were using yeast to make bread, beer, wine, and other fermented beverages. The Egyptians probably didn't know the science behind what made their bread rise, but today we do. The scientific name for yeast is *Saccharomyces cerevisiae*, which means sugar-eating fungus. The name is a straightforward description of what happens when you make bread. The yeast ferments sugars that are in the flour or added to the dough, giving off carbon dioxide (CO_2) and alcohol (ethanol). The CO_2 is trapped as tiny bubbles in the dough, which rises. Why does the yeast do this? To grow and reproduce — to gain energy from the fermentation of carbohydrates. In the yeast experiment, students explore the effect that sugar and water temperature have on yeast growth.

Encourage students to think like ecosystem scientists as they learn more about the world of microorganisms. What are the abiotic and biotic parts of the ecosystem? How do these parts interact? Then encourage students to think about how they can use what they have learned. Challenge them to think about how the scientific data they gather from their experiments can be used in their daily lives.

Growing Microbes Supply List

DO PRESERVATIVES AFFECT MOLD GROWTH? (p. 238)

Supplies

For each group
- 2 sandwich-size resealable plastic bags
- 1 slice bread, made without preservatives; if you cannot find bread made without preservatives, you can substitute leftover pancakes from Lesson 12
- 1 slice bread, made with preservatives
- 2 cotton swabs
- 2 tablespoons water
- 2 labels
- (Optional) Eye dropper

INVESTIGATING MICROBE GROWTH (p. 239)

Supplies

For each group
- 5 eight-ounce plastic cups
- 12 teaspoons active dry yeast (about 6 packages of the yeast used to bake bread; do not use rapid-rise yeast)
- 2 cups warm tap water (H_2O)
- 1 cup cold tap water (H_2O)
- 1/4 cup sucrose (table sugar, $C_{12}H_{22}O_{11}$)
- Measuring spoons
- 4 stirring sticks or spoons
- Towels to clean up spills
- (Optional) Thermometer
- (Optional) Heat source

Do Preservatives Affect Mold Growth?

Students investigate whether or not preservatives in bread have an effect on mold growth. They inoculate two slices of bread — one made with preservatives and one made without preservatives — with mold spores and observe what happens. They record their observations on the ***Mold-Growth Predictions*** activity sheet.

Setup

Make labels for the two plastic bags: "Bread with Preservatives" and "Bread without Preservatives."

Procedure

1. Divide the class into small groups.

2. Distribute the ***Mold-Growth Predictions*** activity sheets to students in each group. Review the sheet with the students.

3. Have each student group carefully observe the bread and record its observations. Allow time for group members to share their observations with each other.

4. Next, have students collect dust.

5. Then have students inoculate each slice of bread with dust and sprinkle the bread with a few drops of water, as described on the activity sheet.

6. Once students have completed preparing the bread slices, have them carefully place each slice of bread inside a plastic bag.

7. Seal the bags and tape a label to each bag. Set the bags aside.

Questions

1. *Did you observe any differences between the two slices of bread when you made your first observations?*

2. *What do you think the dust contains?*

3. *Why do you think you sprinkled water on the bread?*

4. *What do you predict will happen? Do you think the preservatives will have an effect on mold growth? Why do you think this?*

Investigating Microbe Growth

Students investigate what conditions are needed for yeast, a type of microbe, to grow. They record their observations on the **What Makes Microbes Grow?** activity sheet.

Setup

1. Make a label for each cup

– Cup #1: "Yeast"

– Cup #2: "Yeast with Warm H_2O"

– Cup #3: "Yeast with $C_{12}H_{22}O_{11}$"

– Cup #4: "Yeast with Warm H_2O and $C_{12}H_{22}O_{11}$"

– Cup #5: "Yeast with Cold H_2O and $C_{12}H_{22}O_{11}$"

2. Get 2 cups of warm tap water or heat 2 cups of water to between 110° and 115°F.

3. Get 1 cup of cold tap water.

Procedure

1. Divide the class into small groups.

2. Distribute the **What Makes Microbes Grow?** activity sheet to students in each group.

3. Review the sheet with the students.

4. Monitor the experiment for several days, or until students begin to notice changes.

Questions

1. *Which conditions will make the yeast grow? Record your prediction on the* **What Makes Microbes Grow?** *activity sheet. What evidence do you have for your predictions?*

2. *What is the variable in the experiment with Cup #1 and Cup #2?*

3. *What is the variable in the experiment with Cup #3 and Cup #4?*

4. *What is the variable in the experiment with Cup #5?*

5. *Under what conditions did the yeast grow the fastest? Why do you think that happened?*

6. *Compare your results to your prediction. Was your prediction accurate? Were you surprised by anything you learned?*

Name

Date

Putting Food By

People have been inventing ways to preserve food ever since there was enough food for them to have a surplus. It was one way to make sure they did not starve during the times when there was little fresh food. It may be hard for you to imagine life without a refrigerator, but home refrigerators were not invented until the early twentieth century. Mass production of refrigerators did not begin until just after World War II. Before that time, people used ice and snow to cool their food.

Use your imagination. Go back in time to when there were no grocery stores. Think about what your life would be like. You and your family grow or raise all of the food that you eat. You have a huge vegetable garden, fruit trees, and a couple of cows. You also have a field where you grow hay and another field where the cows graze. During the growing season, you have more vegetables and fruit than you can eat. The cows eat the grass in the field and produce lots of milk. You have a surplus of food during the growing season, but will you have enough when winter comes?

You "put it by" or store the surplus food. Long before there were factories to preserve food, people preserved it at home. Think about what you have learned in this unit. To preserve food, you need to use a process that **inhibits,** or limits, the growth of microorganisms. You must use a method that removes the conditions that microorganisms need in order to grow. Think of your experiments with the yeast and the bread. What did you learn from your observations? The yeast and the mold needed moisture and warmth to grow.

Thinking like a scientist, you probably figured out that one method of preserving food is to take away the moisture. To do that, you **dry** the food. What else can you do? You can change the state of water from liquid to solid. You can **freeze** the food. Freezing food also lowers the temperature, which can reduce the growth of populations of microorganisms.

(continued on next page)

LESSON 14: GROWING MICROBES

Farm to Table & Beyond
©2008 Teachers College Columbia University

Name   Date

Putting Food By

Drying is the easiest way to preserve food. You simply remove enough water from the food to inhibit the growth of microorganisms. Some people dry food just by leaving it out in the sun. Of course, you have to make sure you protect the food from insects and dust. It also needs to be a hot day, since the drying should be done as quickly as possible. At the same time, the temperature can't be too hot or it will affect the texture, color, and flavor of the food. Some people air-dry their food. Have you ever seen a string of dried chili peppers or herbs? The food or herb is tied in bunches and hung out of the sun to dry. Sometimes the bunches are placed inside paper bags to protect them from insects and dust. Another drying method is to use an oven, or an electrical appliance called a **dehydrator.** Once the food is dried, it must be stored in airtight packaging to keep out insects and moisture.

Freezing is another simple way to preserve foods. When you freeze food, you keep most of the nutrients in it. The best vegetables to freeze are ones that are usually cooked before eating. Examples are asparagus, beets, cauliflower, broccoli, peas, carrots, corn, spinach, and other green vegetables. Freezing vegetables is easy: clean the vegetables well; cut into bite-sized pieces; **blanch** them (this means you put the vegetables in boiling water for a few minutes — the exception is beets, which need to blanch for up to 50 minutes); cool the vegetables in ice water; dry them; package in freezer containers, leaving space at the top because vegetables expand as they freeze; and place the containers in the freezer. Fruit is frozen the same way, with one exception: fruit does not need to be cooked. Sometimes a sweetened syrup is added to fruit. You can freeze berries by placing a layer of them on a cookie sheet and putting them in the freezer. Once they are hard, put them into freezer bags. With fruit such as peaches, ascorbic acid ($C_6H_8O_6$) is added so the fruit does not turn brown.

Canning is another method for preserving food. Food is canned to remove the air. When you can food, the jars or cans and the food have to be heated to a high temperature to kill all microorganisms. Then the jar is sealed to keep out any new microorganisms.

(continued on next page)

Name   Date

Putting Food By

Pickling and **fermentation** are two of the oldest ways of preserving food. To preserve food by pickling, the food is placed in a **brine** (salt water) or vinegar (a kind of acid). Brine and vinegar inhibit the growth of microorganisms. During pickling, it's important to keep air out so there is no oxygen (O_2). With oxygen present, spoilage microorganisms will grow. Did you know that pickling isn't just for cucumbers? Many vegetables are good pickled, including tomatoes, green beans, peppers, carrots, cabbage, radishes, and corn.

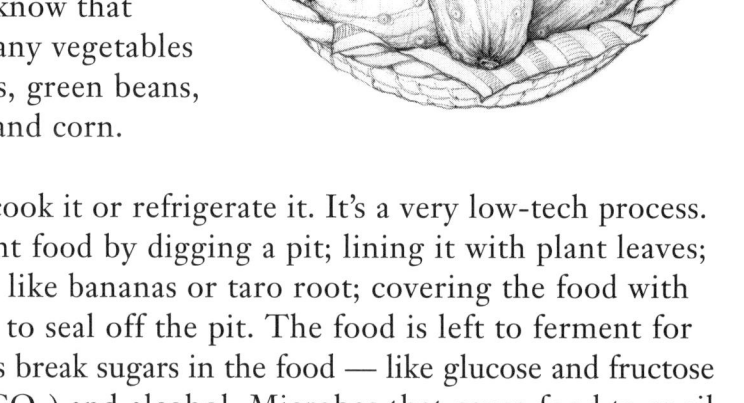

To ferment food, you don't need to cook it or refrigerate it. It's a very low-tech process. Islanders in the South Pacific ferment food by digging a pit; lining it with plant leaves; putting in the food to be fermented, like bananas or taro root; covering the food with plant leaves; and piling rocks on top to seal off the pit. The food is left to ferment for three to six weeks. Beneficial microbes break sugars in the food — like glucose and fructose ($C_6H_{12}O_6$) — into carbon dioxide (CO_2) and alcohol. Microbes that cause food to spoil can't live in this type of environment. When the food is ready to eat, it's taken out of the pit, washed, left to dry in the sun, pounded into a paste, baked or boiled, and eaten.

Drying, freezing, canning, fermenting, and pickling all change the growing conditions, inhibiting the growth of microorganisms. Chemicals called **preservatives** can also be added to food to make it last longer. Preservatives stop microorganisms from growing. This is called **chemical preservation.** Some preservatives are common chemicals like sucrose (table sugar, $C_{12}H_{22}O_{11}$) and salt (NaCl). Sugar and salt preserve foods through a process called **osmosis.** Through osmosis, the added sugar or salt causes water to move out of the cells of the microorganisms. The microorganisms become dehydrated and cannot reproduce. If the microorganisms do not have the water they need to grow and multiply, they cannot cause the food to spoil.

The secret to preserving food is knowing what microorganisms need in order to grow and then changing the conditions so you can inhibit their growth. It's all about science.

Name	Date

Mold-Growth Predictions

In this activity you work in small groups with other students to investigate whether the use of preservatives in bread affects mold growth. Your teacher will provide the materials. Follow the procedure below, fill in the charts, and answer the following questions.

Procedure

1. Use the hand lens to observe the slices of bread. Complete the chart below.

2. Use a cotton swab and collect dust (look on chair legs and in the corner of the classroom).

3. Rub dust from the cotton swab on the bread made without preservatives.

4. Sprinkle the bread with a few drops of water. Use 1–2 teaspoons.

5. Place the bread inside a plastic bag and seal. Tape the "Bread without Preservatives" label to the bag so you can see the whole slice of bread.

6. Repeat steps 2–4 using the slice of bread made with preservatives.

7. Place the bread inside the other plastic bag and seal. Tape the "Bread with Preservatives" label to the bag so you can see the whole slice of bread.

8. Set both bags aside.

OBSERVATIONS		
	Bread Made without Preservatives	**Bread Made with Preservatives**
Color		
Texture		
Smell		
Other Notes		

(continued on next page)

Name   Date

Mold-Growth Predictions

Prediction

1. What do you predict will happen? Write your prediction below.

_____

_____

_____

_____

2. Why do you predict this? What is your evidence?

_____

_____

_____

_____

Record Your Experimental Procedure

Date you began the experiment: _____

Record what you did on the lines below. Add more steps if you need to. Remember to include details such as where you found the dust and how many drops of water you used.

Slice of Bread Made with Preservatives

Step #1 _____

_____

(continued on next page)

BECOMING FOOD SCIENTISTS : INTERACTING PARTS : **FOOD PROCESSING** : ENVIRONMENTAL EFFECTS : WASTE : MAKING CHOICES

Name

Date

Mold-Growth Predictions

Step #2 _____

_____

Step #3 _____

_____

Step #4 _____

_____

Step #5 _____

_____

Slice of Bread Made without Preservatives

Step #1 _____

_____

Step #2 _____

_____

Step #3 _____

_____

Step #4 _____

_____

(continued on next page)

Name   Date

Mold-Growth Predictions

Step #5 _____

_____

Observations

Each day, record your observations on this chart.

MOLD-GROWTH OBSERVATIONS		
	Bread Made without Preservatives	Bread Made with Preservatives
Day 1		
Day 2		
Day 3		
Day 4		
Day 5		

After you have completed your observations and analyzed your data, record your results in your LiFE Log.

Name	Date

What Makes Microbes Grow?

In this activity you will work with a group of students to investigate what conditions are needed for yeast, a type of microbe, to grow. Your teacher will have the materials and setup ready for you. Begin by making your prediction. Then, follow the procedure, and record your observations.

Prediction

1. What do you predict yeast needs to grow?

_____

_____

2. Why do you think this? What evidence do you have?

_____

_____

Procedure

1. Put 2 teaspoons of yeast in each cup (Cups #1 through #5).

2. Put 1/4 cup warm water in Cup #2. If you use a thermometer, the temperature should be between 110° and 115°F. Stir.

3. Record what you observe in Cup #1 and Cup #2 below.

4. Add 1 teaspoon $C_{12}H_{22}O_{11}$ (sucrose) to Cup #3 and 1 teaspoon $C_{12}H_{22}O_{11}$ plus 1/4 cup warm water to Cup #4. Stir.

5. Record what you observe in Cup #3 and Cup #4 below.

6. Put 1 teaspoon $C_{12}H_{22}O_{11}$ in Cup #5. Stir.

7. Put 1/4 cup cold water in Cup #5. Stir.

8. Record what you observe in Cup #5 below.

(continued on next page)

Name

Date

What Makes Microbes Grow?

Observations

What I observed:

_____

_____

_____

What I observed:

_____

_____

_____

What I observed:

_____

_____

_____

What I observed:

_____

_____

_____

(continued on next page)

Farm to Table & Beyond
©2008 Teachers College Columbia University

Name

Date

What Makes Microbes Grow?

Cup #5:
Yeast with
cold H_2O and
$C_{12}H_{22}O_{11}$

What I observed:

_____

_____

_____

1. Compare your results to your prediction.

_____

_____

_____

_____

2. Was your prediction accurate?

_____

3. Were you surprised by anything you learned?

_____

_____

_____

Name   Date

Putting Food By Review Questions

1. Why do people preserve food?

_____

_____

2. What does "putting food by" mean?

_____

_____

3. What is removed when you dry food?

_____

_____

4. What growing condition is changed when you freeze food?

_____

_____

5. What is the process for freezing vegetables?

_____

_____

_____

(continued on next page)

Name   Date

Putting Food By Review Questions

6. How does canning inhibit the growth of microorganisms?

_____

_____

_____

7. How does pickling inhibit the growth of microorganisms?

_____

_____

_____

8. Why is fermenting food considered a low-tech form of preserving food?

_____

_____

_____

9. Based on what you have read, which forms of food preservation do you think have the smallest impact on the environment? Why do you think this?

_____

_____

_____

_____

Investigating Food-Preservation Methods

AIM

To further student understanding of ways in which food changes or is changed.

SCIENTIFIC PROCESSES

- question, explore, observe, gather data

OBJECTIVES

Students will be able to:

- describe why we preserve food;

- discuss the effect of removing moisture from food;

- analyze how different food-preservation methods inhibit microbial growth on food;

- compare and contrast a variety of fresh and dried fruits.

OVERVIEW

Students revisit different kinds of food preservation in this lesson. At the beginning of the lesson, students work with a graphic organizer to review what they have learned about food change. Next, students investigate osmosis by salting cucumbers to draw out water and by soaking dried fruit to draw the moisture back into the food. Note that the emphasis of these investigations is on collecting data to gather evidence of change rather than a detailed understanding of osmosis. Finally, students analyze the effect that drying has on fruit. For homework, students begin to draw together what they have learned about how food changes as it moves from the farm to the table. Help students expand their thinking to include how food-processing technology might affect the environment and what the interacting parts in food-processing techniques are.

MATERIALS

For the teacher:
- *Thinking about Food Change* lesson resource
- *Salt and Cucumbers* experiment sheet
- *Dried Fruit and Water Observations* experiment sheet

For the class:
- Materials from the *Investigating Food-Preservation Methods Supply List* lesson resource

For each group:
- (Optional) *Salt and Cucumbers* and *Dried Fruit and Water* experiment sheets
- Chart paper
- Markers

- Hand lenses
- Metric ruler

For each student:
- *Putting Food By* student reading (pp. 240–242)
- *Mold-Growth Predictions* activity sheet (pp. 243–246)
- *Putting Food By Review Questions* activity sheet (pp. 250–251)
- *Salt and Cucumbers Observations* activity sheet
- *Dried Fruit and Water Observations* activity sheet
- *Taste-and-Compare Fruit Analysis* activity sheet
- LiFE Log

PROCEDURE

Before You Begin

- Gather materials listed on the *Investigating Food-Preservation Methods Supply List* lesson resource.

- Review the *Thinking about Food Change* lesson resource, the *Putting Food By* student reading, and the *Putting Food By Review Questions* activity sheet. Remind students to bring in the *Putting Food By* reading and the *Putting Food By Review Questions* activity sheet from their homework assignment in the previous lesson.

- Check to make sure students have their *Mold-Growth Predictions* activity sheet and mold growth observations that they have recorded in their LiFE Logs from Lesson 14.

- Review the *Salt and Cucumbers* and *Dried Fruit and Water Observations* experiment sheets and the *Salt and Cucumbers Observations* and *Dried Fruit and Water Observations* activity sheets. Make copies of the activity sheets for each student. Decide whether you will do these investigations as a whole-class demonstration or have students work in small groups. Prep materials for the *Salt and Cucumbers Observations* and *Dried Fruit and Water Observations* investigations.

- (Optional) If you decide to do this as small-group investigations, make copies of the *Salt and Cucumbers* and *Dried Fruit and Water Observations* experiment sheets for each group.

- Review the *Taste-and-Compare Fruit Analysis* activity sheet. Prepare materials. Make copies of the activity sheet for each student.

- If you have not already done so, post the Module Question and Unit 3 Question at the front of the classroom.

MODULE QUESTION

What is the system that gets food from farm to table, and how does this system affect the environment?

UNIT QUESTION

What happens to food as it moves from farm to table?

 **QUESTIONING**

1. Review Module and Unit Questions

Remind students of the Module Question and Unit 3 Question. *What have you learned so far in this unit about what happens to food as it moves from farm to table? Why do we preserve foods?* Invite student volunteers to briefly review what they learned from the food-changing and mold-growth experiments. Invite students to share their data from their investigations and the theories they are developing about food changes. Explain that in this lesson we are going to learn about food preservation and how we can prevent food from going bad.

2. Discuss Homework

Invite students to share the list of preserved foods that they found in their homes for their homework assignment from the previous lesson. Record the list on chart paper and post it at the front of the classroom. If students say that they saw the word "preservative" on a food ingredient list, or the words "no preservatives" on the front of a food package, explain that this usually refers to chemicals added to the food to prevent microorganisms from growing. The chemicals can be artificial ones or they can be natural ones like vinegar, sugar, and salt. When chemicals are added to food to make them last longer it is called chemical preservation. Explain to students that throughout this lesson they are going to look at different methods of food preservation and think about what has been done to each food to prevent microorganisms from growing.

Hold up a canned tomato product for the class to see. *Which of the things that microorganisms need to grow has been removed so that they cannot grow on a canned tomato product?* Invite students to share their ideas with the class and encourage them to provide reasons to support their answers. Explain that during canning, food is placed in airtight containers (like cans or jars) with lids. The containers are heated to a temperature that destroys microorganisms. During the heating process, air is driven out of the cans or jars. As the containers cool, a vacuum seal is formed. This seal keeps air and micro-organisms from getting back into the container.

Repeat this procedure for a frozen-food product. Explain that in the process of freezing, both the amount of moisture and the warm temperature change. Water changes state and becomes ice and the temperature is much colder. When these conditions change, the growth rate of the micro-organisms dramatically slows down.

Hold up a dried food product, such as dried fruit or corn. *What happens when food is dried?* (Water is removed and the microorganisms don't have the moisture they need to grow.) Tell students that later in this lesson they will taste dried fruits and explore how they taste compared with fresh fruits of the same type.

Look back at the list of preserved foods on the chart paper. Categorize each of the foods according to the type of food-preservation method that was used.

3. Review Food Change

Have students take out the ***Putting Food By*** student reading and the ***Putting Food By Review Questions*** activity sheet that they com-

pleted as homework. Invite student volunteers to briefly describe how each method helps preserve food. You may wish to use the questions on the activity sheet to guide the discussion.

Have students work in small groups. Distribute chart paper and markers to each group. Remind students that the focus of this unit is on food changes. Tell students that they are going to make a graphic organizer to help them review what they have learned. Remind them that a graphic organizer is a visual tool that they can use to construct knowledge.

Tell students that they will begin to make the graphic organizer as a class, and then complete it within their own groups. Use the ***Thinking about Food Change*** lesson resource as a guide. Draw a bubble on chart paper or the board and label it "food changes." *What are three different ways that food can change?* (Processing, preserving, microorganisms.) Draw a bubble for each one and label each bubble. You may find it helpful to use one method, such as processing, and work through it as a class. It is important for students to direct this part of the activity as much as possible so that they are organizing what they know about food change. These graphic organizers will also help you see areas where students may be struggling with concepts related to food change. Once student groups have completed their work, invite representatives from each group to present their charts to the class. Encourage discussion and debate. Post the graphic organizers at the front of the classroom.

4. Discuss Microorganisms

Refer students to the graphic organizers. *Where do you see evidence of microorganisms?* (Spoiled/ rotten food and fermented food.) *Why do we need to preserve food?* (To inhibit the growth of microorganisms.) Explain that we can understand how to keep microorganisms from growing on food if we understand what they need to grow. *What did we learn about the conditions necessary for microorganism growth from our food-change experiments?* (They need air, water, and warmth.) *What happens if we take away air, water, or warmth?* (We can slow down the rate that microorganisms grow and thus preserve our food.) Explain to students that in the next part of this lesson, the class is going to look at different methods of preservation to learn more about ways to prevent microorganisms from spoiling food.

5. Moisture In and Out

If you have not already done so, while students are working on the graphic organizer, you may wish to set out the materials for the *Salt and Cucumbers* and *Dried Fruit and Water Observations* investigations. Explain that the class is going to investigate moisture in food to learn more about different ways of preserving food. Remind students that microorganisms need moisture to grow and reproduce. In the *Salt and Cucumbers* experiment, the class investigates what happens when salt is added to fresh food. In the *Dried Fruit and Water Observations* experiment, students investigate what happens when moisture is added to dried food. Have students record their results on the *Salt and Cucumbers* and *Dried Fruit and Water Observations* activity sheets. While you begin the investigations in this lesson, it is necessary to leave the experiments for at least 24 hours before collecting the final set of data. The emphasis in both of these investigations is the use of chemicals (sugar and salt) to control moisture content in food.

Salt draws water toward it. Because there is more salt outside the cucumber slices than inside them, the salt coating on the cucumbers draws the water out of the cucumber cells. Water diffuses out of the cucumber in an effort to dilute the salt. The salt dehydrates the cucumber slices.

With the dried fruit, sugar inside the fruit draws in the water. Help students understand that when dried fruit is placed in water, the water penetrates the outer skin and makes the fruit swell. Students can use their measurements of the dried fruit, before and after soaking, as evidence that the change has occurred.

6. Conduct Fruit Tasting

After the students observe the dried and fresh fruit, explain that they will be sampling different dried fruits and comparing them to a fresh version of the same fruit. They will use their senses to complete the *Taste-and-Compare Fruit Analysis* activity sheet. You can do the activity as a whole class, in small groups, or individually. Use whichever approach you think will work best for your class.

Have students begin by tasting dried and fresh versions of three fruits and recording their observations. Remind students to write the name of the fruit on the two charts and their observations below it. Encourage students to take their time, use descriptive words, and be as thoughtful and as thorough as possible as they complete the activity sheet. It may be difficult for students to write unique observations for each fruit since many dried fruits have similar qualities.

After students have completed their observations, have them compare and contrast one fruit in both the fresh and dried form on the last page of the activity sheet. Encourage them to be as detailed and descriptive as possible. When the tasting is completed invite students to share their observations with the class.

7. LiFE Logs

As homework, ask students to answer two questions in their LiFE Logs: *Why do we need to preserve foods?* and *How does preserving food change its taste and other qualities?*

Thinking about Food Change

This sample visual tool is one approach you can use with your students to help them organize what they have been learning about food change in this unit.

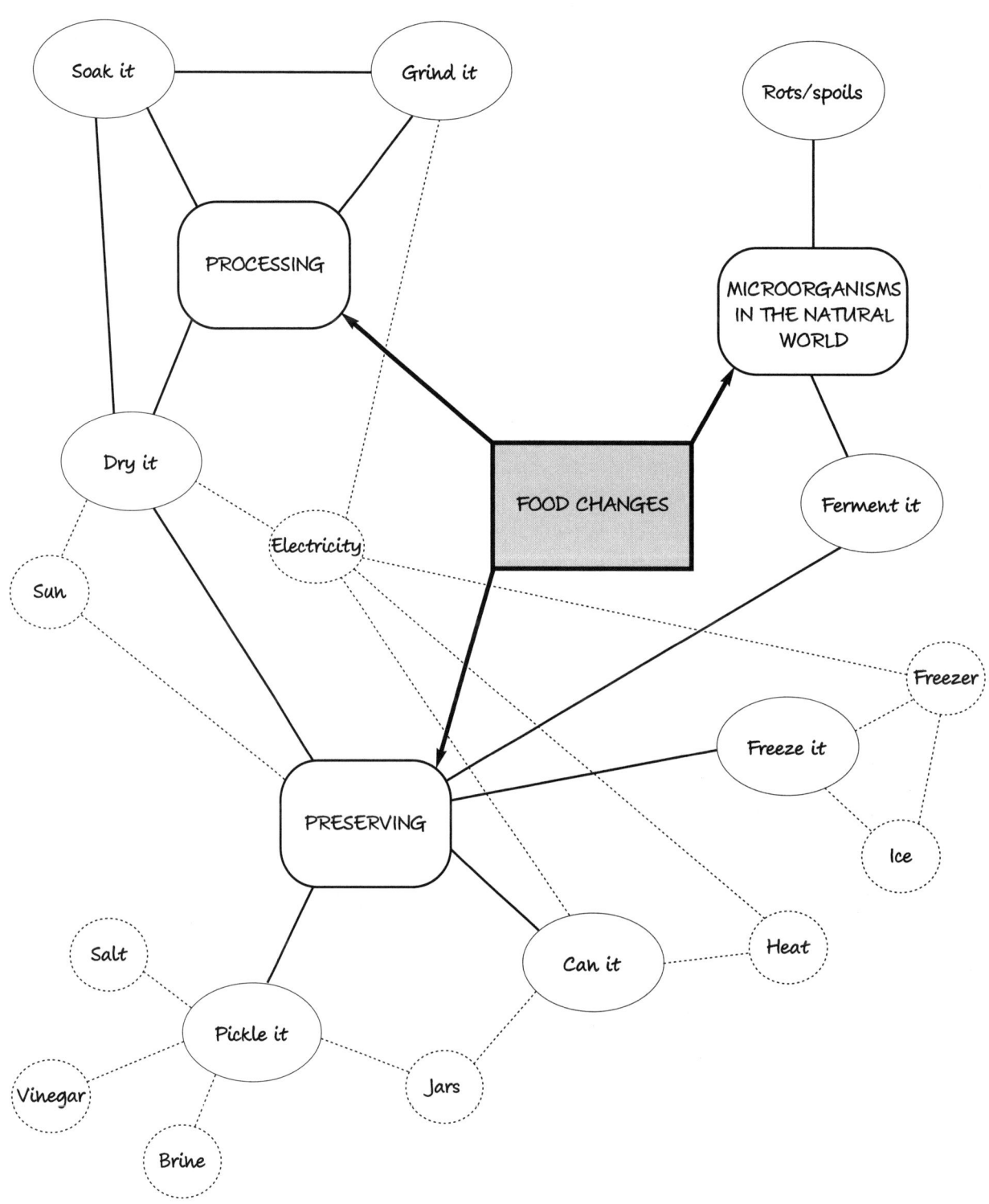

Investigating Food-Preservation Methods Supply List

HOMEWORK DISCUSSION

Supplies
- One canned tomato product
- One frozen-food product
- One bag dried fruit (see below)
- (Optional) Corn from Lesson 11

SALT AND CUCUMBERS (p. 259)

Supplies

For each group
- 1 cucumber
- Scale (a platform scale or electronic scale works well)
- 1-quart jar with cover
- Pickling salt (sodium chloride, NaCl)
- 2-cup measuring cup
- Measuring spoons
- Paper towels

DRIED FRUIT AND WATER OBSERVATIONS (p. 260)

Supplies

For the class
- Towels to clean up spills
- Dried raisins, apricots, or other dried fruits (enough for each student to have two pieces)
- Scale
- Tools for measuring, such as string, thread, metric rule
- 1 small bowl (large enough to hold the water and dried fruit)
- 1 cup fresh water

(continued on next page)

TASTE-AND-COMPARE FRUIT ANALYSIS (p. 265)

Supplies

For the class:
- 1–2 bags of three different kinds of dried fruits*
- fresh fruit versions of the dried fruit**
- paper towels or napkins

* Students will taste and compare three different kinds of dried fruit. The kind of dried fruits you use is not as important as making certain the fruit you use has not been candied and does not have added sugars, like pineapple and cranberries. We have found that any of the following work well: raisins, apples, apricots, bananas, pears, nectarines, peaches, prunes, and mangoes. Try to include a mix of fruits that students are familiar with as well as fruits that may be new to them. Purchase enough raisins for each student to have about four; purchase enough larger fruit for each student to have one or two. Dried apples and fresh apples and raisins and grapes may be the most convenient to use for the part of the activity in which students compare and contrast the dried fruit and the fresh fruit.

**Have samples of fresh apples, grapes, or other fresh-fruit versions of the dried fruit students will be tasting. Provide enough for each student to be able to compare the dried fruit and the corresponding fresh fruit. If possible, use local fruits or fruit that is in season so students have the opportunity to taste just-harvested fruit.

Salt and Cucumbers

Students examine the effect that salt (sodium chloride, NaCl) has on cucumber slices over a period of 24 hours.

Setup

1. Wash the cucumber in clean water.

2. Cut the cucumber into thin slices.

Procedure

1. Weigh the cucumber slices. Record the weight in ounces.

2. Place the cucumber slices in a jar.

3. Add the sodium chloride (NaCl). To determine the amount of sodium chloride, look at the weight of the cucumber in ounces. For every 4 ounces of cucumber, use 1 ounce of sodium chloride.

4. Seal the jar with a lid. Shake it to distribute the salt evenly.

5. Set aside for 24 hours.

6. Remove the cucumber slices. Pour the liquid into the measuring cup.

7. Dry the cucumber slices with paper towels. Weigh them.

Questions

1. *What chemical did you add to the cucumbers?*

2. *What effect did the chemical have on the cucumber?*

3. *Why is salting an effective way to preserve some foods?*

Dried Fruit and Water Observations

Students explore what happens when dried fruit is placed in a bowl of water for 24 hours.

Setup

Set out the paper towels or napkins and the tastings of dried fruit for students.

Procedure

1. Complete the first part of the activity sheet. Have students describe how they will tell whether or not the fruit changes after soaking in the water.

2. Have students taste the dried fruit.

3. Weigh the dried fruit, or use the string or thread and metric ruler and marker to measure the circumference before soaking.

4. Place the dried fruit in the bowl of fresh water.

5. Set the bowl aside for 24 hours.

6. Measure the circumference of the fruit and/or weigh it. Record the data.

Questions

1. *When you tasted the dried fruit before soaking it, could you taste the sugar?*

2. *What happened to the fruit that was soaked in water?*

3. *Why do you think this happened?*

Name	Date

Salt and Cucumbers Observations

In this investigation, you will put salt (sodium chloride, NaCl) on cucumber slices and observe what happens. You will make careful observations, gather data, and analyze your data to draw conclusions.

Day 1 Observations

1. What is the weight of the cucumber in ounces? _____

2. How much sodium chloride (NaCl) did you add to the cucumber slices?

_____

3. Describe what the cucumber slices look like. Do you see any moisture?

_____

_____

Day 2 Observations

1. Describe what the cucumber slices look like. Do you see any moisture?

_____

_____

2. Remove the cucumber slices from the jar. Place them on a paper towel. Measure the liquid left in the jar. How much liquid formed?

_____

3. Dry the cucumber slices with a paper towel. Weigh them. How much do they weigh?

_____

(continued on next page)

Name		Date	

Salt and Cucumbers Observations

4. Is there a difference in how much the cucumber slices weighed in the beginning of this investigation and how much they weigh now?

_____

5. What is the difference in weight?

_____

6. What do you think happened?

_____

_____

_____

7. Why do you think this happened?

_____

_____

_____

8. What evidence do you have that there was an interaction between the sodium chloride (NaCl) and the cucumber slices?

_____

_____

_____

Farm to Table & Beyond
©2008 Teachers College Columbia University

Name	Date

Dried Fruit and Water Observations

Look at the pieces of dried fruit. You are going to soak the fruit in water and observe what happens. You will be looking for ways that the fruit might change. What kinds of data can you collect to help you figure out if the fruit changes? Think about different ways that you can record the size of the fruit. Can you weigh it? Can you measure it?

Day 1 Observations

1. Describe how you will tell whether or not the fruit changes after soaking in the water.

_____

_____

_____

2. Does the dried fruit taste sweet? _____

3. Record the circumference of your dried-fruit sample. _____

4. How much does the dried-fruit sample weigh? _____

5. Describe what you observed when you put the dried fruit in water.

_____

_____

_____

(continued on next page)

Name	Date

Dried Fruit and Water Observations

Day 2 Observations

After the dried fruit soaks for 24 hours, observe it carefully, and answer the following questions.

1. Did you observe a difference in the size of the dried-fruit sample after it was soaked?

_____

2. How much does it weigh after soaking it? _____

3. What is the circumference of the dried-fruit sample after soaking it? _____

_____

4. What do you think happened?

_____

_____

5. Why do you think this happened?

_____

_____

BECOMING FOOD SCIENTISTS : INTERACTING PARTS : FOOD PROCESSING : ENVIRONMENTAL EFFECTS : WASTE : MAKING CHOICES

Name	Date

Taste-and-Compare Fruit Analysis

In this activity, you will taste three different fruits in both dried and fresh forms and compare the two. As you try the fruits, use your scientific-observation skills to clearly describe how each fruit tastes, smells, feels, and looks. Taste the dried fruit and write your observations in the chart below. Use as many adjectives as you can.

DRIED FRUIT OBSERVATIONS		
	Dried Fruit #1 _____	**Dried Fruit #2** _____	**Dried Fruit #3** _____
Appearance			
Taste			
Color			
Texture			
Odor			
Other Observations			

(continued on next page)

Name	Date

Taste-and-Compare Fruit Analysis

Now taste the fresh-fruit version of the dried fruit. Describe how each fruit tastes, smells, feels, and looks. After you taste each fresh fruit, write your observations in the chart below. Use as many adjectives as you can. After you finish making your observations, turn to the next page. Choose one kind of fruit to compare-and-contrast in both the fresh and the dried form. Use your observations to guide your thinking.

FRESH FRUIT OBSERVATIONS			
	Fresh Fruit #1 _____	**Fresh Fruit #2** _____	**Fresh Fruit #3** _____
Appearance			
Taste			
Color			
Texture			
Odor			
Other Observations			

(continued on next page)

LESSON 15: INVESTIGATING FOOD-PRESERVATION METHODS

Farm to Table & Beyond
©2008 Teachers College Columbia University

BECOMING FOOD SCIENTISTS : INTERACTING PARTS : **FOOD PROCESSING** : ENVIRONMENTAL EFFECTS : WASTE : MAKING CHOICES

Name

Date

Taste-and-Compare Fruit Analysis

DRIED FRUIT AND FRESH FRUIT
Compare/contrast thinking-process map

Kind of Dried Fruit:

_____

Kind of Fresh Fruit:

_____

How are they alike?

_____

_____

_____

How are they different?
With regard to:

TASTE?

TEXTURE?

MOISTURE?

WEIGHT?

BECOMING FOOD SCIENTISTS : INTERACTING PARTS : FOOD PROCESSING : ENVIRONMENTAL EFFECTS : WASTE : MAKING CHOICES

Making Pickles

AIM

To further student under-standing of food change and to apply what students have learned.

SCIENTIFIC PROCESSES

- question, explore, observe, apply

OBJECTIVES

Students will be able to:

- summarize why we need to preserve food;

- describe how to inhibit the growth of microorganisms;

- discuss how different food-preservation methods keep microorganisms from growing on food;

- apply what they have learned about preserving food.

OVERVIEW

In this lesson, students apply what they have learned about food preservation by making pickles using cucumbers, sodium chloride (NaCl), and seasonings. First they observe a demonstration of what effect sodium chloride and acetic acid have on microbial growth. Then students apply the principles of food preservation by making pickles. Check for student understanding of technology. Help them understand that processing food and preserving food are examples of technology, which includes methods and procedures involved in changing or modifying the natural environment to satisfy perceived human needs. Using different methods to preserve food, we modify the environment that the microorganisms live in to suit our needs. To complete the lesson, students write in their LiFE Logs about the farm-to-table journey of enriched wheat flour found in frozen pancakes.

MATERIALS

For the teacher:
- *Fermentation* teacher note
- *Inhibiting Growth* experiment sheet
- *Pickles* teacher recipe
- (Optional) *Cooking Tips* lesson resource (p. 213)

For the class:
- Materials from the *Making Pickles Supply List* lesson resource
- Tape
- Labels
- Marker

For each student:
- *Inhibiting Growth Observations* activity sheet
- *Sour Pickles* student recipe
- LiFE Log

PROCEDURE

Before You Begin:

- Gather materials listed on the **Making Pickles Supply List** lesson resource.

- Review the **Fermentation** teacher note, the **Inhibiting Growth** experiment sheet, and the **Pickles** teacher recipe.

- Prepare ingredients for making pickles.

- (Optional) Review the **Cooking Tips** lesson resource.

- Review the **Inhibiting Growth Observations** activity sheet and the **Sour Pickles** student recipe.

- If you have not already done so, post the Module Question and Unit 3 Question at the front of the classroom.

MODULE QUESTION

What is the system that gets food from farm to table, and how does this system affect the environment?

UNIT QUESTION

What happens to food as it moves from farm to table?

 **QUESTIONING**

1. Review Module and Unit Questions

Explain that in this lesson students will be pulling together all they have learned about food preservation and food processing. They will summarize what they have learned and think through the Module and Unit questions.

2. Summarize Unit 3

Review with students what they have learned. *What causes food to change?* (Microorganisms, processing, preservatives.) *What are some of the*

reasons we process food? Why do we change it from one form to another? (It adds variety to our diet; it can make food last longer; people like to eat different kinds of food.) *What are some ways that we preserve food?* (Canning, freezing, salting, drying, fermenting.) *Why do we preserve food?* (To have it available when there isn't any fresh food; to preserve surplus food; it adds variety to our diet.)

What have you learned so far in this unit about what happens to food between farm and table? Explain that in this lesson we are going to learn about fermentation and pickling, other types of technology that humans use. Check for student understanding of technology. It is important for students to understand that technology includes methods and processes involved in changing or modifying the natural environment to satisfy perceived human needs. With this understanding of technology, making pickles is an example of technology.

Point out to students that byproducts, such as brine in the case of pickles, can have an effect on the environment. Consider bringing this up as the class makes pickles. *How does a pickle factory dispose of the brine and vinegar that are so toxic to microorganisms? Do you think brine might have an effect on the environment?* (In fact, there is an effect; in this lesson, however, it is enough to have students think about the brine byproduct and realize that it is one of the trade-offs in the pickle industry.) *Once we eat the pickles, what do we do with the brine?*

 **EXPERIMENTING**

3. Investigate Microbial Growth

Distribute the **Inhibiting Growth Observations** activity sheet to students and prepare the classroom for the **Inhibiting Growth** experiment. In this experiment, sodium chloride (NaCl) is added to one sample of chicken broth, acetic acid (CH_3COOH) is added to another sample of broth, and a third sample of broth is kept as the control. Through observation, students discover that these two chemicals create a toxic environment that inhibits the growth of food-spoiling microorganisms. As you set up the class demonstration, review the activity sheet and have students begin to complete it. Check for student understanding of experimental design. Have students complete the activity sheet through the description of the experimental design. Let the three samples sit at least 24 hours before returning to the experiment.

4. Review Recipe

Turn to the pickle-making setup that you prepared before class. Elicit any questions or ideas that students have about pickling. Remind them that it is a form of preserving food. Have students draw on their experiences from the lessons in this unit and think about what knowledge they have learned that they will apply to this activity. Tell students that, as a class, they will be preparing some cucumbers for pickling. These pickles will be ready to eat in a week or longer, depending on how pickled they want them.

Clarify what students will be doing in each step of the recipe. You may wish to have students work in small groups to make their own jar of pickles or collaborate and work together as a class in an assembly line. Either way you choose, this is a great time to remind students that humans are important "parts" of a food-processing system. You may wish to ask students to think of the steps that are done by machines in factories.

5. Prepare Pickles

Review the *Cooking Tips* lesson resource with the class. If students have not already done so, have them wash their hands. Follow the class recipe and prepare the pickles.

6. Clean Up

Remind students that cleaning up is an important part of food preparation. Point out that not only have they been working with chemicals, they have been working with microorganisms. Have them wipe off any surfaces where they worked, collect any vegetable scraps and put them in a compost box or dispose of the scraps in the appropriate way for your school, pick up anything on the floor, and sweep, if necessary.

7. Finish Observations

Remember to have students complete the *Inhibiting Growth Observations* activity sheet, after the samples have sat for at least 24 hours.

8. LiFE Logs

This is the last lesson in Unit 3, during which students explored the question *What happens to food as it moves from farm to table?* To wrap up the unit, and to assess what students have learned, ask them to write a story in their LiFE Logs that answers this question: *How is the enriched wheat flour in frozen pancakes that you buy in the store preserved, processed, transported, and packaged between the time it leaves the farm as one form, until it arrives, as another form, in the frozen-food section of the store?* Students can begin in class and finish it for homework if needed. You may wish to use this story prompt: "The wheat plant grows on the farm..." and end with "the box of frozen pancakes was stocked in the store's freezer section." Students will share their stories in the next unit.

Fermentation

Any discussion of fermentation wouldn't be complete without mentioning Louis Pasteur (1822–1895). Although fermentation is one of the oldest techniques for preserving food, before Pasteur, people didn't understand how the process worked. It was Pasteur's work on yeast cells that showed how they produce alcohol and carbon dioxide from sugar during fermentation. Pasteur found that fermentation can take place in anaerobic (without oxygen) conditions. It is a form of cellular respiration — a way for yeast cells to get energy when there is no oxygen present. He is quoted as having defined fermentation as "life without air." Pasteur's work arose out of a need to solve a problem that was plaguing the French wine industry — wine that turned to vinegar. Pasteur turned to his microscope and looked for clues to help him figure out how fermentation worked. This lesson offers a wonderful opportunity to bring the history of science to light and show how Louis Pasteur asked questions, experimented, searched for answers, theorized, and applied what he learned to life!

The fermentation process itself is simple. You don't need high-tech equipment or special tools. In fact, as mentioned in the *Putting Food By* student reading, food is fermented in outdoor pits in the South Pacific Islands. While it may seem hard to fathom that fresh food is ever difficult to find in the South Pacific, these tropical islands have unpredictable weather patterns. Destructive storms and tidal waves can destroy crops and leave islanders without food for months. Anthropological research suggests that pit fermentation is one way to respond to these unpredictable weather patterns, providing edible food for indefinite periods of time. The technique is very low-tech, yet effective.

In pit fermentation, the soil on the side of the pits is firmly packed. Islanders line the pit with woven coconut-leaf panels and other plant leaves. Then food, like cassava or banana, is put into the pit and packed to remove any air pockets. More plant leaves are used to cover the pit, sealing it. Finally, clean rocks are piled on top. It takes about six weeks for the food to ferment. As the food ages, the flavor develops. When the food is ready to eat, it is dried in the sun, pounded into a paste, baked or boiled, and eaten.

Making Pickles Supply List

INHIBITING GROWTH (p. 273)

Supplies

For the class demonstration
- 3 eight-ounce plastic cups with plastic wrap or aluminum foil to cover
- 1 cube chicken bouillon
- 1 teaspoon table salt (sodium chloride, NaCl)
- 1 teaspoon vinegar (acetic acid, CH_3COOH)
- 1 cup very hot water
- Measuring cup
- Spoon

PICKLES (p. 274)

Ingredients

Yields 30 pickle halves
- 15 small, very firm Kirby cucumbers, free of bruises or brown spots
- 3–4 cloves garlic, unpeeled but lightly crushed
- 1/2 teaspoon coriander seeds
- 1/2 teaspoon mustard seeds
- 1/2 teaspoon black peppercorns
- (Optional) 3 small bay leaves
- 6–7 sprigs of fresh dill, well washed
- 1/2 teaspoon dried dill seeds
- 2–3 slices of sour rye bread with caraway seeds (one piece for each jar of pickles you make)
- 1 1/2 quarts of water, or as needed
- 1/4 cup plus 2 tablespoons kosher (coarse) salt, or as needed*

Supplies

For the class
- 1 set measuring spoons
- 1 set measuring cups
- 2–3 large, wide-mouthed glass jars with metal canning lids, sterilized
- One small napkin for each student

*It is important that you do not use iodized salt for pickling, as that will leave a bitter aftertaste; if you cannot get kosher (coarse) salt, use non-iodized table salt, substituting about two-thirds of the amount called for.

Inhibiting Growth

In this demonstration, students observe the effects of chemical preservatives on the growth of microorganisms.

Setup

Make three labels: "Sodium Chloride (NaCl)," "Acetic Acid (CH_3COOH)," and "Control." Tape one label to each cup.

Procedure

1. Add the bouillon cube to the hot water. Stir to dissolve. Divide the bouillon equally among the three cups.

2. Add 1 teaspoon salt to the cup labeled Sodium Chloride (NaCl). Do not stir.

3. Add 1 teaspoon vinegar to the cup labeled Acetic Acid (CH_3COOH). Do not stir.

4. Have students observe the three cups and record their observations on the activity sheet.

5. Place the covers on the three cups.

6. Set aside for at least 24 hours.

7. Have students observe the three cups and record their observations on the activity sheet.

Questions

1. *Based on what you have learned in this unit, what do you predict will happen? What evidence do you have to support your prediction?*

2. *What chemicals were added to the bouillon?*

3. *What is the control?*

4. *What are the variables?*

5. *Did you observe any differences the first time you made your observation?*

6. *Were there any differences when you made your second observation?*

7. *What can you conclude from this experiment? What is your evidence?*

Pickles

This recipe does not call for vinegar, which generally is used to make the brine for pickles. Instead, it calls for rye bread. The yeast in the bread creates a mildly fermented brine that lends a subtle flavor to the pickles.

Yields 30 pickle halves

INGREDIENTS

15 small Kirby cucumbers

3–4 cloves garlic, smashed

1/2 teaspoon coriander seeds

1/2 teaspoon mustard seeds

1/2 teaspoon black peppercorns

(Optional) 3 bay leaves

6–7 sprigs fresh dill, washed

1/2 teaspoon dried dill seeds

2–3 slices sour rye bread with caraway seeds

1 1/2 quarts of water

1/4 cup plus 2 tablespoons kosher (coarse) salt

2–3 large, wide-mouthed glass jars and lids, sterilized

DIRECTIONS

1. Carefully wash the cucumbers, rubbing gently with your hands to remove any sand. Slice the cucumbers in half lengthwise.

2. Divide the cucumbers into two or three equal-sized groups and stand them on end in the jars, so that they hold each other in place but not so tightly that they will crush each other. You can add a second upright layer if the jar is tall enough.

3. Add the garlic, herbs, and spices in equal amounts to the jars. Put a slice of bread on top of the cucumbers in each jar.

4. Mix the water with the coarse salt and stir until the salt dissolves.

5. Pour the salt water into the jars to completely cover the pickles. The brine (salt water) should almost overflow so you can be sure no air pockets remain.

6. Put the lids on the jars. Close them tightly.

7. Place the jars on a stain-proof surface in a cool place, with a temperature between 65° and 70°F. Do not refrigerate.

8. Gently shake the jars once a day to mix up the spices.

9. When the pickles are ready* cut into enough pieces so that each student gets a taste, serve on napkins or plates, and enjoy!

* The pickles will be half sour in four or five days, and very sour in about 10 days. If you have only one tasting, students will each get half a pickle. You may decide to have a tasting at five days and again at 10 days. In that case, each student would get a quarter of a pickle at each tasting.

Name	Date

Inhibiting Growth Observations

In this class demonstration, there are three containers of chicken broth. You will test two variables: sodium chloride (NaCl) and acetic acid (CH$_3$COOH). The third container will be the control. You will observe what happens in the three containers.

Label the three containers below: "Sodium Chloride (NaCl)"; "Acetic Acid (CH$_3$COOH)"; "Control."

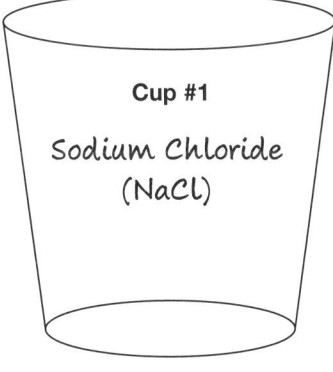
Cup #1
Sodium Chloride
(NaCl)

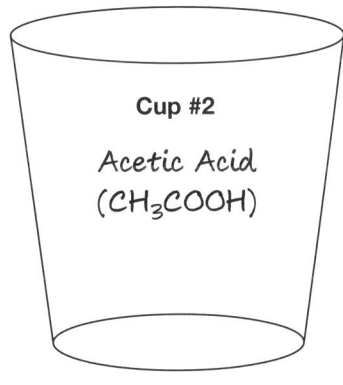
Cup #2
Acetic Acid
(CH$_3$COOH)

Cup #3
Control

In the space below, write a hypothesis about what you think the results of this experiment will be.

1. Cup #1 (NaCl):

_____

_____

2. Cup #2 (CH$_3$COOH):

_____

_____

3. Cup #3 (Control):

_____

_____

(continued on next page)

Name

Date

Inhibiting Growth Observations

Set Up Experiment

Write down the steps in the experiment. These are your methods. Identify the control and the experimental group. Be sure to identify the variable (the thing that will change) for each experimental group. Include the materials the class used in this demonstration.

1. Materials

_____

_____

2. Methods

_____

_____

_____

_____

_____

_____

_____

_____

_____

(continued on next page)

Name   Date

Inhibiting Growth Observations

3. Record your data in the table below.

INHIBITING GROWTH OBSERVATIONS			
	Cup #1 _____	**Cup #2** _____	**Cup #3** _____
Day 1			
Day 2			

(continued on next page)

Name   Date

Inhibiting Growth Observations

4. Examine your results and think about what you have learned. What did the experiment teach you? How can you use this information?

_____

_____

_____

_____

_____

_____

_____

LESSON 16: MAKING PICKLES

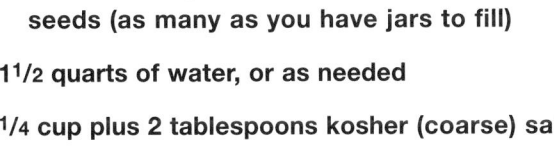

Name	Date

Sour Pickles

Yields 15 pickles

In class, we have been learning about food preservation. Pickling, a form of canning, is a common way to preserve food. To celebrate what we have learned, we made pickles. Take this recipe home and make pickles with your family. Share what you learned about food preservation with your family and enjoy delicious homemade pickles!

INGREDIENTS

15 small, very firm Kirby cucumbers, free of bruises or brown spots

3–4 cloves garlic, smashed

1/2 teaspoon coriander seeds

1/2 teaspoon mustard seeds

1/2 teaspoon black peppercorns

(Optional) 3 bay leaves

6 or 7 sprigs fresh dill, well washed

1/2 teaspoon dried dill seeds

Slices of sour rye bread with caraway seeds (as many as you have jars to fill)

1 1/2 quarts of water, or as needed

1/4 cup plus 2 tablespoons kosher (coarse) salt

2 or 3 large, wide-mouthed glass jars, with lids, sterilized (as many as needed to hold all the pickles)

DIRECTIONS

1. Carefully wash the cucumbers, rubbing gently with your hands to remove all traces of sand.

2. Stand the cucumbers on end in the jars, so that they hold each other in place but not so tightly that they will crush each other. If the jar is tall enough, add a second upright layer.

(continued on next page)

Name   Date

Sour Pickles

3. Add the garlic, herbs, and spices in equal amounts to the jars. Put a slice of bread on top of the cucumbers in each jar.

4. Mix the water with the coarse salt. Stir until the salt dissolves.

5. Pour the salt water (brine) into the jars to completely cover the pickles. The brine should almost overflow. Make sure no air pockets remain.

6. Seal the jars by closing tightly.

7. Place the jars on a stain-proof surface in a cool place, with a temperature between 65° and 70°F. Do not refrigerate.

8. Gently shake the jars once a day to mix up the spices.

9. When the pickles are ready* just reach into the jar, grab a pickle, and enjoy!

*The pickles will be half sour in four or five days, and very sour in about 10 days. When they have reached the degree of sourness you like, store them in the refrigerator. They will keep for about five weeks.

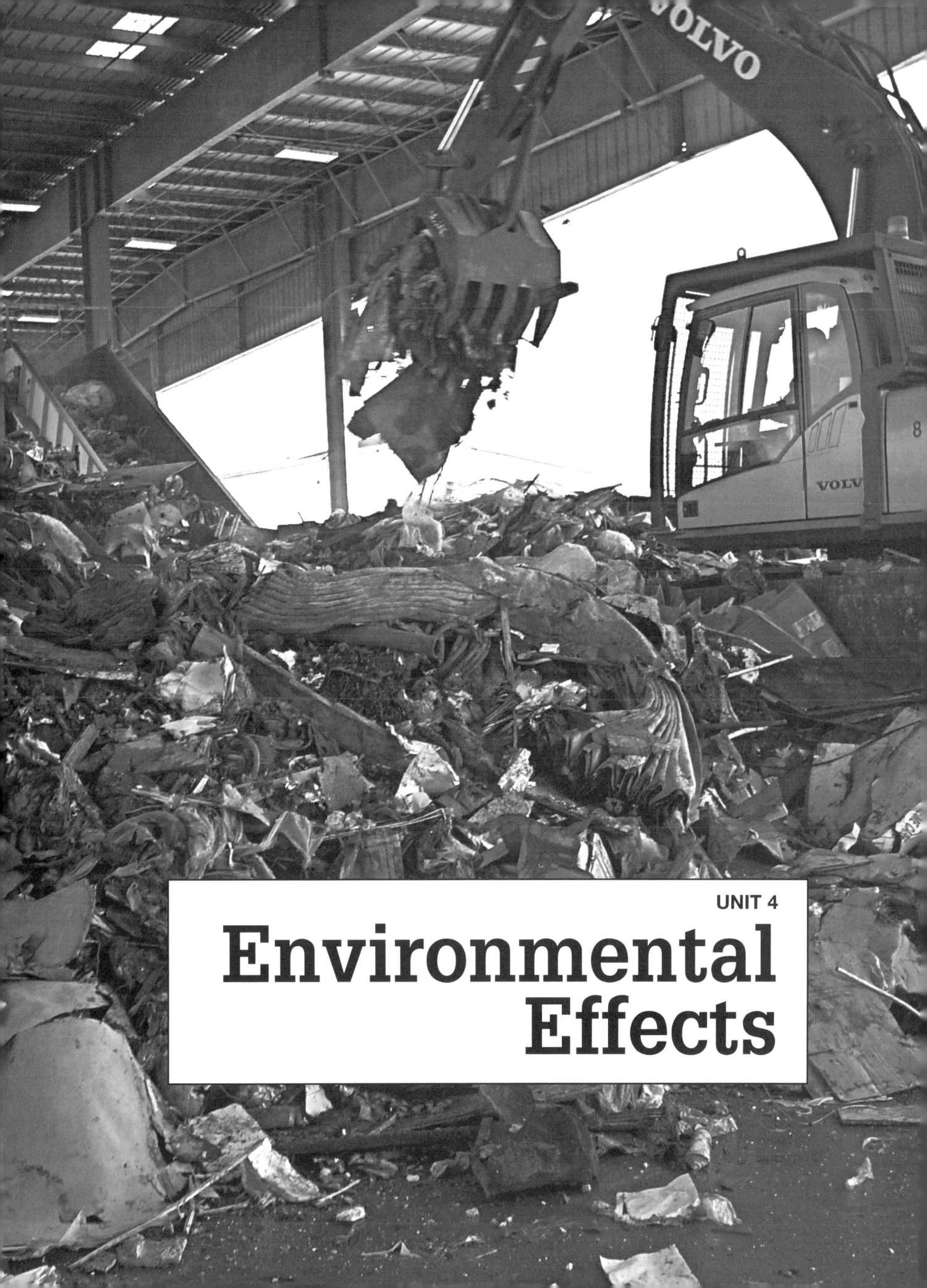

Environmental Effects

Planet Earth

AIM

To gain knowledge about Earth as a system of interacting parts.

SCIENTIFIC PROCESSES

• **research, gain knowledge, summarize, explain**

OBJECTIVES

Students will be able to:

• **name and describe parts of the Earth system;**

• **discuss the water cycle;**

• **describe interactions that take place between parts of the Earth system;**

• **reflect on some of the ways human activities affect the Earth system.**

OVERVIEW

In Units 2 and 3, students explored different parts of our food system and how these parts interact. In investigating packaging, processing, and transportation as systems and subsystems that are part of the larger food system, students began to learn about some of the ways in which human actions affect ecosystems. In this lesson, students are introduced to the study of Earth as a system of interacting cycles and to the Unit 4 Question, *What are the environmental effects of our farm-to-table system?* Working in small groups, students use a jigsaw reading strategy to learn about Earth as a system. They read about the biosphere, the atmosphere, the geosphere, and the hydrosphere. After completing their jigsaw readings, students develop posters that they present to their classmates. Then they read about Earth's cycles, including the water cycle and ocean currents. For homework, they write a brief essay about what they have learned.

MATERIALS

For the teacher:
• *Using the Environment as a Context* teacher note
• *Earth-Systems Approach* lesson resource
• (Optional) *Transportation System Jigsaw* lesson resource (p. 138)

For each group:
• Chart paper
• Markers
• Scissors
• Tape
• (Optional) Magazine with photographs and illustrations to use in presentations

For each student:
• *Dynamic Earth* student reading
• *Earth Cycles* student reading
• *Earth-Systems Worksheet* activity sheet
• LiFE Log

PROCEDURE

Before You Begin:

- Review the *Using the Environment as a Context* teacher note, the *Earth-Systems Approach* lesson resource, the *Dynamic Earth* and *Earth Cycles* student readings, and the *Earth-Systems Worksheet* activity sheet.

- (Optional) You may wish to review the jigsaw strategy described in the *Transportation System Jigsaw* lesson resource.

- Make copies of the student readings and activity sheet for each student.

- If you have not already done so, post the Module Question and Unit 4 Question at the front of the classroom.

MODULE QUESTION

What is the system that gets food from farm to table, and how does this system affect the environment?

UNIT QUESTION

What are the environmental effects of our farm-to-table system?

QUESTIONING

1. Review Module and Unit Questions

Invite students to share their flour-to-frozen pancake stories from Lesson 16. Next, remind them of the Module Question. Briefly discuss some of the human-designed systems students were introduced to in Units 2 and 3 that are part of the food system. Explain that in this unit, they will explore in more detail some of the effects of these human-designed systems. Introduce the Unit 4 Question and tell students that they will explore the environmental effects of our farm-to-table system.

2. Introduce Earth Systems

Have students work in small groups. Distribute the *Dynamic Earth* student reading and the *Earth-Systems Worksheet* activity sheet. Using the jigsaw strategy, assign one topic (biosphere, atmosphere, geosphere, or hydrosphere) to each student in each small group, the "home" group. Have experts move into their topic group. Experts will review the reading and develop a poster. Student experts return to their "home" group and share what they have learned. Alternatively, you may wish to divide the class into four groups and assign one topic to each group. Each group will read about the topic and make a brief presentation to the class.

Review the activity sheet with students. Make sure they understand that this worksheet is to help them organize the information they learn about their topic. It also provides tips to follow when planning the presentation.

SEARCHING

3. Learn about Earth Systems

Allow students time to read about their topic. Tell students that if they have any questions, they should raise their hands and you will come to them to discuss the question. Walk among the groups and check for student understanding. You may wish to discuss the activity sheet with student presenters.

THEORIZING

4. Make Posters

Once students have completed the reading and the activity sheet, have the groups begin work on their poster presentations. If you have time, encourage students to look through magazines or on the Internet for graphics and photographs they can include in their poster presentations.

Remind students that they have about five minutes to make their presentation. Encourage them to outline the main points about their topic and to include interesting details. Make sure they leave enough time for their audience to ask questions.

Before students present, walk among the groups. Listen to the presentations and look over the posters. Check for student understanding of the material. Help students think through any questions they might have.

5. Present Posters

Remind students that they are now the teachers. They will be teaching their classmates about the topic that was assigned to them. Remind the audience to listen carefully during the presentations. Tell students that there will be time to ask questions at the end of each presentation. You may wish to act as moderator. After each presentation, tape the poster at the front of the classroom.

6. Discuss Interactions

Have students turn to the *Dynamic Earth* student reading and read the last paragraph on interacting parts. You may wish to draw the diagram from the *Earth-Systems Approach* lesson resource on the board to help reinforce the interactions. As a class, have a discussion about ways humans interact with the various spheres. *Which sphere is the home to all life?* (The biosphere.) *What are some ways that humans interact with other parts of the biosphere?* (We eat plants and some animals. Farmers grow crops. We use trees to make paper, build houses and furniture.) *What are some ways that humans interact with the atmosphere?* (We take in oxygen and give off carbon dioxide. We fly in planes.) *What are some ways that humans interact with the geosphere?* (Farmers plant crops in soil. They put fertilizer on the soil. We extract minerals from the

Earth.) *What are some ways that humans interact with the hydrosphere?* (We build dams on rivers. We drain swamps. We swim in oceans, lakes, ponds, and rivers. We use water every day. We wash dishes, take baths, wash clothes, brush our teeth, and water the garden. We drink water. Farmers use water for irrigation. Water is used in food processing.) Invite student volunteers to help summarize what they have learned about Earth's subsystems.

7. Discuss Earth Cycles

Distribute the *Earth Cycles* student reading. Have students take turns reading it out loud. Carefully review the hydrologic cycle. *What is another name for the hydrologic cycle?* (The water cycle.) *Where does the energy that powers this cycle come from?* (The sun.) *What happens to water as it moves through the cycle?* (It changes state.) *Does water evaporate from the ocean as fast as it does from the land?* (No. Some water molecules cycle through over a long period of time and some cycle through over a short period.) *Do human activities affect the water cycle?* (Data suggests human activities may affect weather. Irrigation may affect the water cycle.)

8. LiFE Logs

If you do not have time to complete the *Earth Cycles* student reading in class, have students finish it as homework.

Have students write a five-paragraph essay in their LiFE Logs that begins with, "The most interesting thing that I have learned about human interactions with the environment is..."

Using the Environment as a Context

The focus of this unit is environmental effects. As a teacher of science, you may already have faced some of the challenges of bringing environmental issues into the science classroom. As the National Science Teachers Association points out in the preface of *Resources for Environmental Literacy*, "The primary responsibility of teachers of science is to teach science, not to inform their students on environmental issues — and certainly not to influence the stand students may take on those issues."[1] With this in mind, how do you help students gain an understanding of ways to use science to deal with environmental issues and keep personal opinion out of the classroom? This can be particularly challenging if you encourage students to discuss and debate what they are learning about these issues. The good news is that as a teacher, you set the tone for the classroom and the class rules for debating. You can remind students to provide evidence to support their positions.

Introducing environmental issues can be a compelling way to engage students in the application of science to "real-world" problems. In this module, students learn about some of the environmental issues related to the farm-to-table system. They begin to understand that decisions often involve trade-offs. They learn that an action taken in one place may have an effect far away from the original source. They learn that the choices they make as consumers have an effect on the environment. They begin to recognize that they can use the science they learn to make ecologically sound food choices.

This second half of *Farm-to-Table & Beyond* has an emphasis on application and stresses the QuESTA phases of theorizing and applying to life. To support this, we are including examples of science in action. Too often, science is thought of as a body of facts. To help students see science as a process, we have included brief descriptions in student readings of scientific research, highlighting the questions scientists are asking, how they are collecting data, and what their research is suggesting.

---

[1] NSTA, 2007, p. xiii.

Earth-Systems Approach

We continue the systems approach in this unit. This first lesson provides the context for looking at planet Earth as a large, complex network of interacting physical, chemical, and biological interactions. As we have done throughout this module, we look at the system as a whole and some of the subsystems, including the biosphere, the atmosphere, the geosphere, and the hydrosphere. Taking this approach, students can begin to understand how an action that may seem only to affect one subsystem actually reverberates throughout the system. For example, on one level, cutting down acres of rain forest would seem to be an issue that affects only those who live in or near the rain forest or extract products from the forest. It would seem to have a biosphere-specific impact. However, cutting down acres of forest affects much more, including the atmosphere, the hydrosphere, and even the geosphere. As this diagram so clearly shows, all the "spheres" interact, and the energy that powers Earth's systems and interactions comes from the sun.

The unit begins with a student reading that introduces Earth-systems science and some of the interactions that take place on our home planet. This reading is the foundation for the unit's work. It sets the stage for examining the impact of human activity on the Earth system. The second student reading introduces Earth as a system of interacting cycles. Students learn about the hydrologic cycle and patterns of wind and weather. This reading also highlights flotsam science and the work of the oceanographer Curtis Ebbesmeyer in Seattle, Washington. Ebbesmeyer used to collect ocean current data using drift bottles that scientists would drop into the ocean. A chance spill from a container ship in the Pacific changed all that. There are multiple lessons in this story, including how to make lemonade from lemons, as Nike shoes and bathtub toys became data-yielding flotsam. The container spill is also a reminder that the behemoths of the seas that carry society's "goods" are not impervious to the power of Nature. One huge storm that resulted in containers being dumped into the ocean is dramatic evidence of the environmental effects of our global trade system. Cans of Chinese noodles and thousands of bags of chips sealed in water-tight packaging have washed ashore when containers were damaged at sea. The effects of our farm-to-table system are felt around the world.

Name   Date

Dynamic Earth

We live on Earth. That's not a surprise. But did you know that scientists study our planet as a system? They investigate all the parts of the Earth system and how they work together. As a systems scientist, your assignment now is to think about the basic subsystems of planet Earth: the **biosphere,** the **atmosphere,** the **geosphere,** and the **hydrosphere.**

Biosphere

The biosphere is the subsystem where all life exists, including organic matter that has not yet decomposed. To learn about the biosphere, scientists study small parts of it, such as ecosystems. Forests, farms, and lakes are examples of ecosystems.

An ecosystem includes both **abiotic** (nonliving) and **biotic** (living) parts. Temperature, the slope of the land, sunlight, rain, snow, fog, and air are all examples of abiotic factors. These parts can interact, too. Have you ever seen a creek or river flood because of heavy rains? That's one kind of interaction. How about volcanoes? A volcanic eruption can create an island, another interaction.

Biotic factors include all living organisms, which also interact in different ways. Plants and animals compete for resources such as food, air, water, and space. Have you ever heard a gardener or farmer complain about snails or rabbits that eat her plants? Maybe you've seen birds compete for the same food at a bird feeder or in a park. These are all interactions.

Abiotic and biotic parts interact, too. Think about what happens during the process of photosynthesis. Plants use carbon dioxide in the air to help make sugars. They give off oxygen, which is released into the air. The temperature range and availability of water are abiotic factors that affect the kinds of plants and animals that can survive in an environment.

Atmosphere

The atmosphere, another part of the Earth system, is a mixture of gases, including nitrogen, oxygen, water vapor, and carbon dioxide. The atmosphere blankets the Earth, holding in heat to keep the planet warm and absorbing harmful radiation from the sun.

The atmosphere is made up of layers with different temperatures. The layer nearest the Earth is called the **troposphere.** Weather occurs in the troposphere. It contains most of the water vapor, so it's the layer where most of the clouds form. In this layer, air cools gradually as it get farther from Earth. Fast-flowing air currents called jet streams

(continued on next page)

Name   Date

Dynamic Earth

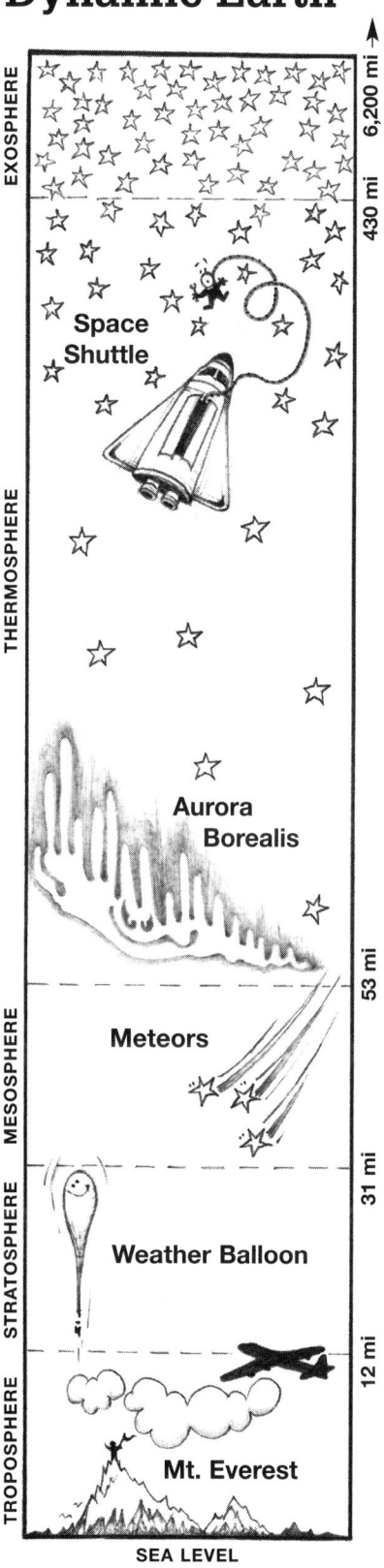

are found here at the top of the troposphere. Many commercial airplanes fly at the boundary of the troposphere and the layer above, called the **stratosphere.** The stratosphere contains the ozone layer, which absorbs the ultraviolet radiation from the sun. This protects life on Earth. In the stratosphere, air is cooler in the lower part and warms up as it gets farther from Earth. Above the stratosphere is the **mesosphere,** which has the coldest temperatures. If you have ever seen a falling star, it's been in the mesosphere. The next layer is the **thermosphere,** where temperatures can be as high as 3,632°F. The space shuttle and space station orbit in this layer. The highest layer of the atmosphere is the **exosphere.** Beyond the exosphere lies outer space.

Geosphere

The geosphere is the solid part of planet Earth. It includes the different layers of the Earth's interior, rocks, minerals, landforms, and the processes that shape our planet's surface. The Earth's interior consists of a series of layers: the inner core, the outer core, the mantle, and the crust. The Earth's crust is the outermost layer. The bottom of the ocean and the continents are parts of the crust. The Earth's crust and the brittle top portion of the mantle make up the rigid shell of the Earth called the **lithosphere.** The lithosphere is broken into plates, which are always moving. Over millions of years, the crust has been shaped into flat lands and mountains.

The Earth's mantle is the middle layer. It's a layer of hot rock that can be as much as 2,000 miles thick. Have you ever seen pictures of red-hot rock erupting from a volcano? That rock comes from the top of the mantle. Scientists have gone into space and peered deep into the ocean, but no one has ever gone to the center of the Earth. Even though no one has traveled there, scientists believe the outer core is liquid and the inner core is

(continued on next page)

Name   Date

Dynamic Earth

solid. If scientists can't see pictures of it and haven't gone there, how can they possibly learn about the Earth's core? They study the Earth's vibrations. These vibrations, or seismic waves, help scientists learn about the inner core.

Hydrosphere

The hydrosphere is the sub-system that includes all of Earth's liquid water. It helps control the temperature and climate on Earth. The hydrosphere includes the oceans, rivers, seas, lakes, ponds, and streams. In fact, almost 75 percent of the Earth is covered by water. About 97 percent

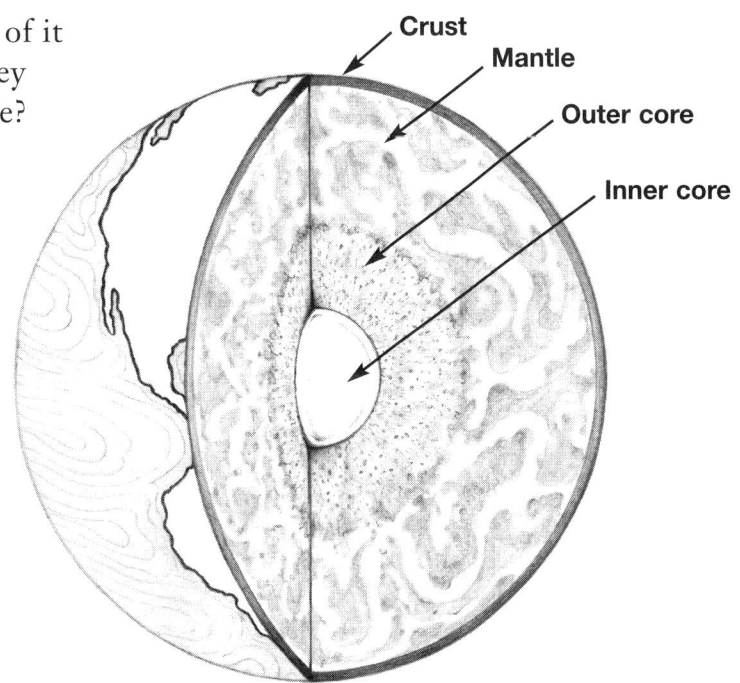

of the water is found in the oceans. The remaining 3 percent is freshwater. But did you know that most of the freshwater is locked up in glaciers and ice caps? What isn't frozen is beneath your feet as groundwater or in rivers, swamps, ponds, and lakes. Like the atmosphere, the hydrosphere is constantly in motion.

There is so much water on Earth that some people call it the blue planet. Water is a substance that is necessary for all forms of life. However, even though most of the planet is cover by water, freshwater is not evenly distributed around the world. Some people have to drill deep wells to reach water. Others have to bring freshwater from far away.

We use freshwater for drinking, farming, and many other things. In the United States, we use huge amounts of water to manufacture products and for irrigation.

Interacting Parts

Imagine wheat plants growing in a field. What interactions take place between the different spheres? During photosynthesis, the wheat plants (part of the biosphere) use carbon dioxide from the atmosphere. Rain (from the atmosphere) falls, landing in streams (part of the hydrosphere), and soaks into the soil (part of the geosphere). The wheat plants take up the water through their roots (part of the biosphere). Where does the energy come from to power all these interactions? It comes from the sun.

BECOMING FOOD SCIENTISTS :: INTERACTING PARTS :: FOOD PROCESSING :: ENVIRONMENTAL EFFECTS :: WASTE :: MAKING CHOICES

Name   Date

Earth Cycles

Earth is a system of interacting cycles. These cycles are the way that matter and energy move from place to place. In this reading, you will learn about several of these cycles.

Hydrologic Cycle

Have you ever been in a desert or seen photographs of one? You may have noticed that compared to a rain forest, there aren't many living organisms. Water is important for life. Plants take it in through their roots and release it through their stomata. Animals, including humans, drink water that came from the atmosphere. How does it return to the atmosphere? We exhale, sweat, and excrete it. The water that we need for life on Earth is constantly cycling between the biosphere, the hydrosphere, and the atmosphere.

Water continually moves between the Earth and the atmosphere in what is called the hydrologic cycle, or water cycle. How does the cycle work? It's powered by energy from the sun. Water molecules (H_2O) change state as they move through the cycle. Liquid water evaporates from a large body of water, like the ocean. It changes to water

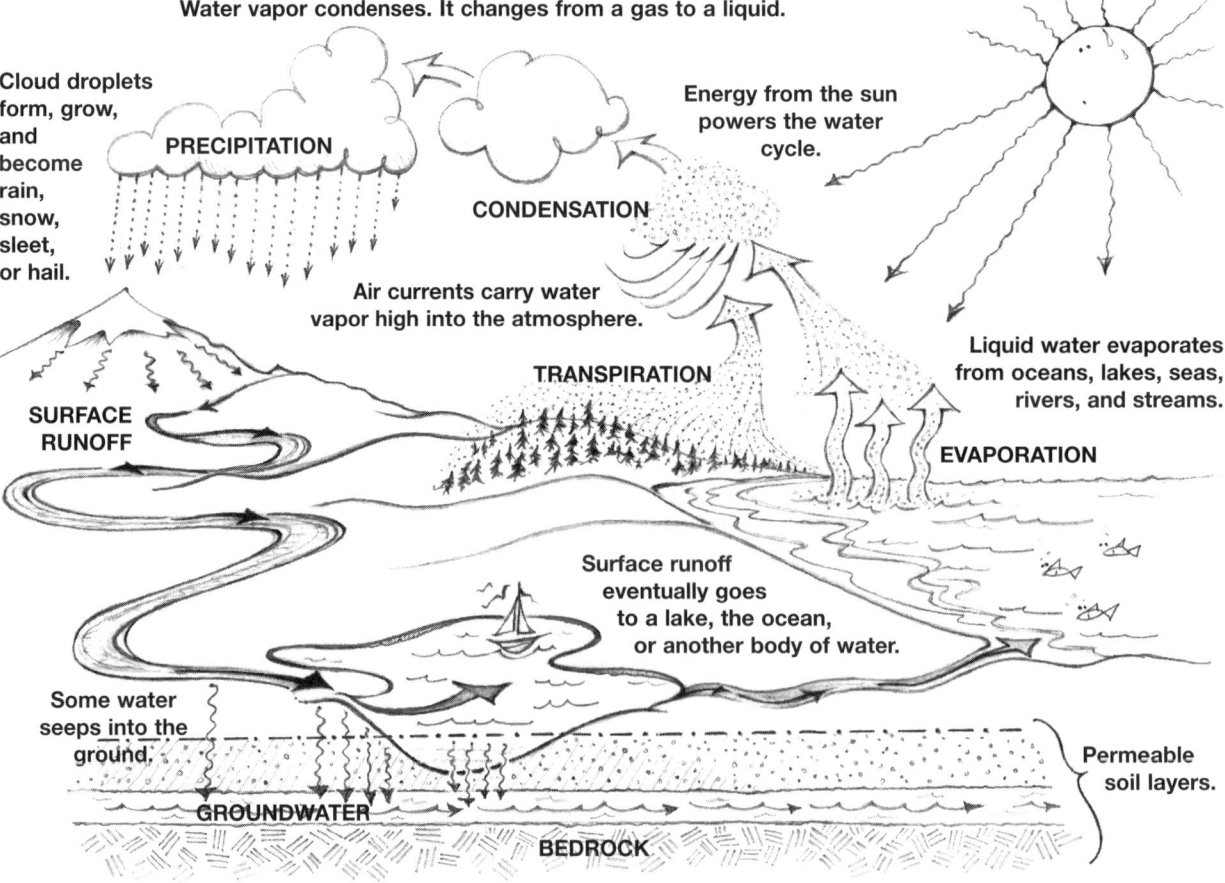

(continued on next page)

Farm to Table & Beyond
©2008 Teachers College Columbia University

Name   Date

Earth Cycles

vapor, a gas, and air currents carry it high into the atmosphere. There, the water vapor condenses. It changes from gas to liquid and forms cloud droplets. These grow and become precipitation (rain, snow, sleet, hail) and fall to Earth. Some of the precipitation seeps into the ground. Some falls on the surface of the land and runs off into rivers and streams. Eventually, it reaches a pond, a lake, or the ocean. The amount of water on Earth is a constant. It is continually recycled.

Water molecules move through the cycle over short- and long-term time scales. For example, scientists estimate that a water molecule might remain in the ocean for a very long time before it evaporates — 3,000 years or more. However, water that evaporates into the atmosphere may return to Earth in about nine days. Scientists who study climate and climate change are interested in the rate of flow and how long water stays in **reservoirs,** like the ocean. About 86 percent of the global evaporation takes place where most of the Earth's water is found — the oceans.

Do human activities affect the water cycle? Scientists at the University of Georgia are using rainfall records and data from satellites to study how large cities, like Atlanta and Houston, can affect their weather. They also want to know whether, if the weather is affected, this could have an effect on the water cycle. Looking at the data, the scientists found unusual patterns that show how human activities are affecting weather in arid regions. What kinds of activities? Irrigation is one example. Based on the data, the researchers think that irrigation and other human activities can change natural systems. No one understands yet how all the variables work, but they are continuing to research these interactions.

Wind and Weather Patterns

What happens when energy from the sun heats the Earth? Think about the shape of the planet. Earth is a sphere. The sun warms the equator more than it does the poles. This creates **air-circulation patterns,** or **winds.** What is wind? It's the movement of air.

In some parts of the world, the direction of the wind changes with the seasons. These winds are called **monsoons.** They cause wet and dry seasons and occur throughout the tropics. About half of the world's population depends on monsoon rains to provide the water they need for agriculture. If you have been in the southwestern United States in the summer and seen thunderstorms approaching, you have experienced the effect of the monsoon. A pattern like this happens near Mexico in summer, which brings about the thunderstorms.

(continued on next page)

Name   Date

Earth Cycles

Ocean Currents

Ocean water has two main parts: the surface layer and the deep waters. The surface layer is the water at the top. It is affected by waves, tides, and weather. This layer is less dense and sits on the surface of the deep waters. Ocean currents are like streams of water, constantly moving. Winds drive these currents, pushing the water along. Ocean currents help move warm water away from the tropics toward the North and South Poles and carry cold water back toward the tropics. Look at the map. Some ocean currents are large and powerful. Land masses, like continents, block the flow of large ocean currents. What happens? These currents loop around. The giant loops that are formed are called **gyres.** North of the equator, the gyres flow clockwise. They flow counterclockwise south of the equator.

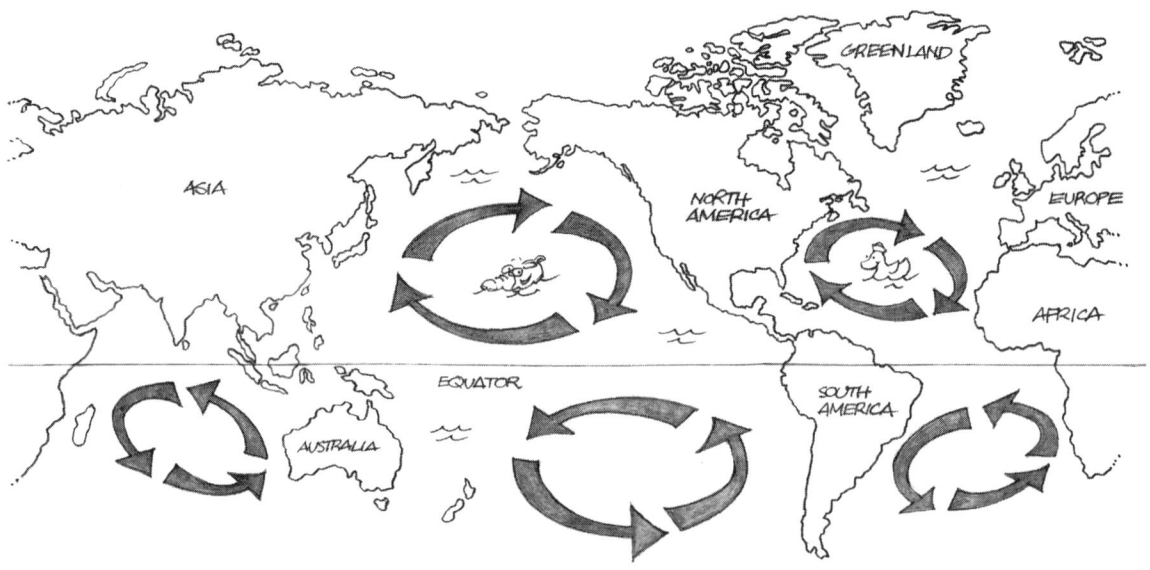

Human Activities and Currents

In the Northern Pacific, a gyre has been making the news. A floating garbage "patch" that has been described as being twice the size of Texas or even as large as the United States is held in place by swirling ocean currents. Charles Moore, a marine scientist, has been studying this garbage patch ever since he stumbled upon it while sailing across the Pacific. He says it's like a soup or stew of plastic and not like an island you can walk on. Debris floating in the ocean is not new. What's new is that this debris is not biodegradable. It's not bits of kelp, wood, or coconut palm. It's plastic and it doesn't go away. However, it does get brittle and break into smaller pieces. These smaller pieces become part of the ocean food chain. Sea birds, turtles, and other marine life think it's food and consume it, but it can harm them.

(continued on next page)

BECOMING FOOD SCIENTISTS : INTERACTING PARTS : FOOD PROCESSING : ENVIRONMENTAL EFFECTS : WASTE : MAKING CHOICES

Name   Date

Earth Cycles

Nike Shoe Spill

Sometimes the floating debris can help scientists. One example is the Nike shoe spill. In May 1990, the container ship Hansa Carrier was caught in a big storm. About 80,000 Nike shoes spilled into the Pacific Ocean. The following winter, the shoes began showing up on the beaches of Washington, Oregon, and Vancouver Island. What happened next is amazing. The shoes had serial numbers that could be traced to the shipment on the Hansa Carrier. With the help of beachcombers, oceanographer Curtis Ebbesmeyer in Seattle, Washington, was able to gather data about where the shoes washed up and use that data to learn more about ocean currents. What happened to the shoes? Beachcombers set up swap meets to try to match lefts and rights of the same size to either wear or sell.

Plastic Duck Armada

The Nike shoe spill was not the only cargo lost at sea. About 10,000 containers fall overboard each year. The cargo that interests research scientists is the cargo that spills along the trade routes between Asia and North America. The data that scientists collect helps them learn more about how the Pacific Subarctic Gyre works. For example, in 1992, a container ship traveling from Hong Kong to Tacoma, Washington, ran into a storm. One of the containers on the ship was full of bathtub toys — ducks, turtles, beavers, and frogs — made in a factory in China. The container went overboard and about 29,000 plastic bathtub toys spilled into the Pacific. Curtis Ebbesmeyer went into action. He began tracking the toys and found them landing on the beaches of Australia, Indonesia, and South America. Some headed north to Alaska, then floated west to Japan, back to Alaska, and north toward the Arctic. Eight years after they began their journey, some of these toys were spotted in the North Atlantic. Using the data from the recent cargo spills and historical data, researchers have figured out that it takes about three years to travel along the Pacific Subarctic Gyre.

Are scientists trying to find out how long it takes a plastic toy to float to the North Atlantic? Not really. What they do want to do is learn more about ocean currents and how they work. How ocean currents move influences climate and living conditions for plants and animals. Tracking the debris helps them do that.

Name	Date

Earth-Systems Worksheet

Your teacher has assigned you a topic to read about and present to your classmates. Use the graphic organizer on this sheet and the presentation tips on the next page to help you with your work.

Topic: _____

Main Idea

Point #1

Point #2

Point #3

Detail

Detail

Detail

Detail

Detail

Detail

(continued on next page)

Name   Date

Earth-Systems Worksheet

Tips to Making a Great Poster Presentation

1. Give your presentation a title.

2. Make sure the information on the poster is well organized.

3. You have about 10 seconds to grab your audience's attention. Keep your information brief and interesting.

4. Use drawings, photographs, and maps.

5. Make sure the type is easy to read. Check for misspellings.

6. The sample below is one way you can set up your poster.

The Hydrosphere   Your Name _____

[Put your artwork here]

Parts of the Hydrosphere

The hydrosphere includes the oceans, seas, rivers, lakes, ponds, and streams.

• 97% of Earth's water is found in the oceans;

[Put your text here: main idea, points, and details]

Natural Resources

AIM

To construct knowledge about the importance and use of natural resources, including fossil fuels.

SCIENTIFIC PROCESSES

- **learn about, think through, construct knowledge, debate**

OBJECTIVES

Students will be able to:

- **define "natural resource";**

- **discuss the use of natural resources;**

- **recognize that people view natural resources in different ways;**

- **discuss uses of fossil fuels.**

OVERVIEW

In this lesson, students learn about Earth's resources. The class discusses what plants and animals need to survive and brainstorms a list of natural resources. Next, students consider all the natural resources involved in making a breakfast of pancakes served with maple syrup. After students gain an understanding of the importance of resources, they participate in a mock town meeting. Town residents are trying to decide about the use of a natural resource. Students play different roles and learn that there can be conflicting viewpoints when it comes to evaluating resource use. They use background information for the differing points of view and a planner to prepare for their roles. This experience provides the foundation for students to investigate our farm-to-table system, resource use, and the system's effects on the environment. The activity closes with a student reading about fossil fuels. For homework, students think through their daily use of fossil fuels.

MATERIALS

For the teacher:
- *Student Ideas about Fossil Fuels* teacher note
- *Point of View* lesson resource
- *Pancake Ingredient Chart* lesson resource (p. 210)
- *Wheat-and-Corn Pancakes* teacher recipe (p. 216)
- Chart paper
- Markers

For each student:
- *Fossil Fuels* student reading
- *Breakfast of Resources* activity sheet
- *Point-of-View Planner* activity sheet
- *Fossil Fuels in Daily Life* activity sheet
- LiFE Log

PROCEDURE

Before You Begin:

- Review the *Student Ideas about Fossil Fuels* teacher note and the *Point of View* lesson resource.

- Read the *Fossil Fuels* student reading and the *Breakfast of Resources, Point-of-View Planner,* and *Fossil Fuels in Daily Life* activity sheets.

- Make copies of the student readings and activity sheets for each student.

- Copy the *Point of View* lesson resource. Cut out each expert point of view to distribute to students so they can prepare for the town meeting.

- Gather the *Pancake Ingredient Chart* lesson resource and the *Wheat-and-Corn Pancakes* teacher recipe.

- If you have not already done so, post the Module Question and Unit 4 Question at the front of the classroom.

MODULE QUESTION

What is the system that gets food from farm to table, and how does this system affect the environment?

UNIT QUESTION

What are the environmental effects of our farm-to-table system?

 QUESTIONING

1. Review Module and Unit Questions

Before this lesson begins, invite students to share their essays from the last lesson's home-work. Remind them of the Module Question. In this lesson, students participate in a mock town meeting and discover that decisions about resource use often involve trade-offs.

2. Introduce Resources

Engage students in a brief discussion. Make two columns on a sheet of chart paper. Label one column Plants and the other Animals. *What do plants need to grow, thrive, and reproduce?* (Air, water, soil, sun.) *What do animals, including humans, need to live and reproduce?* (Light, air, food, water, shelter.) Explain that all these things are called resources and are found in nature.

3. Ecosystem Services

Point out that humans not only get "goods" (soil, water, air) from nature, we also get "services." Ecosystem services are processes, like pollination, nutrient cycling, and waste removal. These processes help provide us with resources. *What resources are provided by pollination?* (Wind pollinates grasses and we eat grains as food; insects pollinate flowers and make honey, which we eat.) *How does nutrient cycling provide us with resources?* (It helps make healthy soils where plants can grow, and we eat the plants.)

 THEORIZING

4. Pancake Resources

Have students work in pairs or small groups. Distribute the *Breakfast of Resources* activity sheet. Go over the *Pancake Ingredient Chart* lesson resource and the *Wheat-and-Corn Pancakes* teacher recipe as a class. Review the instructions and example on the activity sheet. Check for student understanding.

After students have completed the activity sheet, ask volunteers to share their responses with the class. Record them on chart paper. Ask if anyone disagrees with the resources listed or wishes to add more to the list. Invite students to summarize what they have learned about the way humans rely on natural resources.

5. Town Meeting

Tell students that a town meeting is one way that people in a community come together to discuss and debate issues that affect the whole town. The purpose of this town meeting is to discuss and debate whether or not to continue to allow logging in a local forest, to convert it to a tree farm, or to conserve it. Many people have strong feelings about this topic. The town decided to invite different experts to present their points of view. Tell students that they will take turns acting out the roles of experts and town residents. Encourage students to take notes during the presentations so they can provide support for their decision.

Distribute the experts' point-of-view information to students, along with the ***Point-of-View Planner*** activity sheet. Tell students to use the planner to organizer their thoughts. If you have enough time, hold several town meetings so each student has a turn being an expert. Or you may choose to have students work in small groups to prepare the expert's comments and then have each group elect a representative to play the expert role. Allow student experts time to complete the planner.

Set the rules for the town meeting. Remind students to be respectful of differing opinions. Stress that the point of this activity is not to find one right answer. Rather, students will discover that with discussions about the environment and human activity there are often many points of view. It is not always easy to come to agreement.

6. Reach a Decision

After the town meeting has concluded, ask students to think about what they have heard and to make a decision about whether or not logging can continue. Have students write a short paragraph that summarizes the thinking that led to their decision. Invite volunteers to share their thoughts with the class. Remind students that there is not a right or wrong answer.

Tell them that what you are looking for is how they came to their decision. Encourage them to explain why they made this decision. Compliment each student's efforts.

7. LiFE Logs

Have students work in small groups. Distribute the ***Fossil Fuels in Daily Life*** activity sheet. Have students take out their LiFE Logs. Tell them to describe four activities that they have done during the day. Use the following questions as prompts. *How did you wake up this morning? Did an adult wake you or did you use an alarm clock? What did you have for breakfast? How was it prepared? How did you get to school?* Tell students to include at least one food-system activity, such as cooking, eating, or shopping. After they record their daily activities, ask students to read the ***Fossil Fuels*** student reading. Use the guiding questions to engage students in a discussion of fossil fuels. Make sure students understand the difference between the everyday use of words like "oil" and the scientific use. Check for understanding of the difference between renewable and nonrenewable resources.

APPLYING TO LIFE

8. Homework

Review the ***Fossil Fuels in Daily Life*** activity sheet. If students do not complete the reading in class, have them finish it for homework. Tell students to complete the activity sheet after they finish the reading. Encourage them to provide as many details as possible. Ask students to take out their LiFE Logs and write responses to these questions: *How do I rely on natural resources in my everyday life?* and *What would change in my life if there were no more fossil fuels?*

Student Ideas about Fossil Fuels

"Fossil fuels," "natural gas," "petroleum," and "energy resources" are terms that are firmly established as part of everyday language. Energy issues are a central part of our lives, from the price of gasoline to heating oil to climate change to wars being fought over resources. We hear the words so often, it's easy to assume that everyone understands their meaning, including students.

However, research suggests that students hold preexisting ideas and misconceptions when it comes to energy and fossil fuels.[1] In a study of students in first through sixth grades, researchers found that students' preexisting ideas may have come from television, movies, or cartoons. Researchers also found that some students believe that petroleum comes from whale blubber or decaying dinosaur carcasses. They attribute these ideas to historical descriptions of the use of whale blubber as oil. The whale-blubber oil was used for heating and lighting homes, just as petroleum products are today. Popular culture's use of dinosaurs to represent prehistoric times may account for students' idea that oil comes from dinosaurs. This misconception is further supported by a brand of gasoline service stations that uses a dinosaur as part of its logo.

This same study found that many students are not aware of petroleum uses and products. Many students think that petroleum is used as lubrication. They do not realize the origin of manufactured products such as plastics, nylon, asphalt, paints, propane, gasoline, and many waxes. Students also confuse petroleum and vegetable oils. When everyday English has one meaning and scientific terminology has another, it's easy to see how students might get confused.

As you move through this lesson, be sure to check for student understanding. Make the time to carefully review the fossil-fuel readings and activity in this lesson. Engage students in discussions to help uncover any misperceptions that may persist.

---

[1] Rule, 2005, pp. 305–18

Point of View

The town of Forest Mountain has a conflict. Some people in the town want to allow people to cut down as many trees as they want on the nearby mountain for use in manufacturing. They think the forest should be turned into a tree farm. Other people want to either cut down no trees or create limits on how many trees can be cut down, because they think that ongoing logging will hurt the environment. The town leaders decide to hold a meeting so residents can hear the different points of view.

Assign students to play different roles in the town meeting. Make copies of the information below, cut it out, and distribute it to the various role players. Give them time to review the information, complete the *Point-of-View Planner* activity sheet, and prepare for their roles. Have the rest of the class play the role of town residents.

Newspaper Printer

Ashley wants to be able to harvest the trees. She just bought a paper-making business. She used to buy paper from a company in Sweden. To save on transportation costs, she wants to make paper close to her newspaper printing plant. Ashley learned to be a printer from her father. He used to make his own paper, too, and harvested trees from this forest. She argues that it will not harm the environment if they cut down the trees. They will just plant new ones. Trees are plants. They are like a farm crop. While she waits for the new trees to mature, she will harvest another section of the forest. Ashley points out that trees are a resource that she needs so she can stay in business. If Ashley has to close her printing plant because she no longer can afford to buy paper from overseas, where will she get paper? Without paper, there will be no newspapers. How will people get the news? What will happen to all the people in her printing plant? They will lose their jobs.

Ornithologist

Natasha is a raptor specialist — she is a biologist who studies birds of prey, specifically red-tailed hawks. She has been observing a family of red-tailed hawks that are nesting in some of the trees at the edge of the forest. Natasha points out that the hawks need trees for their nests. They prefer tall trees. They build their nests close to the main trunk and about 50 or 60 feet up the tree. Natasha says that she knows that red-tailed hawks can build their nests in lots of different kinds of trees. However, she has been observing that there are not as many tall trees in the area anymore. There just aren't as many trees for the hawks to use. Natasha points out that the hawks play an important part in the local ecosystem. Hawks eat small rodents, like mice. What will happen if the hawks are gone? Will the mouse population grow and create a problem for the town?

Food-Packaging Manufacturer

Chuck owns a large food-packaging company in the area. He says that he needs a paper source to make his food packaging. He wants the source to be close to his factory so he doesn't have to pay lots of money in shipping charges. He says he is helping the environment by using resources close to his factory so he doesn't need to rely on huge trucks that burn lots of fuel. Chuck points out that trees are renewable resources. It will not take millions of years to replace them, the way it does for nonrenewable resources like fossil fuels. Chuck wants to turn part of the forest into a tree farm. He says on a tree farm you can plant the kind of trees you want to grow. In the forest, you have to search for the right kind of tree to cut down to make paper. Chuck thinks the most efficient way to have enough trees is to convert the forest to a tree farm. He points out that his company has been in this area since 1967 and he employs 150 local residents. If he can't find an affordable source of paper to make his products, he may have to shut down his factory. If he does this, many local residents will lose their jobs. What will they do then?

Soil Scientist

Clay is a soil scientist. He knows the soils in the local area very well. He reminds everyone that forests are important to soil conservation, especially in areas like this region. The forests on the mountainside help hold the soil in place and keep it from eroding. Without the forest, the soils would run into the stream. The soil in the stream would change the stream habitat. The soil would cover the stream's gravel bottom, where fish lay their eggs. When this happens, the eggs are smothered and do not survive. The fish can no longer live there. Clay also points out that once the forest is gone, the wind and rain will erode the topsoil. The tree branches and trunks protect the soil from the wind. The tree's roots hold the soil during heavy rains. Without the trees, there could be mudslides. The houses at the bottom of the mountain might get buried in mud. Clay reminds the residents that soil is also a resource. It can be replaced, but it takes a long time.

Ecologist

Juan is an ecologist. He works for a local company that studies ecosystem interactions. He points out that a forest and a tree farm are not the same thing. On a tree farm, one kind of tree is planted, cared for, and harvested in 25–30 year cycles. A forest has more diversity. Forests provide habitat, protection, and food for many plant and animal species. If the forest is converted to a tree farm, what will happen to all these species? The ecosystem will change. Cutting down the trees in the forest will affect human society, too. Juan reminds everyone that forests play an important part in global climate. By taking in carbon dioxide and converting it to oxygen during photosynthesis, trees naturally remove excess carbon from the air. Trees also affect air quality by acting as collection sites for dust and other air particles. Dust particles collect on leaf surfaces and stay there until they are washed to the ground during a rainstorm. Juan reminds residents that a huge factory is on the other side of the mountain. The trees protect the residents from all of the dust and fumes that the factory gives off. If the forest is cut down, there will be no trees between the town and the factory.

Artist

Flora is an artist. She has lived in this area her whole life. As a child, she used to play in the forest. Flora says that her imagination came alive and grew as she played among the trees. She gave them names and remembers when some of them were much smaller. She has watched them grow. Flora shows the residents a series of drawings that she has done over the years. They illustrate how the forest has changed. Flora says the trees are her friends. She is in awe of the wonder and mystery of these green giants. They stand so tall, with roots anchored in the ground. With no help from human beings, trees take water out of the soil and lift it hundreds of feet into the air. Trees remove carbon dioxide from the air and provide the oxygen that people breathe in. Trees make their own food. Flora says these trees are her inspiration. Without them, she will be lost.

Social Scientist

Samuel is a social scientist. He studies human culture and works at a natural history museum near the forest. Samuel has studied civilization's use of wood. He is developing a museum exhibit about the importance of trees to cultures in the past. He points out that long before human society used metals, synthetic fibers, and fossil fuels, people used wood. He says that the development of civilization has been dependent on wood-based technologies. People made tools from wood and used these tools to get food. They prepared food with wood utensils and used wood fires to cook the food. He asks where we would be without fire, agriculture, the wheel, spinning, weaving, water- and land-based transportation, building, and printing. All of these technologies used wood. Samuel says that our technological culture could not have developed without wood. He points out that today there are new materials that are used in place of wood. Instead of cutting down the trees, perhaps the people who need paper and lumber could use something else. If they do need to cut down trees, he would like them to plant new ones to replace the trees they use. He believes some of the forest should be preserved and made part of the museum — an outdoor, living museum.

Medical Researcher/Ethnobotanist

Alicia is a medical researcher and an ethnobotanist. She studies different ways that people use plants for medicine. She asks the town residents if any of them ever take aspirin. She reminds them that aspirin originally came from the bark of willow trees. Alicia points out that many of the medicines that we have today originally came from trees. She explains that one of the first medicines to treat malaria was quinine, which came from the bark of a South American tree. She discusses a cancer drug that comes from the bark of the Pacific yew tree. Alicia explains that Native Americans of the Northwest used yews to make salmon spears, paddles, spoons, bowls, and combs. The yew was also used to treat fever, colds, and stomachaches. Alicia makes the point that researchers are still learning of new medicines that come from trees. If the town allows the forest to be cut down, who knows what potential medical cures might be lost.

Housing Developer

Woody is a housing developer. He wants to develop town property away from the forest. There are no trees on the property. It used to be a factory, but the factory was torn down. Woody wants to build houses for young families. He wants to design a neighborhood with a large playground and picnic area that will be within walking distance of all the homes. Woody also plans to include a community garden space where families can grow some of their own food. Woody explains that he needs trees to use for lumber to build the houses. He wants to use lumber from the forest instead of importing it from somewhere else. Woody points out that it will be less expensive if he uses local wood. He won't have the costs of transportation. He proposes a trade. For every acre of trees he logs on the mountain, he will preserve an acre of tropical rainforest in South America.

Name   Date

Fossil Fuels

Guiding Questions:

- *What are fossil fuels made from?*

- *How long does it take fossil fuels to form?*

- *What are the most common ways fossil fuels are used for energy?*

- *Which household products are made from fossil fuels?*

Fossil fuels are formed from the decayed remains of animals and plants. It takes millions of years for these remains to turn into substances that will burn. There are three main kinds of fossil fuels: coal, petroleum, and natural gas. Each of these is a different state of matter: coal is a solid; petroleum is a liquid; and natural gas is a gas.

Coal

Coal is formed from plants that died millions of years ago when the Earth was covered with swamps. Thick layers of partially decomposed plant matter called **peat** were covered by sediment. Over a long time, heat and pressure changed the peat to coal. It took millions of years to make the coal that we use today. Since we can't make more coal to replace what we use, coal is called a **nonrenewable resource.**

Coal has been burned and used as fuel for hundreds of years. Archaeologists have found evidence in Roman ruins in England that coal was being used before 400 A.D. In the 1300s, the Hopi Indians in the Southwest United States used coal for cooking, heating, and firing their clay pottery.

In the United States, coal is mined in 26 of the 50 states. When it is close to the surface, miners use huge machines to dig it out of the ground. This is called **surface mining.** If the coal is deep in the ground, **mine shafts,** or tunnels, are dug to get to the coal. The mine shafts can be 1,000 feet deep. Miners load the coal into coal cars or onto conveyor belts to get the coal out. At the surface, the coal is loaded onto trucks and taken to be crushed. The crushed coal is shipped by truck, ship, railroad, or barge.

Today, most of the coal in the United States is used to make electricity. It is also used to make steel and cement. In some parts of Asia and Europe, coal is used to heat homes.

(continued on next page)

Name   Date

Fossil Fuels

Petroleum

Petroleum is formed from the remains of marine plants and animals that lived millions of years ago. When the plants and animals died, they sank to the ocean floor. They were covered by sediment. Over a long time, heat and pressure changed the ancient marine life into petroleum. Since petroleum takes millions of years to form and we can't make more to replace what we use, just like coal it is called a nonrenewable resource.

The word "petroleum" comes from Greek and means "rock oil." It is a liquid that is also called **crude oil** or oil. The name "rock oil" gives you a clue about where petroleum comes from. When you see oil above ground, it looks like a black pool, but underground, it's tiny droplets. Oil droplets are trapped inside open spaces inside rocks. You need a microscope to see them. How do oil companies find them, and once they find them, how do they get to the billions of gallons of oil that the world uses each day?

Scientists have different techniques that they can use. Sometimes they use sound waves to explore for oil. Sound waves travel at different speeds through different kinds of rocks. Scientists listen to the sound waves, measure the speed, and figure out where there might be rocks with oil in them. Sometimes scientists send electric current through rock to find oil. Sometimes they drill an exploratory well and look at rock samples. Once they think they have found rocks with oil in them, they begin drilling wells to bring the oil to the surface.

Petroleum is found in many different products. It's used as fuel for transportation. It's also part of asphalt, tar, and wax. Petroleum is found in oil that is used on roads. It's in petroleum jelly. It's in motor oil that is used to keep engines running smoothly. Chemicals from petroleum are found in candles, dyes, fertilizers, fibers, paint, and many other products.

Oil is a fossil fuel just like coal. However, notice the difference. Coal is found where there once were ancient swamps. Petroleum comes from ancient seas.

Natural Gas

When you hear the word "gas," the first thing that comes to mind may be gasoline that you put in a car. However, there is a big difference between gasoline and natural gas. Gasoline is a liquid fuel made from petroleum. Natural gas is another kind of fossil fuel. It is a gas. In fact, it is a mixture of gasses, including methane.

(continued on next page)

Name   Date

Fossil Fuels

Like petroleum, natural gas is formed from the remains of marine plants and animals that lived millions of years ago. When the plants and animals died, they sank to the ocean floor. Over a long time, heat and pressure changed the ancient marine life into natural gas. Since natural gas takes millions of years to form and we can't make more to replace what we use, it is a nonrenewable resource. Oil and natural gas often are found together in sedimentary rocks. Natural gas is odorless. For safety reasons a bad-smelling chemical is added to it to give it an odor.

Natural gas is used to heat homes and for cooking. It is also used when you barbeque over a gas grill. It's used in gas clothes dryers and gas fireplaces. It's used to produce ceramics, foods, glass, iron, paper, and textiles. Natural gas is used to bake the finish on automobiles. Chemicals in natural gas are used in paints, plastics, synthetic rubber, and detergents.

Fossil-Fuel Use

Here are some of the ways we use energy in our homes:

- to power the generators that produce the electricity we use for lights and electronic devices, like computers, televisions, and CD players;

- to heat the water that runs through our faucets, in our washing machines, and in our dishwashers;

- to run our refrigerators, air conditioners, and furnaces to heat our homes.

Here are some of the ways we use energy in factories, schools, libraries, shopping malls, and supermarkets:

- to heat the buildings;

- to produce the electricity for lights;

- to cook food in the cafeteria;

- to power office equipment, like computers, copy machines, and fax machines;

- to heat water;

- to provide the power to make products like processed foods, clothes, books, electronic devices, furniture, dishes, paper, and all the other things we use each day.

(continued on next page)

Name   Date

Fossil Fuels

Products Made from Oil and Gas

Many products that we use all the time come from fossil fuels. Here is a list you can consult as you look for ways that you use fossil fuels in your daily life.

Balloons

Ballpoint pens

Cellular phones

Computer monitors

Crayons

Curtains

Dashboards

Detergent

Dishwashing liquid

Dyes

Eyeglasses

Fertilizers

Fishing rods

Floor wax

Food preservatives

Footballs

Glue

Guitar strings

Hand lotion

House paint

Hula hoops

Ice-cube trays

Ink

Insect repellent

Insecticides

Lipstick

Model cars

Nail polish

Nylon rope

Oil filters

Paintbrushes

Paint rollers

Pajamas

Perfumes

Permanent-press clothing

Petroleum jelly

Pillow stuffing

Plastics

Propane

Purses

Refrigerators

Rubber cement

Rubbing alcohol

Shampoo

Shoe polish

Shoes and sandals

Shower curtains

Skateboards

Skis

Soap dishes

Soft contact lenses

Sunglasses

Surfboards

Synthetic rubber

Tape recorders

Telephones

Tennis rackets

Tents

Tires

Toothbrushes

Toothpaste

Trash bags

Umbrellas

Vitamin capsules

Name   Date

Breakfast of Resources

Look at this stack of wheat-and-corn pancakes. Now think about what you have learned about food processing, ingredients, and natural resources. Brainstorm a list all of the natural resources that were used to make the stack of pancakes. Record them in the space provided. See the example below.
Don't forget the butter and maple syrup!

Product

[Example] Flour for pancakes

Natural Resources Used

[Example] Soil, water, corn and wheat
plants, wind to pollinate
the grasses

_____   _____

_____   _____

_____   _____

_____   _____

_____   _____

_____   _____

_____   _____

_____   _____

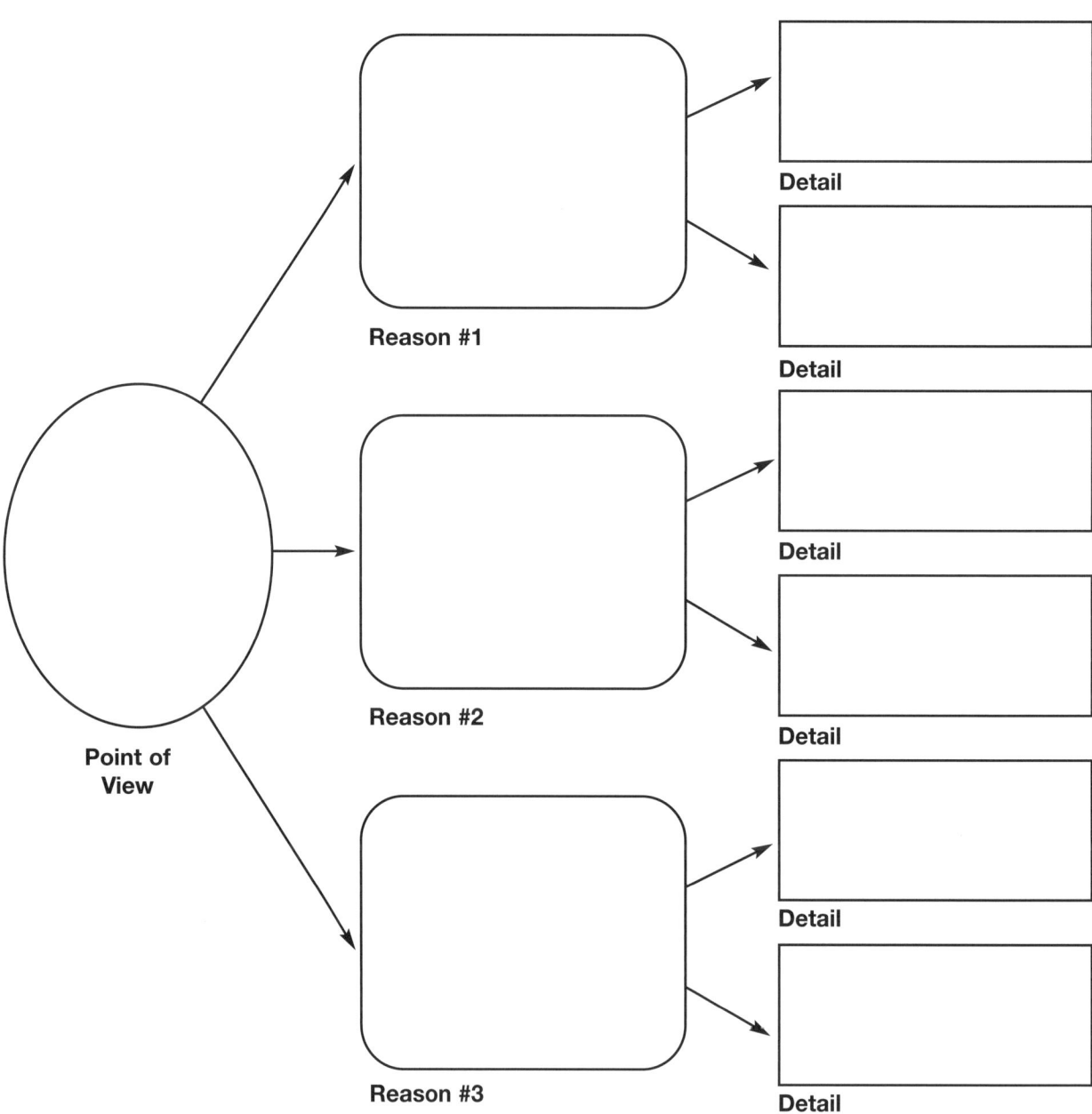

Name   Date

Point-of-View Planner

Your teacher has assigned you a topic to present at a town meeting. Use the graphic organizer on this sheet to help you organize your thoughts.

My Role (name and occupation)_____

Point of View

Reason #1

Detail

Detail

Reason #2

Detail

Detail

Reason #3

Detail

Detail

Farm to Table & Beyond
©2008 Teachers College Columbia University

BECOMING FOOD SCIENTISTS : INTERACTING PARTS : FOOD PROCESSING : **ENVIRONMENTAL EFFECTS** : WASTE : MAKING CHOICES

Name	Date

Fossil Fuels in Daily Life

For this assignment, you are going to examine your own use of fossil fuel. Follow the directions below. Record the activities in your LiFE Log. Try to include at least one food-system activity, such as food shopping, cooking, or eating. Then think about the ways that fossil fuel was used.

1. Describe four different activities that you did during the day. For example, maybe you washed an apple and ate it. Or you washed the dinner dishes with hot water and dish soap. Or you sat at the computer and did your homework.

Example:

Activity One: I washed my hair with shampoo and hot water and dried it with a blow dryer.

2. Draw a circle around each word that describes a product that you used.

Example:

I washed my hair with (shampoo) and (hot water) and dried it with a (blow dryer.)

3. Make two columns in your LiFE Log. List all the words that you circled in the left column. In the right column, describe all the parts of the activity where fossil fuel was used.

Products I Used	How Fossil Fuels Were Used
Shampoo	Shampoo made in a factory powered by fossil fuel. Plastic shampoo bottle was made from fossil fuel. There is fossil fuel in the shampoo.
Hot Water	Water is heated by fossil fuel.
Blow Dryer	The blow dryer is made of plastic, which is made from fossil fuel. It was made in a factory that was powered by fossil fuel. You need electricity to power the blow dryer. The electricity comes from fossil fuel.

The Carbon Cycle

AIM

To gain an understanding of carbon and the carbon cycle.

SCIENTIFIC PROCESSES

- learn about, discover, construct knowledge, think through

OBJECTIVES

Students will be able to:

- describe the composition of the atmosphere;

- outline the carbon cycle;

- explain some of the effects of burning fossil fuels, including the effect on the carbon cycle;

- discuss milestones in our energy use over the last several thousand years.

OVERVIEW

In this lesson, students continue to learn about fossil fuels. First, they review the Lesson 18 homework and discuss their own use of fossil fuels. Next, the class is introduced to carbon and the carbon cycle through a student reading. After reading about and discussing the carbon cycle, students play The Carbon Cycle Game. In this game, students play the roles of air, plants, and animals and pass the Carbon cards, simulating the flow of carbon through the natural environment. They repeat the game, introducing fossil fuels. They "burn" fossil fuels that contain carbon from plants and animals that lived million of years ago and release carbon into the air. Students discuss the effect on the environment when increased amounts of carbon dioxide change the atmosphere's chemical balance. The class adds energy milestones to the time line started earlier in the module and reflects on the increased use of fossil fuels over time. For homework, students review their own use of fossil fuels and look for ways to reduce that use.

MATERIALS

For the teacher:
- *Milestones in Energy History* lesson resource
- *The Carbon Cycle Game* lesson resource
- *Carbon Cycle Game Cards* lesson resource
- Chart paper
- Markers
- (Optional) *Student Ideas about Fossil Fuels* teacher note (p. 299)
- (Optional) *Fossil Fuels* student reading (p. 303)
- (Optional) Large index cards or sentence strips

For each student:
- *Carbon* student reading
- *Carbon Facts* activity sheet
- Carbon cards (copied from the *Carbon Cycle Game Cards* lesson resource)
- LiFE Log

PROCEDURE

Before you begin:

- Review *The Carbon Cycle Game* lesson resource, the *Carbon* student reading, and the *Carbon Facts* activity sheet.

- (Optional) Review the *Student Ideas about Fossil Fuels* teacher note and the *Fossil Fuels* student reading.

- Make sure students have completed the Lesson 18 homework assignment.

- Make copies of the *Carbon Cycle Game Cards* lesson resource, the *Carbon* student reading and the *Carbon Facts* activity sheet.

- Copy the *Milestones in Energy History* lesson resource. Cut out the milestones and glue them onto index cards.

- If you have not already done so, post the Module Question and Unit 4 Question at the front of the classroom.

MODULE QUESTION

What is the system that gets food from farm to table, and how does this system affect the environment?

UNIT QUESTION

What are the environmental effects of our farm-to-table system?

1. Review Module and Unit Questions

Remind students of the Module and Unit questions. Explain that in this lesson the class is going to continue its exploration of fossil fuels.

2. Review Homework

Have students take out their LiFE Logs. Invite student volunteers to share what they learned

about their own use of fossil fuels. Engage the class in a brief discussion of fossil fuels and the food system. *Did anyone record an activity related to the food system? Did anyone cook a meal or make a snack? Did anyone go to the market to buy food? Did anyone work in a garden? How did you use fossil fuels?* (Used heat, gas stove, electricity, hot water, put food leftovers in a plastic container.) *What surprised you the most about your use of fossil fuels?* In this lesson we are going to learn more about fossil fuels.

3. Read about Carbon

Distribute the *Carbon* student reading and the *Carbon Facts* activity sheet. Draw students' attention to the *Carbon Facts* activity sheet. Tell them that each circle represents a topic that is discussed in the reading. The lines are to record details about each topic. They can add more lines if they find they need to. They may also find that they do not need to use all of the lines. The purpose of this activity sheet is to help students organize what they learn about carbon. Tell students to write down details as they read. Have students read silently and complete the activity sheets.

4. Discuss Carbon

Ask a student volunteer to share details that she recorded for matter. *Does anyone else want to add anything? What else did you learn about this topic?* Record the answers on chart paper. Continue with this line of questioning until you have discussed each topic. Conclude with the carbon cycle. Tell students that next they will model how carbon cycles through the environment.

5. Model Carbon Cycle

Distribute the carbon cycle cards to students. Follow the instructions on the *The Carbon Cycle*

Game lesson resource. Modeling the carbon cycle will help students begin to develop an understanding of how carbon moves through the environment. It will also help students gain knowledge about the effect that burning fossil fuels has on the environment.

 THEORIZING

6. Discuss Burning Fossil Fuels

After the class has completed modeling the carbon cycle, pose the following questions for discussion. *What do you think might happen to the extra carbon dioxide in the air that fossil fuels give off when they are burned? Do you think the excess chemicals might have an effect on the environment?*

7. LiFE Logs

Ask students to describe in their own words how burning fossil fuels affects the natural carbon cycle. Tell them that they can refer to the *Carbon Facts* activity sheet for details to support their answer.

 APPLYING TO LIFE

8. Energy Sources

Use the *Milestones in Energy History* lesson resource. If you made cards with the dates, have students add those to the time line you started earlier in this module. If not, have students record dates and events directly on the time line. Once students have added the energy milestones, invite the class to study the time line. Point out that before the Industrial Revolution, most people made their living working in agriculture. However, during the Industrial Revolution, people moved to cities and worked in factories. The machines in the factories burned fossil fuels, like coal. By the nineteenth century, transportation had changed and the steam engine was replacing the horse and carriage. Steam engines used coal, too. All the gases that were released when the coal was burned caused the air quality

to go down in many cities. Some researchers consider the Industrial Revolution the beginning of the time when human beings began to change the chemical composition of the atmosphere.

Ask students to think about the changes in energy sources. Encourage them to think about trade-offs when one form of energy replaced another. *Can you see any changes that might have affected the environment?* (Strip-mining, drilling, using fossil fuels for transportation.) *Do you think there were effects on the environment when energy sources were mined? Do you think that the type of energy resource had an effect on the food system? What other changes influenced the food system?*

9. Effects of Fossil Fuels

Ask students to think about the trade-offs with using fossil fuels as an energy source. Review the uses of fossil fuels from Lesson 18. Invite students to share how the use of fossil fuels may have an impact on the environment. Encourage them to think about fossil fuels from the time they are removed from the ground to the time they are used as an energy source or in products. *Do you think the environment is affected? In what ways might it be affected?* (Drilling or mining to extract the fossil fuels; spills or leaks from transporting the fossil fuels.) Explain that burning fossil fuels generates gases and particulate matter that contaminate the air. The factories that preserve, process, and package food are generally powered by burning fossil fuels, and this also contaminates the air.

Remind students that fossil fuels are nonrenewable resources. *What is a nonrenewable resource?* (A natural resource we cannot replenish. Fossil fuels took millions of years to form.)

10. Homework

Have students review how they use fossil fuels in their daily life. Tell them to describe one change that they can make to reduce their use of fossil fuels. Have them record this in their LiFE Logs.

Milestones in Energy History

5000 B.C.	Wind energy is used to move boats along the Nile River in Egypt
3000 B.C.	Humans begin to use petroleum
1000 B.C.	The Chinese use coal as fuel, one of the first uses of fossil fuels
200 B.C.	The Chinese pump water with simple windmills
500 A.D.	In Iran and Afghanistan, the first horizontal windmills are used in food production
1701 A.D.	Coal is discovered near Richmond, Virginia
1712 A.D.	The first steam engine is developed in England
1751 A.D.	The beginning of the Industrial Revolution in Europe
1800s A.D.	Steam-powered trains use coal as their principle fuel
1816 A.D.	Baltimore, Maryland begins to light its streets with gas made from coal
1830 A.D.	Tom Thumb, the first American-built locomotive, uses coal for fuel
1859 A.D.	Near Titusville, Pennsylvania, Edwin Drake drills the first commercial well and hits oil and natural gas
1860 A.D.	In homes and businesses, wood is the primary fuel for cooking and heating
1862 A.D.	The Union Congress places a $2-per-gallon excise tax on ethanol. The tax money is used to help pay for the Civil War.
1866 A.D.	Strip-mining (mining in strips of land) begins near Danville, Illinois
1900 A.D.	Ethanol competes with gasoline to be the fuel for cars
1910 A.D.	Charles Steinmetz, an engineer, warns of the dangers of air pollution from burning coal
1917 A.D.	The need for fuel during World War I increases the demand for ethanol
1950 A.D.	Due to automobile use, oil starts to become the most-used energy source
1975 A.D.	The U.S. Congress passes a law that requires car manufacturers to build more fuel-efficient cars
1986 A.D.	Consumption of natural gas begins to grow faster than production
1990 A.D.	Clean Air Act Amendments require making changes to gasoline and diesel fuels to make them cleaner

The Carbon Cycle Game

This simulation of the carbon cycle helps students gain an understanding of carbon and how it moves naturally through the environment, as well as how the carbon released from burning fossil fuels affects the carbon balance in the atmosphere.

Distribute Cards

Assign students roles for the game. Give them the cards they will need.

Role: Air. Three students play air. Give each Air student one Air card and one Carbon card.

Role: Plants. Three students are plants. Give each Plant student one Plant card.

Role: People. Three students play people. Give each People student one People card.

Role: Vehicles. One student represents vehicles. Give this student the Vehicle card.

Role: Factories. One student represents factories. Give this student the Factory card.

Role: Fossil Fuels. The rest of the class is part of the Fossil Fuels group. These students represent the carbon in fossil fuels. Give one student the Fossil Fuel card and give each of the remaining students in this group five Carbon cards.

Carbon Cycle Simulation

Tell the Air, Plants, and People to stand up and hold up their Air, Plants, and People cards. These students participate in the first simulation, which models the carbon cycle in nature. In this simulation, carbon cycles through the environment but does not build up in the atmosphere.

Action

1. Plant students take the Carbon cards from the Air students. This represents plants taking in carbon dioxide from the air to use during photosynthesis.

2. Plant students give their Carbon cards to the People students. This action represents people eating plants and taking in the carbon molecules (sugar and starch) that the plants make during photosynthesis.

3. The People students give their Carbon cards to the Air students. This represents people breaking down the carbon molecules (sugar and starch) that they eat and converting them into carbon dioxide. People exhale this carbon dioxide into the air.

4. You may wish to repeat this several times. Engage students in a discussion to check for understanding. Make sure they understand that carbon continuously cycles. There is no beginning or end, since it is a cycle. This simulation starts with plants taking carbon dioxide from the air, but it could have started anywhere. To reinforce this point, repeat the simulation in reverse. Start with people exhaling carbon dioxide.

The Fossil-Fuel Effect

Now that students understand how carbon cycles in nature, add fossil-fuel use to the game. In this version, students simulate activities that burn fossil fuels, which gives off carbon dioxide. The class observes the effect on the environment.

Action

1. Remind students that the carbon in fossil fuels is from plants and animals that lived millions of years ago. This organic matter was buried deep beneath the Earth's surface until people extracted it. The process that converts this organic matter into substances that will burn takes millions of years.

2. The Vehicle person and the Factory person stand at the front of the room and hold up their signs. Three Air students join them.

3. The Fossil Fuels student stands up and holds up his or her sign. All of the remaining students in the Fossil Fuels group line up behind the Fossil Fuels student and hold up their Carbon cards.

4. The Fossil Fuels group, with their Carbon cards, walks by the Vehicles and Factories students. This action represents burning fossil fuels for energy to power vehicles and factories. The Fossil Fuels students give their cards to one of the Air students. This action represents burning fossil fuels, which give off carbon dioxide. The carbon dioxide goes into the air. After all the Carbon cards have been passed from the Fossil Fuels to the Air, have students sit down.

5. Ask the Air students to count their Carbon cards and share how many they have with the class. These Carbon cards represent the buildup of carbon dioxide in the air when large amounts of fossil fuels are burned through human activity.

6. Tell students that plants can absorb some of the excess carbon dioxide, but the Earth cannot support enough plants to absorb all of the carbon dioxide that is given off at the current rate that human beings are burning fossil fuels. Explain that the balance is being upset. Reinforce the fact that human activities move more carbon into the atmosphere than is being removed naturally. This means that the concentration of carbon dioxide in the atmosphere is increasing.

Carbon Cycle Game Cards

Air	**Plants**	**People**
Air	**Plants**	**People**
Air	**Plants**	**People**

Vehicles

Factories

Fossil Fuels

Carbon	Carbon	Carbon	Carbon
Carbon	Carbon	Carbon	Carbon
Carbon	Carbon	Carbon	Carbon
Carbon	Carbon	Carbon	Carbon
Carbon	Carbon	Carbon	Carbon
Carbon	Carbon	Carbon	Carbon

Name   Date

Carbon

What heats our homes, fuels our cars, is in the clothes we wear and the food we eat? Here's a clue: it's also one of the most abundant elements in the universe and its symbol is C. The answer is carbon, a word that comes from *carbo*, the Latin word for charcoal. But before we talk about carbon, let's review some physical science.

Picture a bunch of grapes.
Each grape is made up of matter. In fact, everything in the universe is made up of matter, from the smallest speck of dust to a blue whale to a grape. How can all of these different things be the same thing — matter?

Remember that matter is a general name for something that takes up space and has mass. Matter is made up of **atoms,** small particles that you can't detect without an extremely powerful microscope. Matter changes state. It can be a liquid, a solid, or a gas. Picture an ice cube, which you can think of both as frozen water and as matter. Heat the matter, in this case the frozen water, and it melts and becomes liquid water. This liquid water is still matter, just in a different state. Add more heat to the liquid water and it changes to steam, but it is still matter. Steam is the gas state of matter. Think about the water cycle. The water, or matter, is always present as a solid, a liquid, or a gas.

Consider your food-observations activity. You observed what happened to food, which is also matter and is made up of different kinds of atoms. Think of what you did as watching food change or watching matter change. The physical and chemical properties of the food changed, but it was still matter. You may not have thought of it as food once it started to rot, but it was still matter.

Back to atoms. Elements are substances that have only one kind of atom. Scientists have identified more than 100 elements. Oxygen, hydrogen, nitrogen, carbon, lead, tin, gold, and copper are all elements. Sometimes elements join together and form compounds. For example, two atoms of hydrogen (H_2) plus one atom of oxygen (O) form the compound water (H_2O). Water is matter. It's also a compound, and it's a molecule. When any atoms link, they form a molecule. When atoms from two or more elements join, they form a compound. All compounds are molecules, but not all molecules are compounds.

(continued on next page)

Name   Date

Carbon

Take oxygen. In the atmosphere, oxygen is a molecule with two atoms (O_2). Since both of the atoms come from one element, oxygen, O_2 is not a compound.

Millions of Molecules

The four most common elements found in living things are carbon (C), oxygen (O), hydrogen (H), and nitrogen (N). Although there are just a few elements, they join together in a variety of ways to form millions of different kinds of molecules. Let's look at carbon. Not only is carbon in all living things, it's also found in the sun, the stars, comets, and the atmosphere of many planets, including Earth. Carbon atoms easily link with other elements. There are millions of carbon-containing molecules. A few familiar ones are carbon dioxide (CO_2), which puts the fizz in carbonated water and helps make yeast breads rise. Another is sucrose, or table sugar ($C_{12}H_{22}O_{11}$).

Carbon Cycle

Carbon moves through Earth in a cycle that involves both living and nonliving things. Since it is a cycle, there is no one place where it begins and ends. We'll begin with air. The carbon in air is in the form of carbon dioxide. Plants take in carbon dioxide to make food through the process of **photosynthesis.** Carbon dioxide combines with water and energy from the sun to make sugars and oxygen. The oxygen made during photosynthesis is released back into the atmosphere.

$$6CO_2 + 6H_2O + \text{light energy} \rightarrow C_6H_{12}O_6 + 6O_2$$
(carbon dioxide + water + light energy form sugars + oxygen)

When animals eat plants as food, carbon in the sugars that the plants made is returned to the atmosphere through **cellular respiration.** This is the process of producing energy by breaking down sugar molecules. Plants and animals break down sugar to release energy in a form they can use. Carbon dioxide and water are byproducts of the reaction. In animals, the carbon dioxide goes into the blood and is carried to the lungs. In the lungs, carbon dioxide and oxygen are exchanged during breathing. The carbon dioxide is given off and returns to the atmosphere, where it is available for plants to use again.

$$C_6H_{12}O_6 + 6O_2 \rightarrow 6CO_2 + 6H_2O + \text{energy}$$
(sugar + oxygen react to form carbon dioxide + water + energy)

Carbon is returned to the environment through animal waste and when matter decays. Rocks and water also absorb and give off carbon dioxide as part of the carbon cycle. In the carbon cycle, carbon joins with other elements, it changes physical location, and it changes state. Carbon is all around you and always on the move.

LESSON 19: THE CARBON CYCLE   Farm to Table & Beyond

Name   Date

Carbon Facts

Each circle represents a topic that you just read about. Use the lines to record details that you learned about each topic. See the sample below. Add more lines if you need them.

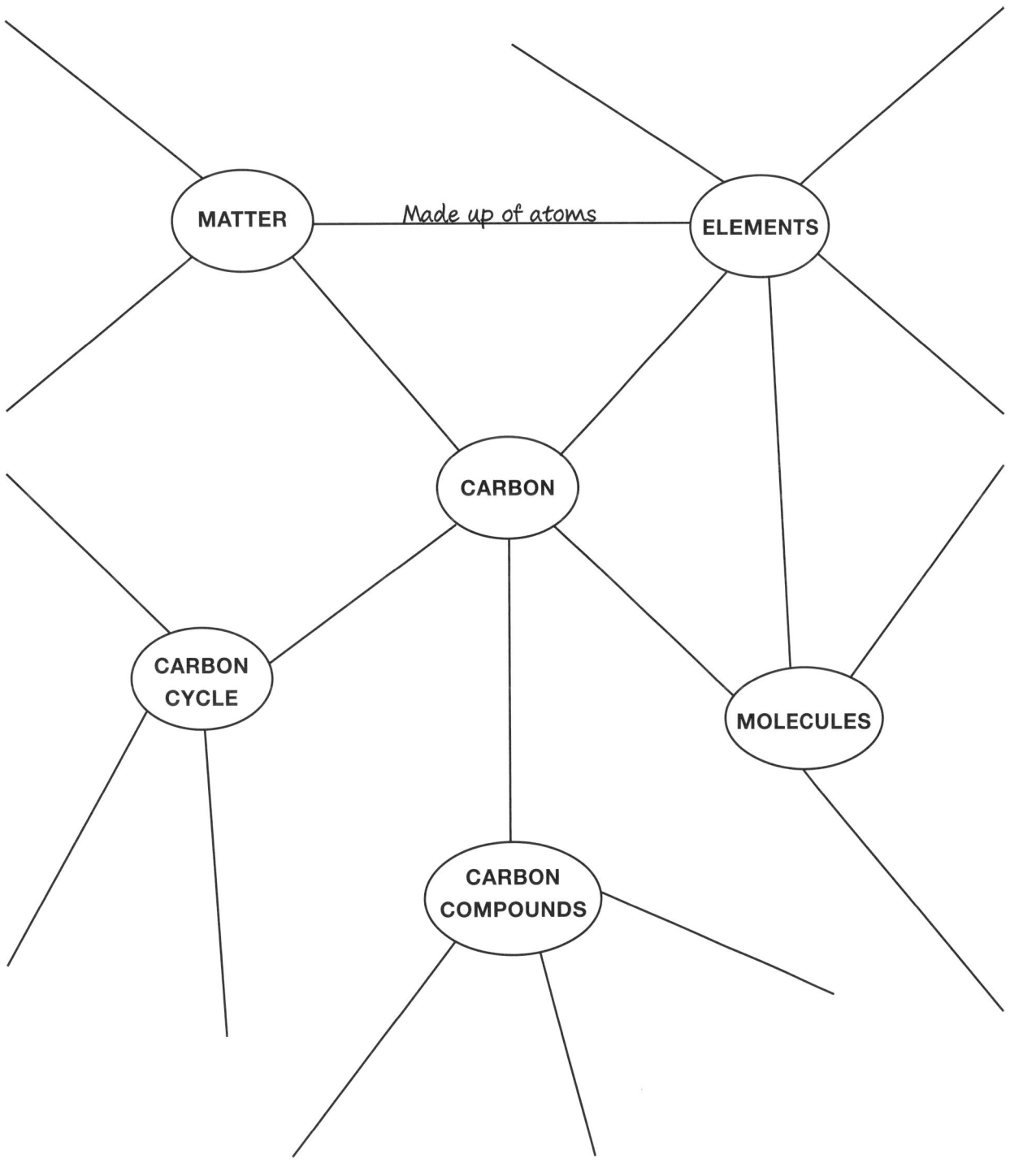

MATTER — *Made up of atoms* — ELEMENTS

CARBON

CARBON CYCLE

MOLECULES

CARBON COMPOUNDS

Earth's Changing Climate

AIM

To develop an understanding of climate change and the effects of human activity.

SCIENTIFIC PROCESSES

- question, gain knowledge about, apply

OBJECTIVES

Students will be able to:

- describe the difference between weather and climate;

- summarize some of the evidence for climate change;

- describe the greenhouse effect;

- identify possible consequences of climate change.

OVERVIEW

This lesson introduces students to Earth's climate system. The lesson begins with a brief assessment of what students already know about climate. This provides an opportunity to check for student understanding of the difference between weather and climate. Next, the class reads about climate change. They learn about the climate system, including a discussion of the cryosphere and greenhouse gases. The reading closes with a discussion of human activities and climate change. Students learn about the enhanced greenhouse effect and the effect of fossil fuels on the concentration of greenhouse gases in the atmosphere. Students are challenged to think about actions they can take in their personal lives to help reduce the use of fossil fuels.

MATERIALS

For the teacher:
- *Teaching about Climate Change* teacher note

For each student:
- *Climate Change* student reading
- *Climate Change Notes* activity sheet
- (Optional) *Carbon* student reading (p. 319)
- (Optional) *Fossil Fuels* student reading (p. 303)
- (Optional) *Fossil Fuels in Daily Life* activity sheet (p. 309)
- LiFE Log

PROCEDURE

Before You Begin:

- Review the ***Teaching about Climate Change*** teacher note, the ***Climate Change*** student reading, and the ***Climate Change Notes*** activity sheet.

- Make copies of the student reading and the activity sheet.

- (Optional) Remind students to bring in the ***Carbon*** and ***Fossil Fuels*** student readings and the ***Fossil Fuels in Daily Life*** activity sheet.

- If you have not already done so, post the Module Question and Unit 4 Question at the front of the classroom.

MODULE QUESTION

What is the system that gets food from farm to table, and how does this system affect the environment?

UNIT QUESTION

What are the environmental effects of our farm-to-table system?

1. Review Module and Unit Questions

Remind students of the Module and Unit questions. Explain that in this lesson the class is going to learn about climate change and Earth's climate system.

2. Discuss Climate

Briefly check for student understanding of climate. You may wish to make a class concept map or do a quick K-W-L chart. Label the first column (K) "What I know," the second (W) "What I want to know," and the third (L) "What I learned." *What's weather? What's climate? How would you describe the difference between climate*

and weather? (Climate is the long-term average weather of a location.) *Does weather change? Do you think climate can change?*

3. Read about Climate Change

Distribute the student reading and the activity sheet. Draw students' attention to the guiding questions at the beginning of the student reading. Invite student volunteers to read them out loud. Review the ***Climate Change Notes*** activity sheet with students. If students are not familiar with this note-taking method, guide them through it. Review the questions on the sheet. As a class, you may wish to do a quick pre-read to see if there are more questions that you would like to add. Students can refer to these notes during the class discussion. Have students silently complete the ***Climate Change*** student reading.

4. Discuss Evidence

Ask a student volunteer for an example of evidence of climate change. *What kind of data do scientists collect?* (Temperature, precipitation, humidity, wind, sunlight.) *What can they learn from ice cores?* (Air bubbles tell them about the atmosphere.) *How about tree rings? What do tree rings tell scientists?* (How much moisture and changes in temperature there were in a certain year.)

5. Discuss the Climate System

Engage students in a discussion of Earth's climate system. Remind students that they can refer to their ***Climate Change Notes*** activity sheet during this discussion. Challenge them to describe some of the parts of the climate system and how they interact. *What are greenhouse gases?* (Carbon dioxide, methane, water vapor.) *How* *do they help keep Earth warm?* (They absorb heat energy and re-emit it directed back to Earth.)

6. Effects of Climate Change

Encourage a whole-class discussion. *What evidence is there that Earth's climate is changing?* (Changes in bird behavior, changes in growing seasons, melting glaciers.) Have students debate the pros and cons of changes in the growing seasons. *If the temperature gets warmer and stays warmer, do you think this would be beneficial for all plants? What would happen to any other species that depend on plants that could no longer live? What do you think would happen if all the glaciers melted? What makes you think this?* Bring the discussion to a close.

APPLYING TO LIFE

7. LiFE Logs

Brainstorm a list of ways to conserve energy with students. For homework, have students write a personal action plan in their LiFE Logs.

Teaching about Climate Change

The media is full of stories about melting sea ice, the threat to polar bears, rising sea levels, and disappearing glaciers. The evidence seems clear. Something is happening. But wait, you read another story, and in it another person is quoted as saying there is no evidence that the climate is changing. Or that, if it is changing, the change is nothing new. Climate changes. Remember, the world hasn't always looked the way it does today. There are so many stories in the media that it can be hard, for adults as well as children, to separate the science from the nonscience. Use this as an opportunity to reinforce the scientific worldview. Remind students that science demands evidence. It explains and it predicts. As a class, you may wish to critique media stories and help students distinguish between fact and opinion.

In discussions of climate change, global warming, and the greenhouse effect, language can obscure understanding. Sometimes this is the result of misunderstandings or factual errors that are perpetuated in the media. Think of the number of times that you have heard people speak of the greenhouse effect and greenhouse gases as something "bad," when just the opposite is true. Energy from the sun heats the Earth. Some of the solar energy is absorbed by land and water on Earth's surface. Some of the energy is reflected back towards space. Greenhouse gases, including water vapor and carbon dioxide, absorb some of this outgoing energy. These gases emit heat energy, which warms Earth. Without this natural warming process, it would be too cold for us to survive on Earth. In other words, Earth's surface receives heat from both the sun and the atmosphere.

Some scientists point out that the "greenhouse effect" is an unfortunate choice of terms. On his Web site, Pennsylvania State University professor Alistair B. Fraser states that "…a real greenhouse does not behave as the atmosphere does. The primary mechanism keeping the air warm in a real greenhouse is the suppression of convection (the exchange of air between the inside and outside). …Indeed, the atmosphere facilitates rather than suppresses convection.[1] Some have even suggested the process should be called "atmospheric effect."[2] However, since the term "greenhouse effect" is so pervasive, it's generally accepted as the term to describe this heating phenomenon. In everyday language, people often equate the greenhouse effect with global warming. This is not the case. Global warming refers to the sustained increase in global average surface temperature, which is one aspect of climate change.

Research on student learning indicates that many students do not distinguish between "weather" and "climate." Even adults use the terms interchangeably. Make sure students understand that climate is the average weather for a specific location for a long period of time — about 30 years.

---

[1] Bad Greenhouse. *www.ems.psu.edu/~fraser/Bad/BadGreenhouse.html* (accessed 9 June 2008)
[2] Nelson, Aron, and Francek, 1992, p. 78.

Name   Date

Climate Change

Guiding Questions

- *What have you learned about climate?*

- *What are the parts of Earth's climate system?*

- *Why are ice and snow important?*

- *Do you think the farm-to-table system affects climate?*

- *What evidence do you have to support your conclusions?*

When you woke up this morning, did you look out the window to see if you needed an umbrella or a jacket? If you did, you were making observations about what the weather would be like for the day. Weather includes rain, snow, heat waves, clear days, and even hurricanes and tornadoes. Weather changes. Sometimes the change is sudden, like an afternoon thunderstorm that comes on quickly and moves on. Sometimes the change takes longer — for example, a heat wave that lasts for several days before the temperature cools. "Weather" describes what's going on in the atmosphere over short periods of time.

"Climate" refers to the average weather conditions that were observed over a long period of time, usually 30 years, for a specific location. Two important features of climate are the average temperature and the amount of precipitation. Other features include wind, humidity, clouds, and fog. A weather change can be from rain one day to blue skies the next. Climate change means there is a change in long-term weather patterns. Climate changes very slowly and over long periods of time.

Climate is an important part of the farm-to-table system. Think about it. Can you grow corn outdoors at the North Pole? How about apples in the tropical rain forest? Farmers use information about their regional climate to make decisions about what crops to grow and when. Information about a growing season includes the time between the last frost in the spring and the first frost in the fall. Climate information also tells farmers how much rainfall they can expect.

Examining the Evidence

What's the evidence for climate change? How do scientists study climate and climate change? Weather data is one source of information. Scientists have been collecting weather data from around the world for a long time. They study records of temperature, precipitation, wind, sunlight, and humidity. Scientists also examine ice cores that

(continued on next page)

Name   Date

Climate Change

they cut out of glaciers. They look for air bubbles in the ice. These bubbles have been buried in the ice for hundreds of years and maybe even longer. The air bubbles help scientists figure out what the climate used to be like. Tree rings are another source of climate information. By looking at tree rings, scientists can tell how much moisture there was and what changes there were in temperature during a particular year. Rocks and fossils provide even more evidence — evidence that goes back billions of years.

Climate System

Think about what you have learned about systems. They are made up of interacting parts. What are the parts of Earth's climate system? The parts include the atmosphere, the hydrosphere, the biosphere, the geosphere, and the **cryosphere.** You have already learned about most of those parts, but what's the cryosphere? It is the regions of Earth where the temperature is zero or below and water exists in solid form as ice, snow cover, or frozen ground. What happens if there is a change to any of the parts of Earth's climate system? What if there is a change to the cryosphere?

Think about what you know about ice and snow. You probably know that you can get a sunburn during the winter if you're out snowboarding or skiing or just playing in the snow. Ice and snow reflect a lot of sunlight rather than absorbing it. What if the snow cover and sea ice melt? Ocean and land surfaces absorb more sunlight as heat than ice and snow do. As more heat is absorbed, more snow and ice melt. Slowly the planet will get warmer. Recently, scientists have observed a decline in the amount of sea ice and have recorded the warmest temperatures in about 400 years.

This is bad news for life in the temperate and tropical areas of the world, not just life in the Arctic. Think about a glacier in Asia or South America. Melting glaciers provides fresh water to rivers. This water is used for agriculture, power, water in households, and manufacturing. Researchers estimate that close to 2 billion people in Asia, Europe, and the Americas depend on river systems that are fed by glaciers. If the glaciers completely disappear, where will the fresh water come from? Think about all the ways that other parts of the climate system are affected by melting ice and snow.

Atmosphere Gases

Remember what you learned about the atmosphere? Oxygen (O) and nitrogen (N) make up about 99 percent of the atmosphere. The rest of the gases, including carbon dioxide (CO_2), water vapor (H_2O), and methane (CH_4) are present in much smaller amounts. For example, carbon dioxide is about 0.03 percent of the atmosphere.

(continued on next page)

Name   Date

Climate Change

Methane, water vapor, and carbon dioxide are called **greenhouse gases.** These heat-absorbing molecules warm the atmosphere in what is called the **greenhouse effect.** The greenhouse gases absorb heat energy and give it off. Here's how it works: when sunlight hits the Earth, soil, plants, and water absorb some of the energy. Some of the energy is reflected back into space. The greenhouse gases in the atmosphere absorb some of this heat energy and then give it off. This way Earth gets heat energy from both the sun and the atmosphere. Scientists estimate that without these gases, the surface temperature of Earth would be below freezing. However, scientists also say that human activities appear to be increasing the greenhouse effect, and this may cause the global climate to change.

Human Activities and Climate Change

What does climate change have to do with you? Earth's climate has been changing for billions of years. At some points it was so cold that ice sheets covered huge regions of the globe. At others, the temperatures were much warmer, and tropical plants grew in what is today the Antarctic. Since there is evidence that there were climate changes long before humans walked on Earth, what does human activity have to do with climate change?

For millions of years, climate changes did happen naturally. However, scientists have been studying data indicating that the world is getting warmer. Over the past 100 years, average global temperatures have gone up by more than 1°F and, in some areas, as much as 4°F. The changes seem small and they happen over a long period of time — it's climate, not weather, right? So, what's the problem? Wouldn't there be changes anyway?

According to scientists, this warming trend has been speeding up. Even though there have been changes in temperature throughout history, scientists are concerned that the rapid changes they are seeing now are not due to natural causes. They say that human activities are part of the problem. Researchers say that since the beginning of the Industrial Revolution, in the mid-eighteenth century, human activity has been releasing more gases into the atmosphere. The Industrial Revolution was a time in history when there was a major change in manufacturing and industry. Before that time, most people worked in jobs related to agriculture. During the Industrial Revolution, people began working in factories and machines were used to help get work done. Fossil fuels, like coal, were burned as a source of energy.

This has continued through to today. Most industries still use fossil fuels as energy. Carbon dioxide comes from burning fossil fuels. We use fossil fuels to make electricity.

(continued on next page)

Name   Date

Climate Change

Our transportation system is dependent on fossil fuels. Landfills, where we send our trash, produce methane, a greenhouse gas. Certain livestock, including cattle, sheep, buffalo, and goats, give off methane. These animals are **ruminants.** They have four stomachs and digest their food in their stomachs instead of their intestines. Bacteria in the stomachs help break down the food and produce methane in the process. The live-stock burp the methane out into the atmosphere. One cow does not give off that much methane. However, according to the U.S. **Environmental Protection Agency** (EPA), there are about 100 million cattle in the United States and they give off about 5.5 million metric tons of methane per year into the atmosphere. In the United States, the largest amounts of methane emissions come from decomposition of waste in landfills, natural gas and oil systems, coal mining, and raising livestock. Rice fields also give off methane. Microorganisms that live in the soil produce methane, which is released into the atmosphere.

Scientists think that **deforestation,** which means changing forested lands to nonforest use, enhances the greenhouse effect in two ways. First, there are no longer trees, which remove carbon dioxide from the atmosphere to use in photosynthesis. Second, when the trees that are cut down are burned or left to decompose, they release carbon dioxide.

Effects of Climate Change

You may wonder if you can observe any of the effects of climate change now or if this is something that might happen hundreds of years from now. Scientists report that cli-mate change is affecting us right now. They say that increasing temperatures, increasing amounts of CO_2, and changes in patterns of precipitation are already affecting agricul-ture, water resources, and biodiversity.

Some scientists report that birds are already telling us that climate change will have an affect on ecosystems. The Cornell Lab of Ornithology reports that tufted puffins that live near Vancouver Island in Canada are being affected. It seems that their main food source, a small fish, is affected by changes in water temperature. When the puffins' food supply moves away, the birds don't have food. Some bird species are changing their nest-ing dates. They need to adjust so that they have food resources when their young hatch.

If you have a garden, you may have been noticing changes in the growing season. The Arbor Day Foundation updated its hardiness zone maps, which help people select the right kind of tree to grow in the place where they live. Researchers used 15 years of data to update their maps. When they did this, they found out that some states have

(continued on next page)

Name   Date

Climate Change

shifted one full hardiness zone. For example, Washington, D.C., is now in the same zone as parts of North Carolina and Texas. This means that some of the cold-loving tree species will find it harder to thrive in Washington.

Taking Action

What can you do if climate change is already happening? Think back to what you learned about fossil fuels. You probably discovered that in simply living your daily life you were using products made from petroleum or using natural gas to heat your home or coal to generate electricity. Even if you don't drive a car, there are plenty of ways to burn fossil fuels and contribute greenhouse gases to the atmosphere. The question is, what can you do about it? Many people around the world and close to home are trying to reduce carbon emissions. The action is called reducing your **carbon footprint.**

"Carbon footprint" is a shorthand way of referring to the amount of greenhouse-gas emissions caused by human activities. The footprint is measured in units of carbon dioxide over a period of time, often for one year. Are there ways that you can reduce your own use of fossil fuels? Think about what you have learned about the global farm-to-table system. Are there changes that you could make in your food choices that might help reduce our dependence on fossil fuels? Look around you — you may even find that your local community or your school is trying to reduce its carbon footprint.

Name   Date

Climate Change Notes

Use this page to take notes while you are reading. Review the questions. Record your notes in the column on the right. Remember, you are taking notes. You do not need to write complete sentences. If you need more space, write in your LiFE Log.

Questions	What I Have Learned
1. What is climate?	**a.** _____
	b. _____
	c. _____
2. What are the parts of the climate system?	**a.** _____
	b. _____
	c. _____
3. Why are ice and snow important?	**a.** _____
	b. _____
	c. _____
4. Does the farm-to-table system affect climate?	**a.** _____
	b. _____
	c. _____
5. What is the evidence for climate change?	**a.** _____
	b. _____
	c. _____

Transportation Effects

AIM

To develop an understanding of the difference between environmental problems and environmental issues.

SCIENTIFIC PROCESSES

• **question, learn about, construct knowledge, think through**

OBJECTIVES

Students will be able to:

• **discuss ways that human-designed systems have an impact on the natural environment;**

• **understand that technology can have unintended consequences;**

• **develop their own opinions concerning the environmental effect of human activities;**

• **discuss and debate the advantages and disadvantages of transportation.**

OVERVIEW

In this lesson, students continue their transportation studies. The lesson begins with a brief review of how food is moved by ocean freighter. Ocean-transportation-system experts from Lesson 8 share their posters and diagrams with the class. Next, students read about an environmental problem: the North Atlantic right whale is being affected by human activities. Students learn that whales and ships collide and some whales die as a result. After learning about scientific research and new technology that can help the whale, students learn that the shipping industry does not believe the evidence is strong enough to make changes. In the course of the reading, students learn that environmental problems can become issues when people do not agree on what should be done. Through class discussion, students gain an understanding of different perspectives and how they influence decisions. For homework, students contemplate the effect on their own lives if the ship that collides with a whale is carrying tropical fruit to their market. Students make a personal choice and explain their decision.

MATERIALS

For the teacher:
• *Environmental Problems and Issues* teacher note

For each student:
• *Whale Traffic* student reading
• *Whale Traffic Notes* activity sheet
• (Optional) *Moving Food by Ocean Freighter* student reading (pp. 155–158)

• (Optional) Ocean transportation student posters or diagrams
• LiFE Log

PROCEDURE

Before You Begin:

- Review the *Environmental Problems and Issues* teacher note and the *Whale Traffic* student reading and the *Whale Traffic Notes* activity sheet.

- Make copies of the *Whale Traffic* student reading and the *Whale Traffic Notes* activity sheet to distribute to each student.

- (Optional) Review the *Moving Food by Ocean Freighter* student reading.

- (Optional) You may wish to have the ocean-freighter experts from Lesson 8 bring in their posters or diagrams to share with the class.

- If you have not already done so, post the Module Question and Unit 4 Question at the front of the classroom.

MODULE QUESTION

What is the system that gets food from farm to table, and how does this system affect the environment?

UNIT QUESTION

What are the environmental effects of our farm-to-table system?

 **QUESTIONING**

1. Review Module and Unit Questions

Review the Module and Unit questions with students. Explain that in this lesson they are going to continue to investigate transportation. Tell them they are going to learn about an environmental problem that is linked to transportation.

2. Review Ocean Transportation

If the ocean-freighter experts from Lesson 8 have their posters or diagrams, invite them to post them at the front of the class. Ask student volunteers to briefly review the information with the class. *What have you learned about the ocean transportation system's effect on the environment? Is ocean transportation a form of human activity? Why do you think that?* Accept all answers. If students do not think of transportation as a human activity, ask questions that will help them understand the connection between humans and transportation. *What are some activities in the ocean transportation system that have an impact on the environment?* (Building a port; dredging; releasing ballast water; transporting goods.)

Are plants and animals affected by the ocean transportation system? What animals are affected? If students do not mention the North Atlantic right whale, prompt them. Tell students they are going to read about an environmental problem that involves whales and ships.

 **SEARCHING**

3. Read about Whales

Distribute the *Whale Traffic* student reading and the *Whale Traffic Notes* activity sheet to each student. Draw students' attention to the guiding questions at the beginning of the reading. Invite student volunteers to read them out loud. Review the *Whale Traffic Notes* activity sheet with students. If students are not familiar with this note-taking method, guide them through it. Review the questions on the sheet. As a class, you may wish to do a quick pre-read to see if there are more questions that you would like to add. Students can refer to these notes during the class discussion. Have students silently read the story about whales and the shipping industry.

4. Discuss Problem

Use the guiding questions to direct class discussion. *Why do scientists think the North Atlantic right whale is in trouble?* (The population is not increasing.) *Why do scientists say that human activity is harming the whale? What evidence do they have?* (Dead whales; injured whales.) *Why is it difficult for ships to avoid hitting the whales?* (They can't see them. If they do see them, the ships can't stop in time and hit the whales.) *What is the North Atlantic right whale's habitat?* (Coastal waters from Florida to Canada.)

What research are scientists doing to help people find out where the whales are? (They are studying whale sounds. They are developing new sonar technology.) *What research are scientists doing to discover whether or not reducing a ship's speed will help reduce ship-whale collisions?* (They are studying the structure of whale bones.)

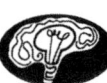 **THEORIZING**

5. Discuss Solutions

Ask a student volunteer to describe the solution that is being used in Canada. *What is being done to help the whales?* (Changes were made to the shipping lanes to keep the ships away from the whales.) *What was the plan for Massachusetts?* (Changes were made to the shipping lanes.) *Where are the whales located?* (Stellwagen Bank National Marine Sanctuary.)

Do you think the shipping-industry leaders in the story were making their decisions based on scientific evidence? What kind of information did they use to make their decision? (Financial information.) Tell students that an environmental problem can become an issue when people disagree on the problem. The shipping-industry leaders in this story disagreed with the whale researchers. They did not believe that reducing the speed of the ships would help save the whales.

 **APPLYING TO LIFE**

6. LiFE Logs

For homework, pose this scenario to students: You love tropical fruit, especially bananas, pineapples, and mangoes. You have just found out that this fruit is shipped to the United States on a huge container ship. You live in New Jersey. You know that the ship travels from South America to Florida. Then it continues along the Atlantic coastline until it reaches New Jersey. You heard that the ship killed a North Atlantic right whale on its last voyage. *Will you stop buying tropical fruit? Will you want to learn more about what can be done to protect the whales? Will you write a letter to the shipping company? Will you just keep eating tropical fruit?* Write a short essay in your LiFE Log that demonstrates how you will apply what you have learned about container ships and North Atlantic right whales to your tropical-fruit-eating practices. Use information from the ***Whale Traffic*** student reading to help you give details and supporting evidence for your actions.

Environmental Problems and Issues

Today's headlines provide an almost endless opportunity to introduce students to important issues that involve science and technology. In this lesson, we take a real-life environmental problem and use it to introduce students to some of the different ways in which scientists are investigating it. We encourage you to use this lesson to introduce your students to the difference between an environmental problem and an environmental issue. "An environmental problem is a threat or difficulty; an issue arises when individuals or groups disagree on the problem. Differing beliefs and values are what drive issues. The issue may pit economic interests against ecological interests or political interests against health and safety concerns. Of course, many times an array of differing beliefs and values are involved within any one issue."[1]

Through the student reading we set up the problem: Although protected, the North Atlantic right whale population is not increasing. What is happening to the whale? We walk students through some of the scientific research that is being conducted, as well as some of the applications of what the scientists have learned. We also introduce the issue: Some people do not accept the scientific evidence. They say the changes will cost too much money. Will people change their behavior to protect the whale?

This issue links to the global food system in several ways. Ship-whale collisions have killed whales. Some of the ships that collide with whales transport crude oil, natural gas, and food. If changing the shipping lanes or reducing the speed of the ship might help the whales, will the public and industry agree to these changes? What if the changes mean an increase in costs to the consumer?

Exploring a topic like this helps students develop critical thinking, creative thinking, and problem-solving skills. The reading closes with several examples of innovative solutions that were arrived at through collaboration among scientists, industry, and government officials. If you and your students would like to learn more about North Atlantic right whale research, visit the Right Whale Listening Network (*www.listenforwhales.org*). Christopher W. Clark, one of the scientists featured in the reading, launched this Web site so the public can listen to whale calls and see where researchers are finding the whales in the shipping lanes that run through Stellwagen Bank National Marine Sanctuary.

---

[1] Volk, 1993, p. 49

Name   Date

Whale Traffic

Guiding Questions

- *Why do scientists think the North Atlantic right whale is in trouble?*

- *Why do scientists say that human activity is harming the whale?*

- *How did the right whale get its name?*

- *Why is it difficult for ships to avoid hitting the right whale?*

- *What is the North Atlantic right whale's habitat?*

- *What research are scientists doing to help people find out where the whales are?*

- *What research are scientists doing to discover whether or not reducing a ship's speed will help reduce ship-whale collisions?*

Imagine you are 45 to 55 feet long and weigh about 70 tons (140,000 pounds). At that size, you may think nothing can touch you. There are not many things in nature that are that big. However, as scientists have found out, being big does not mean you are untouchable, especially if you are a right whale.

The Problem

Although the North Atlantic right whale is protected as an endangered species, the population of the world's most endangered species of whale is not increasing. Scientists are concerned that this species may become extinct. If this whale is being protected, why are there so few of them? What's going on with the right whale?

Scientists wanted to figure out why this protected species is still struggling. They started out by researching what is already known about the North Atlantic right whale. Here's what they learned.

For centuries, the right whale was hunted for its blubber and baleen. The blubber was boiled to make whale oil that was used as lamp fuel and as a lubricant. The baleen was used to make fishing rods, hoops for women's skirts, umbrella ribs, and buggy whips. The right whale got its name because it stayed close to shore, making it easier to hunt, and its thick layer of blubber made it float after it was killed. Whalers said it was the "right" whale to kill. Commercial whalers continued to hunt the right whale into the twentieth century. Today, however, there are several laws that protect the whale, including the Endangered Species Act. With all of this legal protection, it seemed as though the right whale would be saved, and safe. However, this is not the case.

(continued on next page)

Name   Date

Whale Traffic

Scientists estimate that there are about 350 of the whales left and the population is not increasing. There must still be something affecting this marine mammal. Researchers wondered whether human activity might be a problem. How could scientists find out what was going on?

Researchers began to look for evidence. They read what other scientists had learned about the right whale. They looked at data, analyzed records, and studied photographs of right whales. They analyzed dead whales that washed up onshore. They figured out that human activity was harming the whales. Based on the information they found, it appeared that some of the whale deaths were due to whales colliding with ships. The next question was, Are there any solutions?

Urban Ocean

The North Atlantic right whale's habitat is the coastal waters of the Atlantic Ocean, from Florida to Nova Scotia, Canada. The pregnant female whales winter and give birth to calves off the coast of Florida and Georgia. No one knows where the rest of the whale population goes during winter. In the springtime, the right whales head to New England and feed in the waters of Cape Cod Bay. These whales stay close to shore. During the summer, the North Atlantic right whales move to Canada and the Bay of Fundy for summer feeding and their nursery grounds.

Cargo ships and freighters travel through these same coastal waters on their way to ports up and down the Eastern Seaboard and east and west across the Atlantic. These are busy waters. While it may seem hard not to see a 70-ton, 55-foot whale, remember how big some of the cargo ships are. Remember, too, that there is a lot of ship traffic. For example, researchers found that Jacksonville, Florida, has 2,000 ships entering and departing each year. That's a lot of ships for whales to avoid, and Jacksonville is just one port. Scientists also found out that ships of all sizes injured the whales, not just massive container ships.

You may wonder why whales don't stay below the surface to avoid the ships. One reason may be the way right whales feed. They

(continued on next page)

BECOMING FOOD SCIENTISTS : INTERACTING PARTS : FOOD PROCESSING : ENVIRONMENTAL EFFECTS : WASTE : MAKING CHOICES

Name   Date

Whale Traffic

don't gulp, like some other whales. Right whales skim feed. They feed while moving with their mouths open through a patch of zooplankton. This means the whales stay on the surface for a long time. Some researchers think this may be one reason so many right whales are hit.

Researchers began to think about some actions that could reduce injury to the right whale from ship strikes. Would it help to have ships change their routes so they went around critical areas? Would reducing the ship's speed help? Could new technology help ships spot whales more effectively? There were lots of questions. Scientists began to gather data.

Whale Watching

Scientists knew that there already were systems in place to watch for whales and alert ships. When the whales are in their winter habitat in Florida, daily flights go out and report whale sightings. Of course, the flights only go out if the weather is good and they don't fly at night. That means there are a lot of hours when no one is watching for whales. Sometimes ships that are in the area also report whale sightings. In the Cape Cod area, a similar whale-warning system is in place. However, people can only report what they can see. Was there a way to improve this warning system?

James Miller, an ocean engineer at the University of Rhode Island, thought so. He wondered if it would be possible to use sonar imaging to see the whales. He knew that most sonar devices look straight down. This wouldn't help with avoiding whales. Ship captains need to see ahead. Miller designed a sonar device that emits signals that send back real-time images of what's in front of the ship. This means that ship captains would know when there was a whale in front of them no matter what the weather was like or whether it was day or night.

Another group of scientists were interested in whale sounds. Christopher W. Clark, a scientist at the Cornell Lab of Ornithology, wondered if it would be possible to listen for the whales. Dr. Clark specializes in bioacoustics, which is a field of science that studies the sounds animals use to communicate. He worked with researchers at Woods Hole Oceanographic Institution in Massachusetts to develop the technology that would let researchers listen for the whales. The scientists developed an underwater listening system that lets them track the right whales when they are in Massachusetts Bay. Using this technology, researchers can alert ships and warn them to slow down when a whale is nearby.

(continued on next page)

Name   Date

Whale Traffic

Listening to the underwater sound devices, researchers also discovered that the whales' habitat is very noisy. Some of the noise is not from the whales. It comes from underwater pipelines, cruise ships, whale-watching boats, tankers, and construction projects. Scientists are learning more about the effect of all of this man-made noise on whale behavior.

Changing Speeds

Cargo ships can weigh as much as 500,000 tons and travel at 20 knots (about 23 miles per hour). Any whale that was in the way of such a ship would be harmed. Some people said ships should slow down when they were in the whale habitat and that would reduce the number of collisions. The shipping industry protested. It said slowing down would cost it as much as $150 million per year in lost time and extra fuel costs. In addition, members of the shipping industry said there was no evidence that slowing down would help the whales.

A team of researchers decided to find out if there was a link between whale deaths due to ship-whale collisions and the speed of the ship. They decided to investigate what it would take to break a whale bone. Before they could begin their work, they needed some whale bones.

As you can probably guess, it's not easy to find whale bones — first you have to have a dead whale. When a dead female North Atlantic right whale washed ashore in North Carolina, the scientists removed the jawbone. They picked the jawbone for a good reason. They knew from existing data that some whales had died after their jaws were broken. They also knew that they had not seen a whale skeleton with a healed jawbone. This told them that an injury to the jaw might kill a whale. Now what they wanted to find out was what it would take to break the bone.

The scientists found out that whale bones are not like any other mammal bones. Since whales float in the water, they don't need rigid bones to support them. Once the scientists knew what the structure of the bone was, they could begin to figure out exactly what would happen when a ship hit the whale. One of the researchers, Regina Campbell-Malone, worked with small pieces of the whale jaw. She applied different amounts of force to the bone to find out when it would break. The data the scientists are collecting will be used to predict whether reducing a ship's speed will reduce the number of whale deaths as a result of collision.

(continued on next page)

Name   Date

Whale Traffic

Developing a Solution

What happens after scientists complete their investigations and present their data? Imagine you are a politician and you have to decide what to do. Do you protect the whale and tell ships to reduce their speed when they are in the whale habitat? Or do you tell the public that it will cost too much for the ships to slow down? If the shipping industry's costs go up, your costs will go up, too. Some of these ships bring in food from other countries. Some of them bring crude oil and natural gas. You use these fossil fuels every day. What will happen if oil becomes too expensive to buy?

Deciding the answer to a problem like this is challenging, but it can be done. For example, in Canada, government officials moved ships away from the whale habitat. Whale researchers and officials from a shipping company worked together. They analyzed data to find out where the whales would be. Then they moved the shipping lanes. If the ships aren't where the whales are, there should be fewer collisions.

The National Oceanic and Atmospheric Administration (NOAA) and the U.S. Coast Guard worked together on a solution for whales in Massachusetts waters and ships traveling in and out of Boston Harbor. NOAA researchers studied right-whale data that had been collected for 25 years. They looked at the number of whales that had been sighted in the Stellwagen Bank National Marine Sanctuary. Researchers wanted to find out if they could reduce the risk of a whale being hit by a ship if they moved the shipping lanes. The Coast Guard worked with NOAA to see what the effect would be on the shipping industry. About 3,500 ships pass through the Stellwagen Bank National Marine Sanctuary each year. By having the ships take a slightly different path, they can avoid the whales.

Through collaboration, it appears that the whales and the shipping industry can coexist.

Name   Date

Whale Traffic Notes

Use this page to take notes while you are reading. Review the questions. Record your notes in the column on the right. Remember, you are taking notes. You do not need to write complete sentences. If you need more space, write in your LiFE Log.

Questions	What I Have Learned
1. Why do scientists think the North Atlantic right whale is in trouble?	**a.** _____ **b.** _____ **c.** _____
2. Why do scientists say that human activity is harming the whale?	**a.** _____ **b.** _____ **c.** _____
3. Why is it difficult for ships to avoid hitting the right whale?	**a.** _____ **b.** _____ **c.** _____
4. What research are scientists doing to help people find out where the whales are?	**a.** _____ **b.** _____ **c.** _____
5. What is the North Atlantic right whale's habitat?	**a.** _____ **b.** _____ **c.** _____

BECOMING FOOD SCIENTISTS : INTERACTING PARTS : FOOD PROCESSING : ENVIRONMENTAL EFFECTS : WASTE : MAKING CHOICES

Packaging Effects

AIM

To research and think through the effects of packaging on the environment.

SCIENTIFIC PROCESSES

- construct knowledge, think through, explain, summarize

OBJECTIVES

Students will be able to:

- identify interactions between human systems and natural systems;

- describe in words and through images some of the environmental effects of packaging;

- summarize what they have learned to create an informational display about the environmental impact of packaging.

OVERVIEW

In this lesson, students conclude the packaging research projects that they began in Lesson 7. Students review their research, summarize what they found out, and develop a presentation. They think through the natural resources that are used as raw materials and the energy sources that are used in the manufacturing process, focusing on the impact of food packaging on the environment. Students communicate what they have learned by making informational displays, such as multimedia presentations or posters. After they present what they have learned, they consider how they can use this new knowledge. For homework, students prepare for Lesson 23. They read about the applesauce-making process and use a flowchart organizer to think through what they have learned.

MATERIALS

For the teacher:
- *Packaging Presentations* lesson resource
- *Product Research Checklist* lesson resource (p. 114)

For each group:
- Poster board, chart paper, or three-panel display board
- Markers
- Colored pencils or pens
- Scissors
- Glue
- (Optional) Examples of different kinds of food packaging
- (Optional) Digital camera
- (Optional) Magazines and brochures with images of food packaging

For each student:
- *Environmental-Impact Analysis* student reading
- Research projects started in Lesson 7
- *Sand to Glass, Plant to Paper, Petroleum to Plastic,* and *Ore to Aluminum Can* activity sheets (pp. 130–133)
- *Packaging Effects Guiding Questions* activity sheet
- *Packaging Flowchart Organizer* activity sheet
- *Processing Flowchart Organizer* activity sheet
- LiFE Log

PROCEDURE

Before You Begin:

- Review the *Packaging Presentations* and *Product Research Checklist* lesson resources.

- Make copies of the *Packaging Effects Guiding Questions* and *Packaging Flowchart Organizer* activity sheets for each student.

- Make copies of the *Environmental-Impact Analysis* student reading and the *Processing Flowchart Organizer* activity sheet. This is homework for students to complete before you begin the next lesson.

- Bring in old magazines and brochures with images of different kinds of food packaging.

- Sign on to the LiFE Web site (*www.tc.edu/life*) for more information about packaging, including links to videos that illustrate the manufacturing process for several types of food packaging.

- If you have not already done so, post the Module Question and the Unit 4 Question at the front of the classroom.

MODULE QUESTION

What is the system that gets food from farm to table, and how does this system affect the environment?

UNIT QUESTION

What are the environmental effects of our farm-to-table system?

 QUESTIONING

1. Review Module and Unit Questions

Before you begin this lesson, invite students to share their essays from Lesson 21. Review the Module and Unit questions with students. Explain that in this lesson the class continues its investigation of the environmental effects of different parts of the farm-to-table system. In this lesson, students conclude their food-packaging research projects.

 SEARCHING

2. Discuss Presentations

Brainstorm with students a list of possible presentation styles. Refer to the *Packaging Presentations* lesson resource for some examples. Tell students that the audience for the presentation is the students' peers. This is an opportunity for your class to share what they have learned with others at your school.

3. Organize Information

Have students work in the four project teams from Lesson 7: glass, paper, aluminum, and plastic. Distribute the *Packaging Effects Guiding Questions* and *Packaging Flowchart Organizer* activity sheets. Tell students to use these activity sheets to help them organize their thinking and the information that they have gathered. Remind students to include information on the environmental impact of manufacturing the packaging. Brainstorm a list of different kinds of environmental effects. *Does the manufacturing process used to make your product contaminate water or air? Does harvesting or extracting the raw materials for your product cause loss of wildlife habitat? Are there other effects on the environment?* (Increase in noise levels; loss of scenic views.)

Have each project team elect a team leader, if they have not already done so. The leader's role is to guide the development of the presentation. Help team leaders identify tasks and make assignments. Depending upon the size of each project team, you may wish to divide the team into smaller groups or have students work in pairs to produce different parts of the presentation or multiple presentations.

4. Develop Presentations

Remind students to begin with an outline of what they want to include in their presentation. Next, have them elaborate on the information in the outline. Tell them to critique what they have written and make any changes that they think are necessary. If students in groups are working on different parts of the presentation, have them exchange their outlines with another member of their group so they can critique each other's work. Once the text is complete, have students put together their informational displays.

5. Present Research

Invite each product team to display its presentation. Ask each team leader to lead the discussion of the team's work. After team members have made their presentations, moderate a question-and-answer session with the audience.

At the end of the presentations, congratulate students on their work. Consider keeping the displays to use at the *Farm-to-Table & Beyond* Expo described in Lesson 29.

6. LiFE Logs

Ask students to write in their LiFE Logs. Have them pick one type of packaging and use these questions to focus their writing: *What was the most interesting thing that you learned about packaging? Did you learn anything that changed your thinking about packaging? How can you use what*

you have learned? If students do not complete this assignment during class, have them complete it as part of their homework.

7. Homework

Distribute the ***Environmental-Impact Analysis*** student reading and the ***Processing Flowchart Organizer*** activity sheet. Tell students to use the questions to guide their reading and to use the flowchart to organize their thoughts. They will refer to the flowchart in Lesson 23.

Packaging Presentations

Have students use the flowchart and the guiding questions on the activity sheet to organize their work. This activity offers students an opportunity to reflect on what they have learned and to summarize the information to present to an audience of their peers. Remind students to outline their presentation before they begin.

Here are some suggestions.

Life Cycle of a Glass Jar

From a speck of sand to a jar of spaghetti sauce to the recycling bin

Overview or introduction: What is glass?

Step 1: What raw materials used to make this product come from the natural environment? What natural resources were used? What kind of energy was used to mine the resources?

Step 2: How is glass made from these raw materials? What process is used? What energy sources are used to process the raw materials?

Step 3: After the glass jars are made, how are they transported from the manufacturing plant? What are the main uses of glass in food packaging? What kinds of foods are packaged in glass?

Step 4: After the glass jar has food in it, how is it transported to stores? What kind of energy is used? How does the glass jar get to a consumer's home?

Step 5: After the jar is empty, what happens to it? Is it recycled? Can it be reused? This is the end of the life of the product.

This same life-cycle approach can be done for a paper milk carton, an aluminum can, a paper cereal box, or a plastic juice bottle.

Three-Panel Display

Using this approach, students can work individually or in pairs to develop a panel topic.

Panel 1: Natural Resources. What raw materials are used to make the product? What energy sources were used to harvest or extract the resource?

Panel 2: Product. How is this product used in food packaging? What kinds of food are packaged in this product? Can the finished product be recycled or reused?

Panel 3: Environmental Effects. What are the environmental effects of using this product in our farm-to-table system?

Name   Date

Environmental-Impact Analysis

You are an environmental scientist. You want to know if fruit processing has an effect on the environment. You decide to investigate apples. To begin, you gather some data about the number of apples consumed each year. You find that according to the U.S. Apple Association, in 2004 the average American consumed a total of about 50.4 pounds of apples and apple products. Of the 50.4 pounds, an estimated 31.83 pounds were processed apples. If the average American consumer eats about 32 pounds of processed apples and in 2004 the estimated population of the United States was 293,655,404 — well, that's a lot of processed apples. You decide to research applesauce.

First, you define the focus of your study. The processing plant — what happens from the time the apples are unloaded until they are ready to be shipped as applesauce — looks like a good place to begin.

At the Processing Plant

You watch as the apples are unloaded. The first thing you notice is that the apples are always moving. They are placed on a conveyor belt and moved from one location to another inside the plant. You make a note that the conveyor belt is run by a machine. You follow the apples to a tank where the different varieties are mixed together. Another machine called a deleafer removes any leaves left on the apples. Next, the conveyor belt moves the apples to a place where they are graded and inspected. Factory workers stand next to the conveyor belt and look for apples that are rotten. They pull the rotting fruit off the conveyor belt so it doesn't get mixed in with the applesauce. You make a note about the waste products and the machines.

Next, the apples are rinsed with water to wash off any dust or dirt. You make a note to find out what happens to all this dirty water after the apples are removed. The rinsed apples are sprayed with clean water. Any apples with decay or leaves still attached are removed. You make a note about more water and more waste. The apples move to a machine that peels them as another machine rotates them. Most of the apple peel is removed at this stage, along with the core — more waste. Another machine chops the apples and they move into a steam cooker. The steam heats them so they become soft. You make a note about the chopper and the steam heat.

At the Finisher

The soft apples move into a finisher. This machine has rotating paddles that push the apple through a screen, which acts like a separation device. Here, any leftover apple peel or core is removed. All the defects (peel, core, seeds) are kept behind while the

(continued on next page)

Farm to Table & Beyond
©2008 Teachers College Columbia University

Name   Date

Environmental-Impact Analysis

screened apple continues along its way. You note more machines and more waste and wonder how the screens get cleaned. Next the screened apple "mush" goes into a blending kettle, where the applesauce ingredients are blended. Then the finished applesauce moves to a filler where it is poured into glass jars. The filled jars move along the conveyor belt, the lid is sterilized with heat, and then the jars are transferred to a cooler. Here they are cooled down before being put in the warehouse to be stored. You notice some bent lids and broken glass in containers in the corner. You make a note: more waste and energy used to cool the applesauce. Before the jars are shipped to a store, labels are added. You make a note about more machines and paper. Finally, the jars are placed in cardboard boxes, the boxes are stacked on pallets, and a machine wraps the pallets in plastic. They are ready to ship. You note the forklifts and small trucks at the plant and the machine that wraps the plastic.

Apple Waste

Before leaving the processing plant, you ask about all of the apple peels, cores, stems, leaves, and seeds. What are they going to do with the waste? They can't leave it in a pile next to the factory. It will ferment, attract insects, and smell. The manager tells you that they reuse some of the waste. The apple peels and cores are pressed to get out all of the juice, which they use to make vinegar. Of course, there is a trade-off. Making vinegar takes more energy and more processing, so there will continue to be an impact on the environment, but the impact is less than if the waste were just thrown away. The manager tells you they used to send the leftover apple waste to a landfill, but now they sell it for animal feed. You make some final notes and head back to the office to review your data.

Innovation: Apple Paper

If you lived in Italy, you might try to think of something else to do with the apple peels, cores, and seeds. In Italy, the waste produced by the apple industry is classified as "special waste." This means it has to be disposed of using a very expensive process. An Italian engineer by the name of Dr. Albert Volcan did research to see if he could find a way to handle the apple waste that would not require the expensive process. He spent several years doing experiments. He developed something called the Volcan process, which includes drying, cooling, and grinding the waste. It is changed into something that looks like flour and can be used as a raw material to make paper. Dr. Volcan's product is called Cartamela, which means "apple paper" in Italian.

(continued on next page)

BECOMING FOOD SCIENTISTS : INTERACTING PARTS : FOOD PROCESSING : **ENVIRONMENTAL EFFECTS** : WASTE : MAKING CHOICES

Name   Date

Environmental-Impact Analysis

Using the Volcan process, one region of Italy that generates about 30,000 tons of moist waste can make about 5,000 tons of paper. Researchers estimate that it takes about 15 trees to produce one ton of this paper. To make the equivalent amount of regular paper out of trees would take about 75,000 trees.

Analysis

1. Go back over the reading and circle or highlight each step in the process where you think there is an impact on the environment.

2. Use the ***Processing Flowchart Organizer*** activity sheet to make a flowchart that illustrates the applesauce-making process. At each step, write down any effect on the environment and what that effect is. For example: A truck that delivers apples to the factory uses gasoline, which is a fossil fuel and a nonrenewable resource. The exhaust gives off particulate matter, which goes into the air.

Here are some questions to guide you:

- *Was water used? Where?*

- *Was heat used? Where?*

- *Were there machines in the factory? What powered the machines?*

- *Was any waste generated? What kind of waste?*

- *Do you think that there are lights in the factory? What powers the lights?*

- *Did you see any use of paper or paper products? What is used to make paper?*

In your LiFE Log, write at least two paragraphs that respond to the following questions: "If you had to decide between making apple waste into paper and feeding apple waste to animals, which would you choose?" "Why did you make that decision?"

BECOMING FOOD SCIENTISTS : INTERACTING PARTS : FOOD PROCESSING : ENVIRONMENTAL EFFECTS : WASTE : MAKING CHOICES

Name   Date

Packaging Effects Guiding Questions

Think about the natural resources used in the process of extracting or harvesting (cutting down) the raw materials used in your product.

- *What machines are used? What fuels the machines? Is the air affected?*

- *What raw material or materials are used to manufacture your product?*

- *Is the raw material mined, cut down, or drilled?*

- *Is energy used in the process of getting the raw material? Where does the energy come from? Is there any effect on the environment?*

- *How does the raw material get to the manufacturing plant? Is any energy used?*

- *Describe the process used to make the product. Be sure to include a discussion of energy. (machines, human labor, electricity, natural gas, hydropower, etc.)*

- *If water is used in the manufacturing process, what happens to the wastewater?*

- *Is heat part of the process? Is the product cooled during the manufacturing process?*

- *How is the product packaged (pallets, plastic warp, cartons, etc.)?*

- *Are recycled materials part of the process?*

- *What can be done with your product after it is used?*

Name   Date

Packaging Flowchart Organizer

Use this flowchart to organize your thoughts about what happens as raw materials are made into packaging for our farm-to-table system.

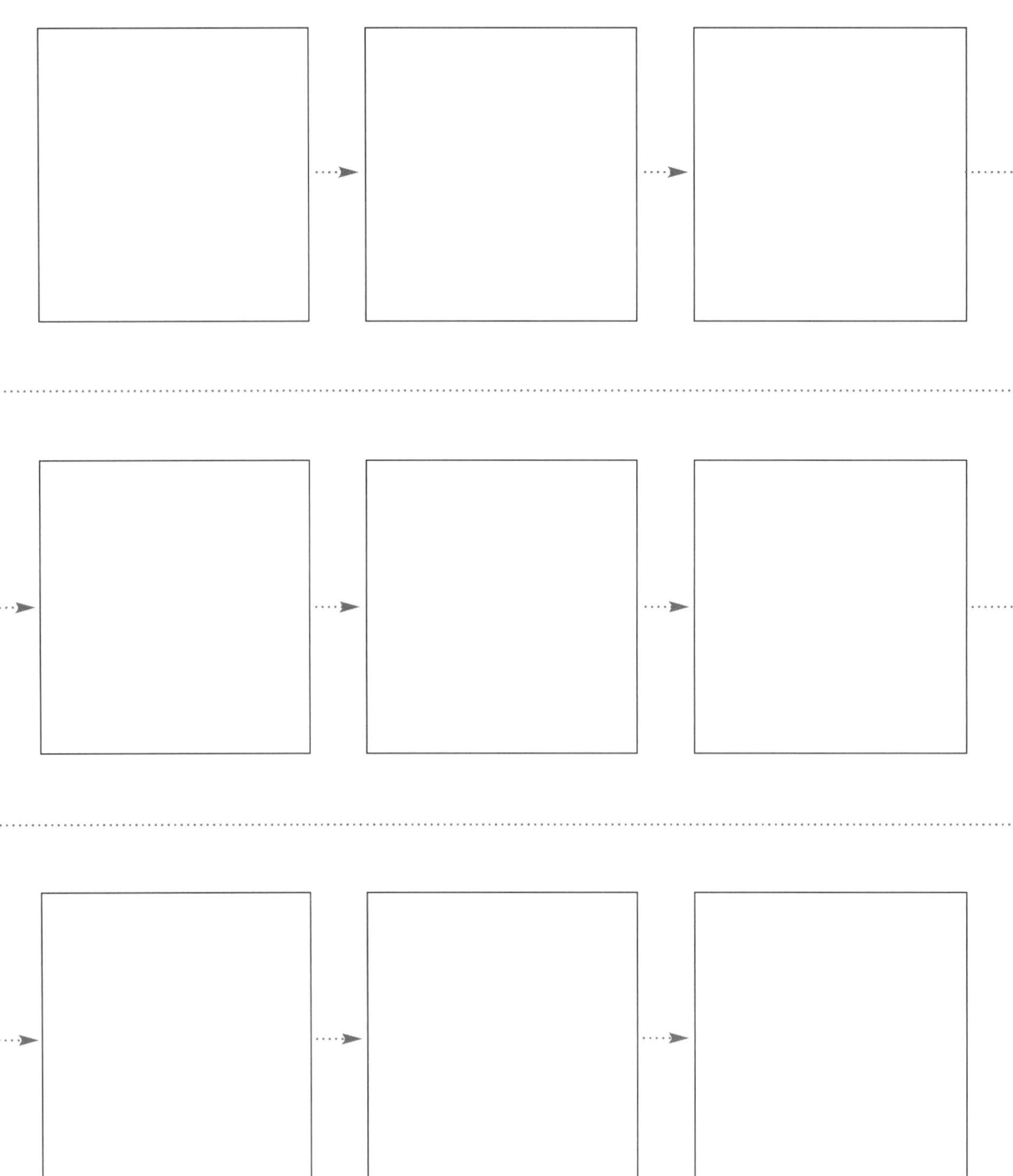

Name   Date

Processing Flowchart Organizer

Use this flowchart to organize your thoughts about what happens as raw ingredients are made into processed foods for our farm-to-table system. Start with the unprocessed food, or raw ingredients (apple). End with applesauce.

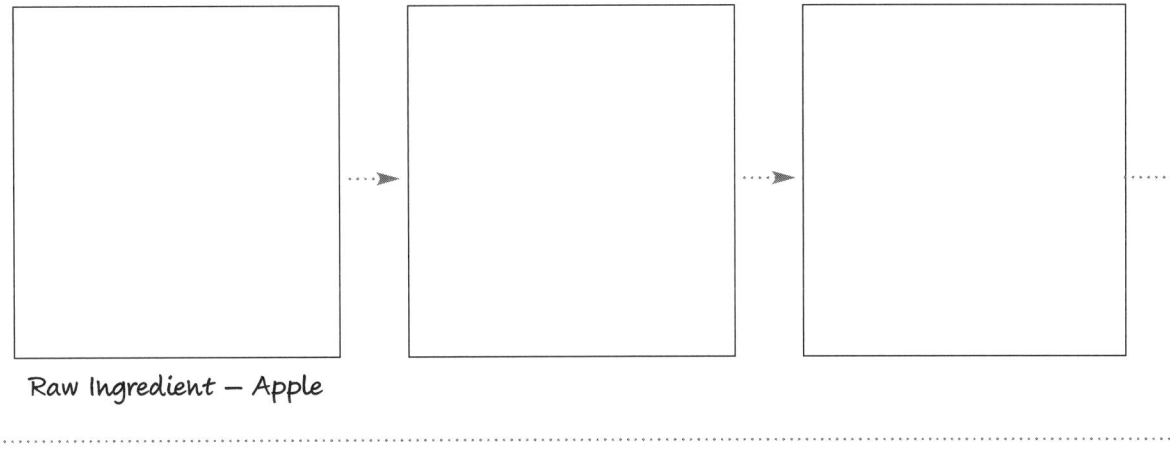

Raw Ingredient — Apple

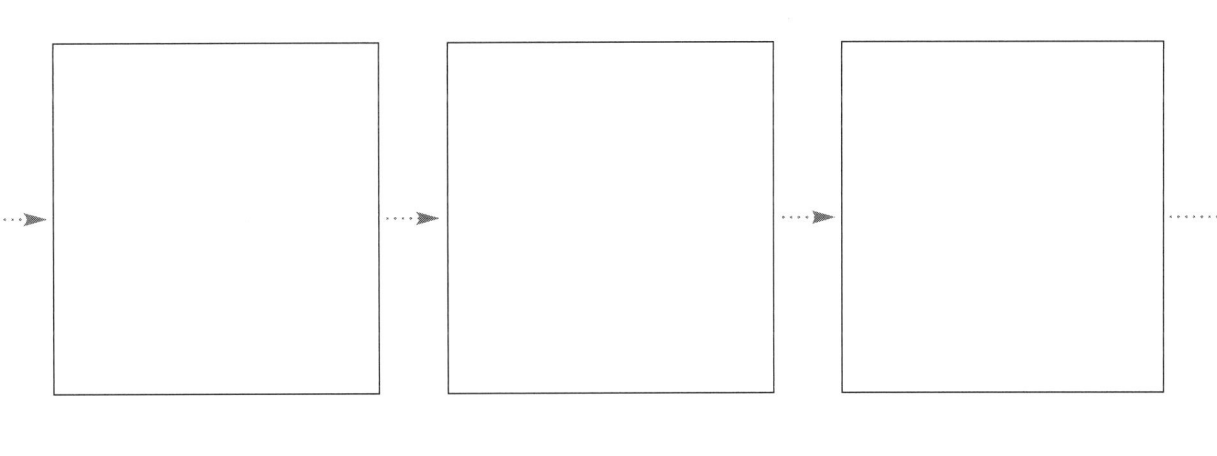

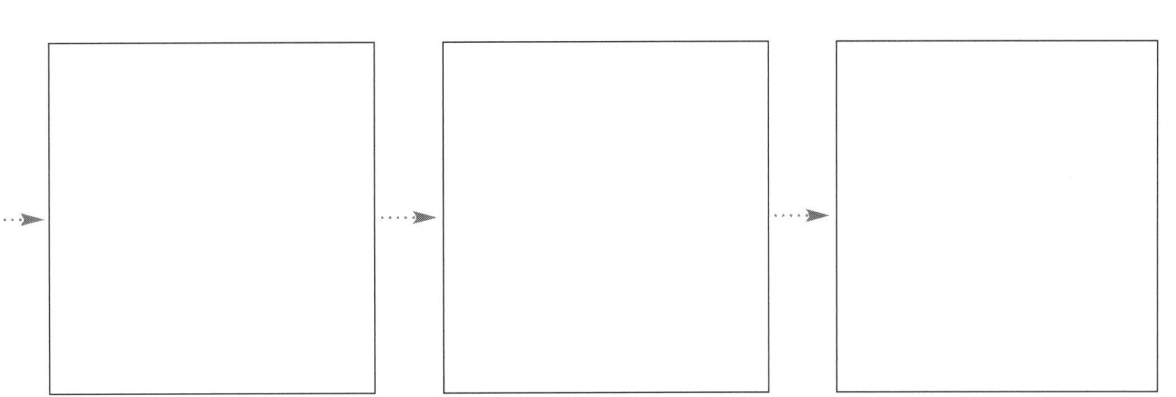

Processing Effects

AIM

To construct knowledge about the environmental effects of our food-processing system.

SCIENTIFIC PROCESSES

- discuss, consider, think through, apply

OBJECTIVES

Students will be able to:

- discuss the process of making applesauce;

- list and explain the trade-offs of processing food;

- describe energy sources and byproducts of a snack-manufacturing system;

- deliver an informational presentation about a method for food processing with minimal effect on the environment.

OVERVIEW

In this lesson, students continue to gain knowledge about the environmental effects of our farm-to-table system. The lesson begins with a review of the Lesson 22 homework, including a discussion of the apples-to-applesauce flowchart. Next, the class takes a more detailed look at one byproduct of the pickle industry — the brine. The class reads about how a scientist is trying to solve a problem with pickle brine and considers the trade-offs that might be involved. Then, students take on the role of food processor. Working in small groups, students play the role of executives of a snack company, X-tremely Healthy Snacks. The company has an environmental mission. The executives' job is to run a profitable company, yet minimize its impact on the environment. The company wants to launch a new snack. They are in the prototype phase and need to design a system to manufacture the snack with a minimal effect on the environment. The CEO of the group presents this design to the board of directors.

MATERIALS

For the class:
- (Optional) *Pickles* teacher recipe (p. 274)

For each group:
- Chart paper
- Markers

For each student:
- *Environmental-Impact Analysis* student reading (p. 346)
- *Processing Flowchart Organizer* activity sheet (p. 351)

- *In a Pickle* student reading
- *Snack Assessment* student reading
- *Snack-System Design* activity sheet
- LiFE Log

PROCEDURE

Before You Begin:

- Remind students to bring in their homework from Lesson 22, including the **Environmental-Impact Analysis** student reading and the completed **Processing Flowchart Organizer** activity sheet.

- Review and make copies of the **In a Pickle** and **Snack Assessment** student readings and the **Snack-System Design** activity sheet.

- If you have not already done so, post the Module Question and the Unit 4 Question at the front of the classroom.

MODULE QUESTION

What is the system that gets food from farm to table, and how does this system affect the environment?

UNIT QUESTION

What are the environmental effects of our farm-to-table system?

 QUESTIONING

1. Review the Module and Unit Questions

Review the Module and Unit questions with students. Explain that in this lesson the class investigates environmental effects of food processing.

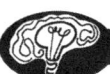 THEORIZING

2. Discuss Processing Homework

Have students take out their LiFE Logs and their homework from Lesson 22. *What are some facts you learned about fruit processing? Does it have an effect on the environment? How is energy used in the process?* Compile a list on the board.

What did you learn about food waste? What happens to all of the apple peels, cores, and seeds? What trade-offs are there when the food waste is used to make vinegar? What about making paper from apple waste? What do you think the trade-offs are? (Energy to make the paper. It doesn't take many trees, but it still takes energy.) Invite student volunteers to share with the class whether they would make apple waste into paper or feed it to animals. Have them read what they wrote in their LiFE Logs and explain their decisions.

Tell students that now they are going to learn about one example of scientists working with food manufacturers to try to solve a food-processing problem.

 SEARCHING

3. Read about Pickle Brine

Distribute the **In a Pickle** student reading. Engage students in a brief discussion of the class pickle-making experience. *What ingredients did we use to make pickles?* If students need to be reminded, have one student read the ingredient list of the **Pickles** teacher recipe. *Do you think that any of these ingredients could have an impact on the environment?*

Have students turn to the student reading. Invite volunteers to take turns reading paragraphs aloud. After students have finished reading, ask them if they can think of any trade-offs in using the enzyme. *Do you think that getting the clay from the riverbed would have an impact on the environment? How important is it to pack pickles in brine?* (Very important. People don't like soggy pickles.) *If you were a pickle manufacturer, what would you do? Would you want to use the clay? What if you lived near the river on the Georgia/ Florida border? Would you want someone to take the clay away?* Have students explain their answers.

Becoming Food Scientists : Interacting Parts : Food Processing : ENVIRONMENTAL EFFECTS : Waste : Making Choices

4. Design a Snack

Distribute the *Snack Assessment* student reading and the *Snack System Design* activity sheet to students. Have students work in small groups. Tell each group to select a CEO. The other group members are part of the design team. The CEO is responsible for presenting the group's design to the class. The class plays the role of members of the board of directors of X-tremely Healthy Snacks.

As a class, review the questions on the activity sheet. Tell students to use these questions to guide their design and presentation. Have students outline their presentations on chart paper. Review and discuss each outline with the group members. Allow students time to revise their outlines.

5. Finalize Presentation

Encourage students to consider the best verbal and visual ways to present their information. They may wish to include drawings as well as text. Ask them to think about what will make their presentation interesting, lively, and engaging. Remind them that the board of directors wants evidence that their design is "green."

Brainstorm a list of presentation tips. Record them on the board. Tips might include: organize what you are going to say; organize your presentation materials; explain the problem you solved; explain your solution.

6. Present to Board

Outline the presentation procedure. Tell students that the CEO of each group will present to the board. At the end of each CEO's presentation, the board can ask questions. You will be moderator and monitor the time.

7. Homework

As homework, ask students to think about the presentations. Have them write in their LiFE Logs and describe the design system they believe would minimize the environmental effects of making the snack. Tell students to explain why they came to this conclusion.

Name

Date

In a Pickle

Commercial pickle packers had a problem — pickle brine can harm the environment. Seriously. Think about the ingredients in the brine. Salt, or salt water, can get pickle producers in a pickle.

You may be thinking about the amount of brine you made when you made pickles and wonder how such a small amount can cause harm. Remember, though, you made about 15 pickles. What if you made 20 billion pickles? That's how many pickles Americans consume each year, according to Pickle Packers International. Imagine how much brine you would need to make 20 billion pickles!

Commercial pickle packers process pickles on a large scale. They put cucumbers in a large tank with salt, water, and spices. The pickles stay there while they ferment, which can take up to three months. After the pickles are removed, they are rinsed and put into jars. What happens to all that leftover brine?

Pickle packers can't just dump it into the soil or pour it down the drain. It might contaminate the water. Think of all the plants and animals that would be affected. Removing all the chemicals and purifying the waste is very expensive. What else can pickle packers do? They could recycle the brine and use it to store pickles. However, sometimes there is a problem when they reuse it. Sometimes the brine becomes contaminated with an enzyme that makes the pickles get soggy instead of staying crisp. No one wants a soggy pickle. The pickle packers had to figure out what to do about that enzyme.

A food-science professor at the University of Arkansas, Ron Buescher, had an idea. He looked at different materials that were already being used by the food industry. He wondered if any of them would get rid of the enzyme. He tested different substances that absorb enzymes and found one that he thought would work. It was a special kind of clay that is found in a dry riverbed on the border between Georgia and Florida.

Here's the process. After the pickles are removed from the tanks, the clay is mixed in. It settles and attracts the enzymes. Then the clay is removed from the brine, which is now free of the enzymes. The enzymes can be removed from the clay, and it can be reused on the next batch of brine. And the consumer — you — can have crisp, crunchy pickles.

Name   Date

Snack Assessment

As executives of X-tremely Healthy Snacks, your goal is to run a company that is profitable, yet minimizes its impact on the environment. You understand that making products creates waste and uses energy. You want to keep the impact low. To do this, you are looking at each part of your farm-to-market system to make sure it has a low impact, too. For example, you will buy your ingredients from farmers who use sustainable practices. Your local transportation system uses hybrid vehicles to reduce the use of fossil fuels. The cargo ships are trying out a new way of using wind power along with their engines.

You also made a promise to your customers to use the latest "green" innovations in packaging. You will work with sustainable-design consultants who develop products that can be reused or that will last a long time. They also work with 100-percent renewable resources, recycled resources, or nontoxic materials. Whenever possible, the designers choose local materials that are manufactured with renewable energy.

Food scientists in your test kitchen have been developing new energy snacks. The food scientists have developed a great-tasting dried-fruit-and-nut mix, but before you take it to market, you need to think through what you will do with all of the food waste and the energy resources you will need to produce it.

You study the ingredient list: raisins, dried apples (chopped), dried apricots (chopped), sunflower kernels, roasted peanuts, and cashews. You can buy the raisins, apples, and apricots already dried and chopped. It costs less to buy the sunflower kernels still in their hulls, but that means you need to remove the kernels before you can use them. You buy the cashews out of the shell and ready to use. It costs much less to buy roasted peanuts in the shell, but you will have to remove the shells.

You need to design a system for removing the shells from the peanuts and the hulls from the sunflower kernels. Next, think about the energy resources you will need. Here are some of the factory requirements to consider: lights, ventilation, machines for mixing, and forklifts to move the containers. You also have to consider how to keep the machines clean, how to fill the packages with the snack mix, and what to do with the shells from the peanuts and the hulls from the sunflower kernels.

Your assignment is to prepare a presentation that your CEO will give to your board of directors. Your goal is to convince them that this new snack will have a low impact on the environment. They want to hear all the details.

Farm to Table & Beyond
©2008 Teachers College Columbia University

Name   Date

Snack-System Design

Here are some guiding questions to think about as you develop a food-processing system for X-tremely Healthy Snacks. Remember to think about energy resources you will need and how you will handle any byproducts from the manufacturing process. Be sure to include transportation, food processing, and packaging in your presentation to the board.

Transportation

- You can purchase raisins, dried apples, dried apricots, sunflower kernels, and roasted peanuts from suppliers in the United States. Cashews come from Brazil. *How will you transport these ingredients to your factory, which is located on the Mississippi River?*

- *Will there be any impact on the environment? Describe the environmental effects.*

- *After you manufacture the snack, you want to ship it to grocery stores and vending machines across the United States. How will you do this?*

Food Processing

- You buy all of the ingredients ready to use, except the sunflower seeds and roasted peanuts. *What process will you use to remove the sunflower-seed hulls and peanut shells?*

- *What energy resources will you need? Will there be an impact on the environment?*

- *What will you do with the byproducts, the empty hulls and shells?*

- *How will you mix all the ingredients to make the snack? What energy resources will you need?*

- *How will you prevent the ingredients from spilling onto the floor? What will you do with any ingredients that do spill?*

- *How will you dispose of the food waste?*

- *What resources will you use to clean the factory after the workers make this snack? Will the machines need to be washed before a different kind of snack food is made?*

Packaging

- You plan to sell this snack in three different sizes. The smallest size (2 oz.) is for vending machines. You plan to sell two other sizes (8 oz. and 16 oz.) in grocery stores. *What kind of food packaging will you use?*

- *How will you fill the packages? What energy resources and raw materials will you need to use?*

- *Will there be a label for the snack? What resources are used to make the label?*

- *How did the labels get to your factory? What energy resources were used?*

Effects, Effects, Effects

AIM

To think through how the farm-to-table system affects the natural environment.

SCIENTIFIC PROCESSES

- discuss, investigate, gather data, construct knowledge

OBJECTIVES

Students will be able to:

- demonstrate understanding of ways that human-designed systems have an effect on the natural environment;

- discuss advantages and disadvantages of an industrial food system;

- explore what happens to garbage.

OVERVIEW

Throughout this unit, students have been learning about the environmental effects of our farm-to-table system. In this culminating lesson, students synthesize what they have learned about the impact of the transportation, food-processing, and packaging systems on the natural environment. The lesson begins with a review of the previous lesson's homework. Then, students work in small groups to make posters that trace the path of applesauce from orchard to table, making note of environmental impacts along the way. The class also considers what happens to all of the solid waste that society generates. As a class or in small groups, students build a model landfill and make observations of the contents. They will make additional observations throughout Unit 5. Finally, students make food logs to use to collect data on their own food habits. For homework, students write responses to the Unit 4 Question in their LiFE Logs.

MATERIALS

For the teacher:
- *How Do We Generate Waste with Our Food System?* sample conversation
- *Burying Our Trash* teacher note
- *Keeping a Food Log* lesson resource
- *Investigating Landfills* experiment sheet
- (Optional) *Apples to Applesauce Concept Map* lesson resource (p. 76)

For the class:
- Materials from the *Effects, Effects, Effects Supply List* lesson resource
- Stapler

For each group:
- Chart paper
- Markers
- Scissors

For each student:
- (Optional) *Environmental-Impact Analysis* student reading (p. 346)
- (Optional) *Processing Flowchart Organizer* activity sheet (p. 351)
- *Landfill Observations* activity sheet
- Personal Three-Day Food Log from the *Keeping a Food Log* lesson resource
- LiFE Log

PROCEDURE

Before You Begin:

- Review the *Effects, Effects, Effects Supply List* lesson resource. Gather materials.

- (Optional) Have the *Apples to Applesauce Concept Map* lesson resource, the *Environmental-Impact Analysis* student reading, and the *Processing Flowchart Organizer* activity sheet available for students to use for reference.

- Review the *How Do We Generate Waste with Our Food System?* sample conversation, the *Burying Our Trash* teacher note, the *Investigating Landfills* experiment sheet, and the *Landfill Observations* activity sheet. Cut the two bottles and prepare them to set up the model landfill. Prepare more bottles if each group is doing a model.

- Review the *Keeping a Food Log* lesson resource. Make copies of the *Personal Three-Day Food Log* mini-book template for students. Assemble the mini-books before class, or plan to have students make them during class.

- Make copies of the *Landfill Observations* activity sheet for students.

- If you have not already done so, post the Module Question and Unit 4 Question at the front of the classroom.

MODULE QUESTION

What is the system that gets food from farm to table, and how does this system affect the environment?

UNIT QUESTION

What are the environmental effects of our farm-to-table system?

 **QUESTIONING**

1. Review the Module and Unit Questions

Tell students that this is the final lesson of Unit 4. In this lesson, they think through what they have learned and write their ideas about answers to the Module and Unit questions. They also begin to explore what happens to garbage.

 **THEORIZING**

2. Snack-System Review

Review the homework assignment. Invite students to describe the design system they believe would minimize the environmental effects of making the snack. Remind them to explain why they came to this conclusion. Encourage other students to ask questions and to discuss their classmates' decisions.

3. Farm-to-Table Posters

In this section of the lesson, students make posters of the farm-to-table system, indicate all the places in the system where there are environmental effects, and briefly describe the effects. If your class needs support with this work, you may wish to have them sketch out the apples-to-applesauce system, using the *Environmental-Impact Analysis* student reading and the *Processing Flowchart Organizer* activity sheet. To start them off, refer to the *Apples to Applesauce Concept Map* lesson resource. However, if your students are ready to venture out and explore a new food, feel free to do so. This activity serves as a post-assessment of what students have learned thus far.

The goal is to take a farm-fresh food and follow it as it moves through the system — from harvesting to transportation to a processing factory, where it is made into a different form (applesauce, jam, pasta) and packaged and shipped to a grocery store. The food is purchased at the grocery store and served at home. Students think through what they have learned and summarize the environmental effects of bringing this food to the consumer.

4. Display Posters

After student groups have completed their posters, choose a representative from one group to present its work. After the presentation, ask students if they have any additions they would suggest. *Did anyone see any other environmental effects to add to this poster?* Remind students to explain why they would add this. *Do you disagree with anything? Why do you disagree?* Invite the other groups to show their posters. Tape them at the front of the class. Consider inviting another class to view the posters and listen to the explanations. Save the posters for the expo in Unit 6. Close this part of the lesson.

5. What Happens to Garbage

Remind students that one of the environmental effects they have been learning about is waste. Use the sample conversation to guide students through a discussion of waste. Introduce landfills. *What is a landfill?* (A place where garbage is buried between layers of soil.) *Do you think landfills might have an effect on the environment?* (Byproducts might leak into the ground and contaminate groundwater.) *How can we investigate landfills without going to one?* (Build a model.)

EXPERIMENTING

6. Model Landfill

Tell students that landfills are the most common method of disposing of waste. We send about two-thirds of our waste to landfills. Explain that landfills have a liner that helps prevent toxic materials from getting into the surrounding water and soil. They are also covered with a layer of soil at the end of each day to control the odors and keep animals out.

Distribute the *Landfill Observations* activity sheet to students. If you are making the landfill model as a class demonstration, have student volunteers help you assemble the landfill. Use the *Investigating Landfills* experiment sheet to guide the discussion of setting up the model. Describe each section of the landfill as you assemble it. Ask students to make initial observations about their landfills and record them in the Observations #1 column on their *Landfill Observations* activity sheet. Tell students they will be returning to this experiment over the next several days (in Lessons 26 and 27) to make additional observations about the landfill, which they will also record on the activity sheet.

7. Introduce Food Logs

Tell students that they are going to begin to collect data about their own food habits. Explain that over a three-day period, students will record everything that they eat. Use the *Keeping a Food Log* lesson resource to guide the discussion. If you have made the food logs ahead of time, distribute one to each student. Otherwise, distribute the Personal Three-Day Food Log mini-book template to students and demonstrate how to assemble the books. Draw the sample food log on the board. Review the directions for filling in the chart. Remind students that it is very important for them to fill it in as completely as possible. Tell students that they will be using this information at the end of the module, in Lesson 30.

APPLYING TO LIFE

8. Homework

Ask students to write an answer in their LiFE Logs to the Unit 4 Question, *What are the environmental effects of our farm-to-table system?* Then have students use that answer as a springboard and come up with ideas for changes that could be made to the food system that might help reduce the environmental effects of our farm-to-table system.

How Do We Generate Waste with Our Food System?

This sample conversation in the **Theorizing** phase of the QuESTA cycle will help you guide your students through a conversation that will enable them to construct new knowledge. As you engage students in your own conversation, encourage them to bring together what they know about waste and our food system and to develop new theories about the environmental effects of our farm-to-table system. Encourage students to recognize how they might use this new knowledge in their daily lives. This is a guide. Feel free to adjust your questioning to the needs of your class.

MR. G: Let's have a discussion to summarize what we have learned about the environmental effects of our farm-to-table system. *Who can start by telling me about how food packaging is made?*

MARIE: Food packaging is made in factories out of materials that are made from sand, metals, trees, and oil. I forget what they're called. Most of the materials come from the earth.

MR. G: Right, food packages are made from things that come from the environment. We call these materials natural resources. *Is it easy to make packages from natural resources?*

MARIE: No, it takes a lot of steps and a lot of energy.

MR. G: *Are there any environmental effects when the packages are made? Is there any waste?*

JEFF: I think so. When I helped my dad make a box and lid out of cardboard pieces, there were little pieces of cardboard that fell out when we folded the box together.

RITA: When my mother sews a dress, she cuts it out of fabric and there are always pieces of fabric left over.

MR. G: That's good thinking, Jeff and Rita. Those are two good examples of waste created during the process of making something. *Can anyone think of another example of waste?*

MARIE: After we eat food, we throw the packages away. Or sometimes we recycle them. Most of the time we throw them away.

MR. G: *And what happens to the waste when we throw it away?*

JEFF: It goes to a landfill or gets burned. In a landfill the waste just sits there. It's gross.

MR. G: *Can anyone think of an example of an environmental effect from transportation?*

ALEJANDRO: We use fuel in engines. The exhaust can pollute the air. Sometimes there can be oil spills or leaks, and that pollutes, too.

Continue the conversation to include environmental effects of different modes of transportation and the natural resources that are used in food packaging and processing. Encourage students to think about food waste when food with bruises is thrown out during food processing.

Burying Our Trash

We are a "throw-away" nation. What we don't recycle or reuse ultimately ends up in a landfill. For as long as human societies have had surplus goods, they have dumped, buried, or burned what they no longer wanted or needed. In *Moveable Feasts*, the writer Sarah Murray describes an archaeological dig in Italy, at the top of Monte Testaccio, a short distance from Rome. The site is a garbage dump. For more than two centuries, olive-oil amphorae were dumped there after their contents had been unloaded and distributed. The result is a hill made up of broken bits of pottery.[1]

Today, we hide our garbage from sight. We bury it in landfills, but not just anywhere. Potential landfill sites need to meet certain criteria. Engineers inspect the geology of the landscape, looking for sites that will minimize any possibility that the waste will contaminate the groundwater. Landfills are designed to hold trash and prevent it from creating environmental problems in the future. They are built into or on top of the ground and are designed to prevent any contamination between the waste and the surrounding environment.

In addition to an appropriate geological setting, a secure landfill has three critical elements: a bottom liner, a leachate-collection system, and a cover. The bottom liner prevents the trash from coming in contact with the outside soil and groundwater. Landfills use one or more layers of clay, a tough plastic liner, or a combination of these. A plastic liner is sometimes also surrounded on either side by a textile mat that helps to keep the plastic liner from being punctured by nearby rock and gravel layers. The bottom liner forms a barrier between the liquid waste, called leachate, and the ground. If the liner fails, the leachate will permeate the soil.

The landfill is divided into cells. At any given time, only a few cells are open and filled with trash. This minimizes the trash's exposure to rain. It is important to keep the landfill as dry as possible to reduce the amount of leachate. One way to prevent leachate is to keep rainwater out of the landfill.

Leachate seeps to the bottom of a landfill and is collected by a system of pipes. It is pumped out and treated at the surface. At some landfills, ground wells are drilled around the landfill. These wells are used to monitor groundwater quality. This is one way to ensure that the local community's drinking water stays pure.

At the end of each day, workers spread soil over the waste. This cover reduces the odor and helps keep animals out. When a section of the landfill is finished, it is sealed permanently with a cap, which forms a barrier between the waste and the surface to prevent precipitation from seeping into the waste. The cap includes a barrier, such as clay, covered by a layer of soil and plants.

Landfills can also be sources of energy. Methane is produced as organic wastes in the landfill decompose. Some landfills burn the methane to get rid of it. Others collect it, treat it, and sell it as a commercial fuel. They can even use it to generate electricity.

---

[1] Murray, 2007, pp. 1-5

Keeping a Food Log

A food log is one way to study what we eat each day. Looking at this information, we can decide what food choices we might want to change. In this lesson, students begin a food log, which they will use to record what they eat over the course of three days. Students will use this data in Lesson 30, so there is ample time for them to complete their food logs.

When you introduce students to the food-log activity, emphasize that they will need to be diligent about recording what they eat. Explain that food logs can teach us a lot about what we eat, when we eat it, and with whom. Tell students to make the food log as accurate as possible. Encourage them to carry the log with them at all times so they can record the data at the time they eat rather than have to rely on their memories. Tell students to be precise and to record as much detail as they can. For example, if students eat chips, tell them to record the brand name, the type of chip, and the size of the bag or serving. Use the chart below to model several examples for students. Point out that the more details they write on the food log, the more they will be able to learn about their eating habits. With this information, students will be able to decide which habits they may wish to change in the future. Be sure to remind students often to keep up with their food logs, especially several days before you plan to begin Lesson 30.

Date: 18 September	Day of the Week: Monday

Time	Foods I Ate	Notes
8:00 a.m.	Large bowl of corn flakes with 2% milk, 10 oz. glass of orange juice	At home in the kitchen with my little brother Juan.
11:30 a.m.	3 chicken fingers, 1 bite of green beans, 1 roll, 8 oz. carton of 1% milk, oatmeal cookie	At school, in the cafeteria with my class. It was really loud. I had a hard time concentrating on eating because of all the noise.
3:00 p.m.	12 oz. can of ginger ale, apple	At my friend Oscar's house after school.

Consider keeping a food log of your own eating habits. You may be surprised at what you learn!

How to Make a Mini-Book Food Log

Make a double-sided copy of the Personal Three-Day Food Log (pp. 364-365). Using the solid outer rule as a guide, cut off the page margins. Then cut the resulting page in half along the horizontal solid line. Fold the pages in half along the vertical dotted lines. Make sure the pages are in numerical order. Staple the pages together along the spine.

BECOMING FOOD SCIENTISTS : INTERACTING PARTS : FOOD PROCESSING : **ENVIRONMENTAL EFFECTS** : WASTE : MAKING CHOICES

LINKING FOOD AND THE ENVIRONMENT

Personal Three-Day Food Log

Name: _____

Use this food log to gather data about what you eat every day. For the next three days, record everything you eat in your food log. Carry it with you all the time. Fill it in as completely as possible.

PAGE 8

PAGE 1

Date: _____

Time   **Foods I Ate**

_____ _____
_____ _____
_____ _____
_____ _____
_____ _____
_____ _____
_____ _____
_____ _____
_____ _____
_____ _____

PAGE 6

Day of the Week: _____

Notes

_____
_____
_____
_____
_____
_____
_____
_____
_____
_____

PAGE 3

Date: _____

Time   **Foods I Ate**

_____   _____

_____   _____

_____   _____

_____   _____

_____   _____

_____   _____

_____   _____

_____   _____

Day of the Week: _____

Notes

_____

_____

_____

_____

_____

_____

_____

_____

_____

Date: _____

Time   **Foods I Ate**

_____   _____

_____   _____

_____   _____

_____   _____

_____   _____

_____   _____

_____   _____

_____   _____

Day of the Week: _____

Notes

_____

_____

_____

_____

_____

_____

_____

_____

_____

Effects, Effects, Effects Supply List

INVESTIGATING LANDFILLS (pp. 367–368)

Supplies

For each group of 4 students
- 2 clean two-liter beverage containers with caps, preferably clear bottles
- 1 utility knife
- Masking tape
- One 4-inch-square piece of mesh fabric (cheesecloth or stocking)
- 1 small plastic bag
- 1 rubber band
- 1 pair of plastic or garden gloves
- 2 cups of paper trash ripped or cut into small pieces (classroom waste works fine)
- 2 cups food scraps cut into small pieces (from the cafeteria or bring in from home)
- 1 1/2 cups gravel
- 1 1/2 cups garden soil (do not use sterilized potting soil)
- 1/2 cup clay or clay soil
- 1 cup water

Investigating Landfills

Students observe what happens to "garbage" in this simulated landfill. These instructions are for one model landfill. You can do this investigation as a class or have students work in groups and have each group make its own landfill.

Setup

1. Gather the materials. Remove the labels from the bottles.

2. Bottle #1. Cut the top off, as shown in the diagram. Use a utility knife to make an incision about 8 1/2" from the bottom of the bottle. Use scissors to cut around the bottle. Trim any jagged edges. Place masking tape over the edges to prevent cuts. Use the base of Bottle #1 as the base of the landfill. Place the top of Bottle #1, including the cap, off to one side.

3. Bottle #2. Use the utility knife to make an incision about 1 1/2" from the bottom, as shown in the diagram. Remove the bottle cap. Discard the cap and the bottom of Bottle #2. Recycle if your local recycling laws permit.

4. Use the cut edge of one of the bottles and trace a circle onto a piece of mesh fabric, such as cheesecloth or a stocking. Cut out the circle.

5. Use the cut edge of one of the bottles and trace a circle onto the plastic bag. Cut it out a little larger than the circle so it will completely cover the layer of clay soil.

Procedure

1. Assemble the landfill. Use a rubber band to hold the piece of mesh fabric over the neck of Bottle #2.

2. Invert Bottle #2 into the bottom of Bottle #1, as shown. Keep the top of Bottle #1 to use later, as the landfill cover.

3. Put on the gloves and show students the paper and food waste that represent the garbage. Have them record the types of waste on their *Landfill Observations* activity sheet.

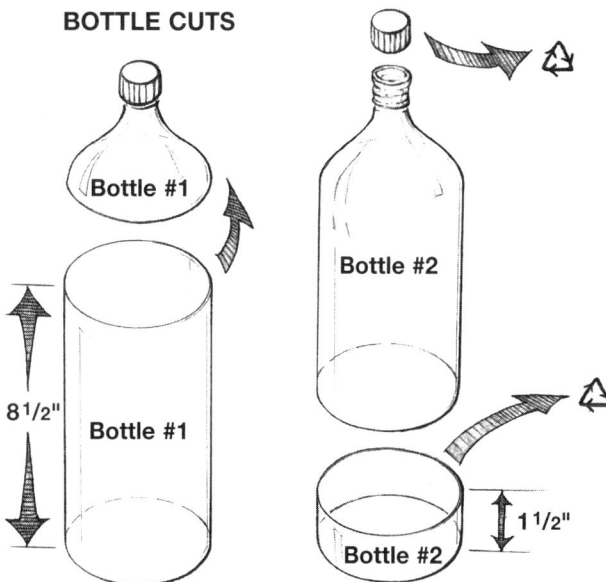

BOTTLE CUTS

Bottle #1

8 1/2" — Bottle #1

Bottle #2

Bottle #2 — 1 1/2"

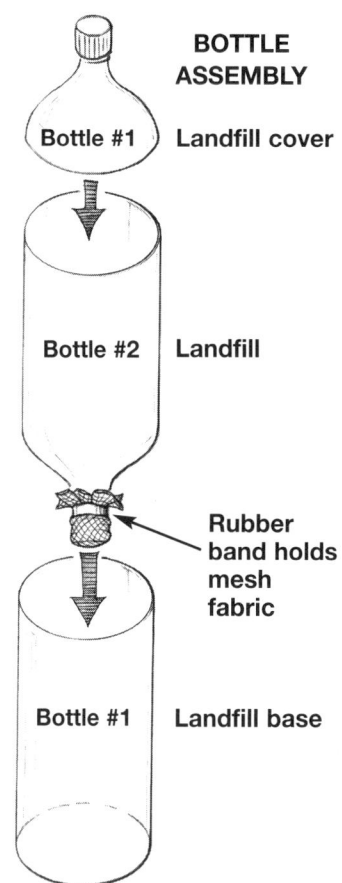

BOTTLE ASSEMBLY

Bottle #1 — Landfill cover

Bottle #2 — Landfill

Rubber band holds mesh fabric

Bottle #1 — Landfill base

4. Assemble the layers. If you run out of room, stop after you cover the first layer of garbage with soil.

a. 1 cup gravel

b. 1/2 cup soil

c. 1/2 cup clay or clay soil

d. Plastic circle

e. 1/2 cup gravel

f. 1/2 cup soil

g. 1 cup garbage

h. 1/2 cup soil

i. 1 cup garbage

j. 1/2 cup soil

5. Sprinkle 3/4 cup of water over the top layer of the landfill to simulate rain.

6. Use the top of Bottle #1 as the landfill cover.

7. Place the model landfill in a spot where it will have indirect light. Choose a location where students can observe it, but where it will not be disturbed.

Questions

1. *Why are there layers of gravel, soil, clay soil, and a plastic liner before the garbage layer?*

2. *Why did we cover the garbage with a layer of soil?*

3. *What do you predict will happen to the garbage?*

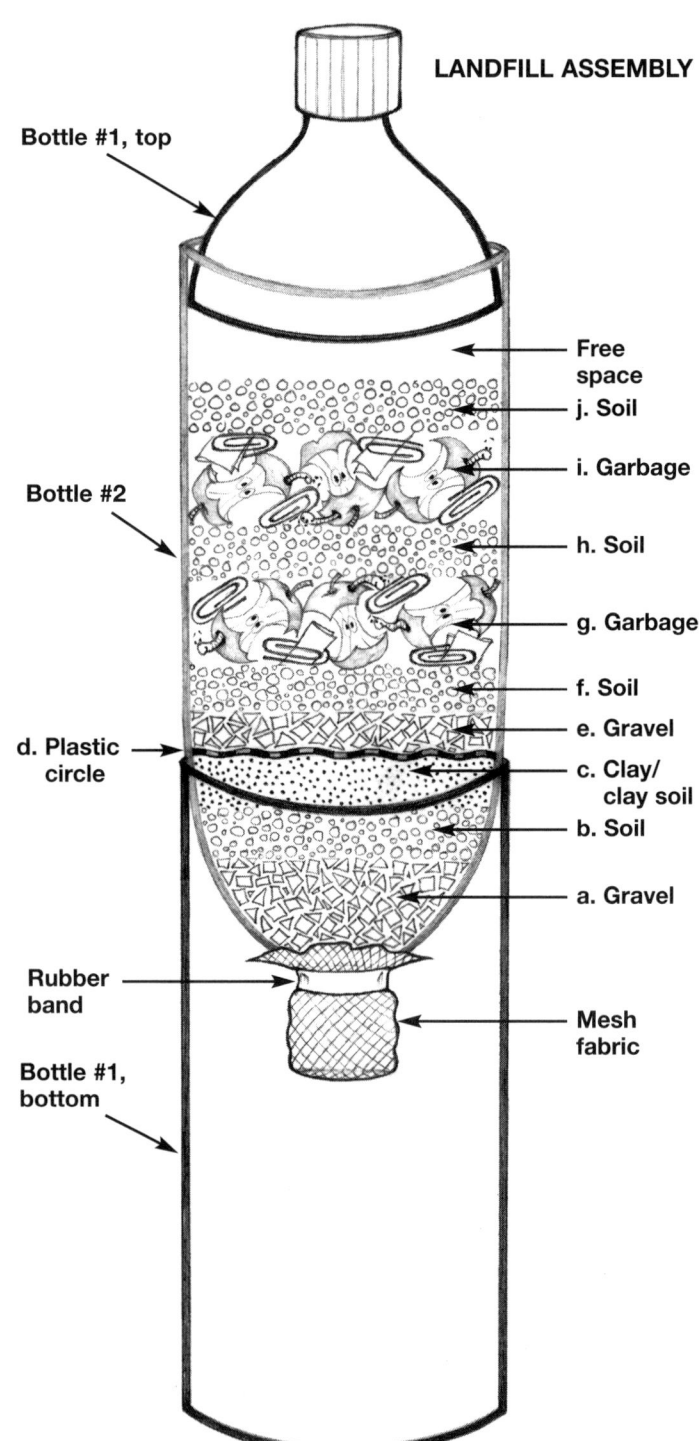

LANDFILL ASSEMBLY

Bottle #1, top

Free space

j. Soil

i. Garbage

Bottle #2

h. Soil

g. Garbage

f. Soil

e. Gravel

d. Plastic circle

c. Clay/ clay soil

b. Soil

a. Gravel

Rubber band

Mesh fabric

Bottle #1, bottom

Name	Date

Landfill Observations

In this experiment, we are investigating what happens to garbage when it is buried in a landfill. To find out, we made a model landfill. The model has layers of gravel, soil, and garbage. We also simulated rainfall by sprinkling water over the top of the landfill. Finally, we used a bottle top as the landfill cover. We buried garbage in the landfill.

Your assignment is to observe the model landfill and record your observations. First make a prediction. Then observe the landfill and record your data using pictures or words, or both. Use the chart on the next page to record how everything looks and smells. After you have completed your observations, look at your results and think about what you have learned. Then write your conclusion. Finally, compare your conclusion to your prediction.

Prediction: What do you predict will happen to the garbage that is buried in the model landfill?

_____

_____

_____

Conclusion: After you make your last observation, what conclusions can you draw?

_____

_____

_____

Look at your prediction. Compare your conclusion to your prediction. Have your ideas changed? Explain your answer.

_____

_____

_____

(continued on next page)

Name

Date

Landfill Observations

Landfill Observations			
	Observation #1 Date:_____	**Observation #2** Date:_____	**Observation #3** Date:_____
Paper Trash			
Food Scraps			
What happens to the water that was sprinkled on the landfill?			

Farm to Table & Beyond
©2008 Teachers College Columbia University

UNIT 5
Waste

Beyond the Trash Can

AIM

To begin to investigate food-related waste.

SCIENTIFIC PROCESSES

- investigate, gather data, construct knowledge, apply

OBJECTIVES

Students will be able to:

- describe different methods of waste management;

- discuss reducing, reusing, and recycling;

- gather data about their own food practices.

OVERVIEW

In this lesson, students begin to explore the Unit 5 Question, *How can we reduce the food-related waste that we produce?* In prior lessons, they investigated different ways that byproducts are generated through our global food system. Now students begin to consider different ways to reduce that waste. Students review the three R's — reducing, reusing, and recycling — and reflect on the advantages of each. Through readings and discussion, students learn about waste management in their own community. They refer to their manufacturing research projects from Lesson 7 and share what they have learned about paper, glass, plastic, and aluminum cans. Students also check the model landfill and record their observations. For homework, each student keeps a food log for one day and records all of the packaging associated with each food item.

MATERIALS

For the teacher:
- *Milestones in Garbage History* lesson resource
- *Waste Hierarchy* lesson resource
- *Waste Watchers Web-Based Resources* lesson resource
- Chart paper
- Markers
- (Optional) Large index cards or sentence strips

For the class:
- Recycling laws, guidelines, and related educational materials from your community
- (Optional) *Where Does the Garbage Go?* by Paul Showers

For each student:
- *Waste Not!* student reading
- *Time-Line Analysis* activity sheet
- *Food-Related-Waste Data* activity sheet
- *Sand to Glass, Plant to Paper, Petroleum to Plastic,* or *Ore to Aluminum Can* research project from Lesson 7
- LiFE Log

PROCEDURE

Before You Begin:

- Gather recycling laws, guidelines, and related educational materials from your community. Place these materials in a Waste Watchers reference center in the classroom. Review the *Waste Watchers Web-Based Resources* lesson resource for online information.

- Copy and review the *Milestones in Garbage History* lesson resource. Cut out the milestones and glue them onto index cards. Or copy the milestones onto sentence strips.

- Read the *Waste Hierarchy* lesson resource.

- Set up a *Waste Watchers* reference center in the classroom. Place copies of your community's waste-management laws and guidelines here, as well as any educational materials that might be available.

- Review the *Waste Not!* student reading and the *Time-Line Analysis* and *Food-Related-Waste Data* activity sheets.

- Make copies of the *Waste Not!* student reading and the *Time-Line Analysis* and *Food-Related-Waste Data* activity sheets for students.

- If you have not already done so, post the Module Question and Unit 5 Question at the front of the classroom.

- (Optional) Make an appointment for a field trip to a local recycling facility.

- (Optional) Discuss this unit's activities with your school's custodial staff and food-service staff. Invite them to meet with your students to discuss food-related waste at your school. You may wish to ask them to serve as informal advisors.

MODULE QUESTION

What is the system that gets food from farm to table, and how does this system affect the environment?

UNIT QUESTION

How can we reduce the food-related waste that we produce?

QUESTIONING

1. Review Module and Unit Questions

Remind students of the Module Question. Introduce the new unit and the new Unit Question. Explain that this first lesson is about learning the options we have for managing waste. Tell students that throughout this unit, they will collect and analyze data about the food-related waste generated at their school.

SEARCHING

2. Discuss Waste

Review with students what they already know about waste. *Remember what we discussed about garbage and where it goes? What are some of the ways that garbage is managed?* (It's recycled or composted or it goes to the landfill.) *Do you think that garbage has always been handled this way?* Use the *Milestones in Garbage History* lesson resource to guide the conversation. You may wish to read Paul Showers's *Where Does the Garbage Go?* as a class to add to the discussion. If you made cards or sentence strips with the dates, have students add them to the time line. If not, have students record dates and events directly on the time line. Once they have added these milestones, invite the class to study the time line. Ask them to think about changes in the ways people have handled waste.

Distribute the *Time-Line Analysis* activity sheet. Have students work in small groups or pairs. Point out the events related to changes in the way that food was packaged and processed. *Do you see any developments or changes that might have resulted in an increase in waste?* (Food packaging, new materials.) Tell students to use the activity sheet to record at least two events that may have caused an increase in waste, and an effect each event may have brought about. For example, after the refrigerated car, fresh produce was shipped by train from coast to coast. *What kind of waste may have resulted from this?* (Pollution from burning fuel for transportation and cooling, containers manufactured for shipping, paper used to make labels.) Invite student pairs to share their ideas with the class.

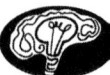 **THEORIZING**

3. Compare Waste-Management Methods

Distribute the *Waste Not!* student reading to students and allow them time to read it in class. Engage them in a brief discussion. *What are some ways that we manage waste? What do we mean by the three R's?* (Reducing, reusing, and recycling.) Tell students that they are going to be learning more about waste management.

Review the *Waste Hierarchy* lesson resource and use it to guide the class discussion. Draw or reproduce the diagram on the board. Label it as shown. Engage students in a discussion. *Why is "reduce" the most preferred method?* (If we reduce the amount of waste produced, there is less to dispose of. It means using fewer raw materials and less energy during the manufacturing of packaging or in transporting products.) *Based on what you have learned about landfills, why do you think they are the least preferred method of managing waste?* (The waste remains. It's just buried or out of sight. The landfill may leak.) *Why does "reuse" follow "reduce" as an option?* (To reuse waste, it has to exist in the first place.)

4. Discuss Reducing

Ask a student volunteer to discuss what "reduce" means in terms of waste management. Prompt students to discuss reduction both in terms of raw materials and energy used and in terms of the number of products that consumers purchase. Explain that as consumers, we have the power to refuse to purchase products if they do not meet our standards. For example, if there is excessive packaging, consumers can write letters of complaint and refuse to buy the product. We can refuse to purchase food that is transported long distances. This approach reduces waste at the source. *Can you think of any products you or your family might not buy so you can reduce waste? What products might you buy in their place?* (Buy products in bulk; buy from a local farmer.)

5. Discuss Reusing

Elicit ideas about the meaning of "reuse." *Does anyone have any ideas about what it means to reuse something like a food package?* Students may have several different ideas. Explain that reuse is when we return reusable packages to the food manufacturer to be cleaned and reused to package more food. *Any ideas about what kinds of food packages are considered reusable? Think back to our lessons on food packaging and transportation. Has anyone ever taken a bottle or crate back to the manufacturer or farmer for reusing? What are some of the effects of reusing food packages?* (It reduces the amount of raw materials that have to be used to make new containers. The containers have to be sterilized before being used again, and that process takes lots of heat.)

6. Discuss Recycling

Ask students to describe recycling. *When you recycle waste, what happens to it?* (It is turned into a valuable resource. It can be manufactured into the same product it was before — recycling glass to make a glass bottle. Or it can be made into a new product, like turning recycled plastic into plastic lumber.)

Ask students to share their ideas about composting. *Has anyone ever had a compost bin? What do you know about composting? Why do you think composting is considered recycling?* (Kitchen waste is turned into another product.) If your school has a compost bin, use it to give students an idea of what composting entails. Explain that composting is nature's system of recycling. When we compost, we are mimicking nature. It helps us reduce waste and produce something useful.

7. Review Local Guidelines

Have students work in small groups or as a whole class to briefly review these resources. Have them look through the local information you have collected. Invite students to share information about waste management that they did not know before. If possible, have your school custodian discuss with the class where the items that are recycled in the school go after they get picked up.

8. LiFE Logs

Ask students to choose one of the waste-reduction methods (reducing, reusing, recycling) and write a paragraph that describes how this method decreases food-related waste. Tell them to discuss the pros and cons of using the method they've selected. If time permits, invite a few students to read what they wrote to the entire class. Use this as a way to review how the methods reduce waste and to clarify any misconceptions.

9. Homework

Have students record everything they eat for one day on their ***Food-Related-Waste Data*** activity sheet. Point out that this is separate from their three-day food log assignment. Remind them to bring the log with them everywhere they go so they won't forget anything

that they eat. Explain that the data they collect in this food log will be used in the next lesson. It is very important for them to complete the log so the class can use their data. Explain that they do not need to complete the "Factory Points," "Transportation Points," "My Data," or "My Group's Data" portions of the activity sheet. These will be completed during the next lesson.

Milestones in Garbage History

500 B.C.	Early records show that one of the first city dumps was in Athens, Greece. By law, the dump had to be located at least 1 mile from the city limits.
1354 A.D.	In London, "rakers" rake trash, load it in carts, and remove it once a week
1388 A.D.	In England, it is illegal to dump waste into public waterways and ditches
1690 A.D.	At Rittenhouse Mill in Philadelphia, Pennsylvania, paper is made from recycled fibers (waste paper and rags)
1757 A.D.	Instead of throwing garbage out windows and doors, American households dig refuse pits
1834 A.D.	In Charleston, West Virginia, a law protects vultures from hunters. The birds help eat the city's garbage.
1842 A.D.	In England, a report makes the connection between filth and disease
1872 A.D.	New York City stops dumping its garbage from a platform built out over the East River
1890 A.D.	In England, the British Paper Company is established to make paper and board from recycled materials
1896 A.D.	New York City requires residents to separate household waste. Food waste is placed in one container, ash in another, and dry trash is placed in a bag or bundle.
1900 A.D.	In the Unites States, pigs are used to help get rid of garbage. Towns build "piggeries," where the animals are fed fresh or cooked garbage. It is estimated that 75 pigs can eat one ton of garbage per day.
1904 A.D.	The nation's first aluminum-recycling plants open in Chicago, Illinois, and Cleveland, Ohio
1908 A.D.	Paper cups replace tin cups at water-vending machines on trains and in public buildings
1916 A.D.	Waxed paper is commonly used to wrap bread
1920s A.D.	Landfills become a popular way to reclaim swampland and get rid of trash. Cities begin to fill in wetlands with garbage, ash, and dirt as a way to dispose of waste.
1934 A.D.	The United States Supreme Court bans dumping waste into the ocean
1960 A.D.	Americans generate 88.1 million tons of municipal solid waste

1960s A.D.	Store-bought bread is sold wrapped in plastic bags instead of waxed paper
1965 A.D.	The first federal solid-waste-management laws are enacted
1970 A.D.	The first Earth Day is celebrated. The U.S. Environmental Protection Agency (EPA) is created.
1984 A.D.	During the Olympic Games in Los Angeles, athletes, trainers, coaches, and spectators generate 6.5 million pounds of trash in 22 days, more than six pounds per person per day.
1986 A.D.	Rhode Island enacts the nation's first statewide mandatory-recycling law
1986 A.D.	Fresh Kills, in Staten Island, New York, becomes the largest landfill in the world
1997 A.D.	Americans generate about 22 million tons of food waste
2006 A.D.	Americans generate 251 million tons of municipal solid waste

Waste Hierarchy

Before you begin discussing solid-waste management, check for student understanding of these commonly used terms: "solid waste," "waste management," "reduce," "reuse," "recycle," and "landfill" or "dump." In your discussions with students, you may find that some children assume that garbagemen do the recycling after they pick up the garbage.

Draw or project this diagram on the board. Use it to guide the class discussion about different methods of waste management.

The Waste Hierarchy

The Most Preferred Environmental Option

REDUCE

What this means is buy less and use less. Reduce the amount of raw materials and energy being used to make products. Buy what you need and then use it or pass it on to someone else who can use it. This method has the smallest impact on the environment. Reducing also means to buy wisely when it comes to packaging. You can reduce waste by only buying products packaged with the minimum amount of packaging needed to keep a product clean and damage free. Avoid single-serving containers — buy in bulk. Avoid packaging that exists only to sell the product. Bring your own bag to the grocery store. Compost at home to reduce the food waste that you send to the landfill. This is the most preferred form of waste management because it prevents waste being generated in the first place.

REUSE

In this method of waste management, you reuse materials in their original form. You give unwanted toys or clothes to a charity. You return an empty egg carton to the farmer to have it refilled with eggs. You use washable cloth napkins and if you use disposable plastic utensils, you wash them and use them again. Reusing a product is preferable to recycling it. When you reuse a product, it does not have to be broken down and manufactured again before it can be used again.

RECYCLE

To recycle, you take a product to a center where it is made into the same kind of product or a new product. For example, paper, aluminum cans, glass bottles, and plastic bottles can all be recycled. They are collected, separated, and sent to places where they can be processed into new materials or products. Plastic bottles can become jackets or even playground lumber. Paper can be recycled and used to make paper again. This method helps cut down on the energy and the amount of raw materials needed to make new paper, cans, and bottles. Composting is also recycling. Composting takes kitchen and garden scraps and turns them into a new product — soil amendment. Compostable items include eggshells, newspapers, raw fruit and vegetable scraps, coffee grounds, and tea bags.

DISPOSE

This is the least preferred option for waste management. If you can't reduce, reuse, or recycle something, then you need to dispose of it and send it to the landfill. This method has the greatest impact on the environment.

The Least Preferred Environmental Option

Waste Watchers Web-Based Resources

- **Earth 911** — *http://earth911.org*

  Earth 911 looks at environmental conservation from every angle: recycling, consumer choices, water, air quality, composting, household items, and climate change. Feature articles give readers practical tips for a green lifestyle, and a student section lists national contests and other readings.

- **Environmental Literacy Council** — *www.enviroliteracy.org/subcategory.php/41.html*

  This Environmental Literacy Council page discusses effects of waste on the natural environment. Read here about how our waste affects the air, the land, the water, and Earth's ecosystems, and research how our energy use and food are linked to the environment.

- **EPA's Regional Web Sites and Waste Programs** — *www.epa.gov/epaoswer/osw/regions.htm*

  Investigate the differences in waste processing by region of the United States. Click on your state and region on the interactive map to learn more.

- **EPA's Wastes Web Page** — *www.epa.gov/epaoswer/osw/index.htm*

  What are the effects of wood waste versus plastic waste? How can we recycle our cell phones? Find out on this site's extensive list of waste materials with links to other resources.

- **EPA's Student Center: Waste and Recycling** — *www.epa.gov/students/waste.htm*

  This student section of the EPA Web site links to games, informational sites, and programs working to keep waste out of our natural environment.

- **National Institute of Environmental Health Sciences (NIEHS)**
  National Institutes of Health (NIH)
  Department of Health and Human Services (DHHS)
  — *http://kids.niehs.nih.gov/recycle.htm*

  This site from the NIEHS educates students about practical steps they can take to reduce, reuse, and recycle. Through this site, link to other schools' recycling initiatives and find classroom activities and ideas.

- **National Recycling Coalition** — *www.nrc-recycle.org*

  Use the site's recycling calculator to measure the impact you have on the environment, watch computer graphics that teach about the effects of recycling, and learn more about recycling initiatives nationwide.

- **Rotten Truth (About Garbage)** — *www.astc.org/exhibitions/rotten/rthome.htm*

  Read here for straight talk on garbage: how we create it, where it goes, and what our alternatives may be. An exhibition informs viewers of the "rotten truth," and discusses actions others have taken. Look for a resource section on waste and practical suggestions for taking action in your community.

- **National Recycling Coalition** — *www.nrc-recycle.org/localresources.aspx*

  What is your state doing to promote recycling efforts? Find out at this site, where you can click on your state on the map and learn more.

- **EPA's Waste Prevention and Recycling Clip Art and Other Visual Resources** — *www.epa.gov/epaoswer/education/photos.htm*

  This clip-art section of the EPA Web site is a rich source of illustrations and photos that might be of interest when students are making visual presentations.

Name   Date

Waste Not!

When you hear the word "waste," what image comes to mind? Do you think of a trash can? Maybe you think of a recycling bin or a compost bin. If you're thinking like this, you're thinking about ways to manage waste. But before we talk about ways to manage waste, let's think about what the word "waste" means.

"Waste" means things that can't be used or reused. According to the U.S. Environmental Protection Agency (EPA), Americans generate millions of tons of waste — garbage — each year. In fact, the EPA reports that the average American generates 4.5 pounds of trash every day. What do we do with all of that waste?

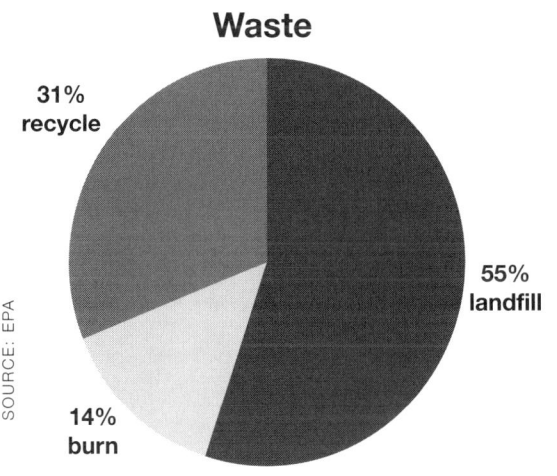

Waste

- 31% recycle
- 55% landfill
- 14% burn

SOURCE: EPA

Look at the pie chart on the left. We recycle some of it, burn some of it, and send the rest to landfills.

You may already know the three R's: reduce, reuse, and recycle. Which one do you think is the most preferred way to manage waste and the most beneficial for the environment? If you said "reduce," you were right. One of the most effective ways to cut down on the amount of waste is to prevent it from the beginning. Reducing means to consume and throw away less.

Ways to Reduce

 What are some ways to reduce?

 Buy goods that will last a long time.

 Look for products that use fewer raw materials.

Look for products that have materials that can be used again after their original use.

(continued on next page)

Name   Date

Waste Not!

Ways to Reuse

Reusing items is the next best way. How can you reuse items?

 Repair items instead of throwing them away.

 Donate items to a charity or give them to someone else.

 Collect empty glass bottles and return them to the store.

Ways to Recycle

Recycling means you turn waste into a valuable resource. Did you know that composting is a kind of recycling? How do you recycle waste?

 Return aluminum cans to recycling centers.

 Recycle your unwanted CDs. Put them in a box labeled "Recycle CDs Here." When the box is full, send them to a CD recycling center.

 Compost kitchen waste.

Now think about the three R's and what you have learned about the food system. What actions can you take that will help manage waste?

Becoming Food Scientists : Interacting Parts : Food Processing : Environmental Effects : **WASTE** : Making Choices

Name   Date

Time-Line Analysis

Look at the food-systems-history time line. Find at least two events that you think are related. For example, you may see that plastic is invented and then notice that bread begins to be sold in plastic bags. The plastic bags became part of the waste that people threw away. Use this activity sheet to organize your thoughts.

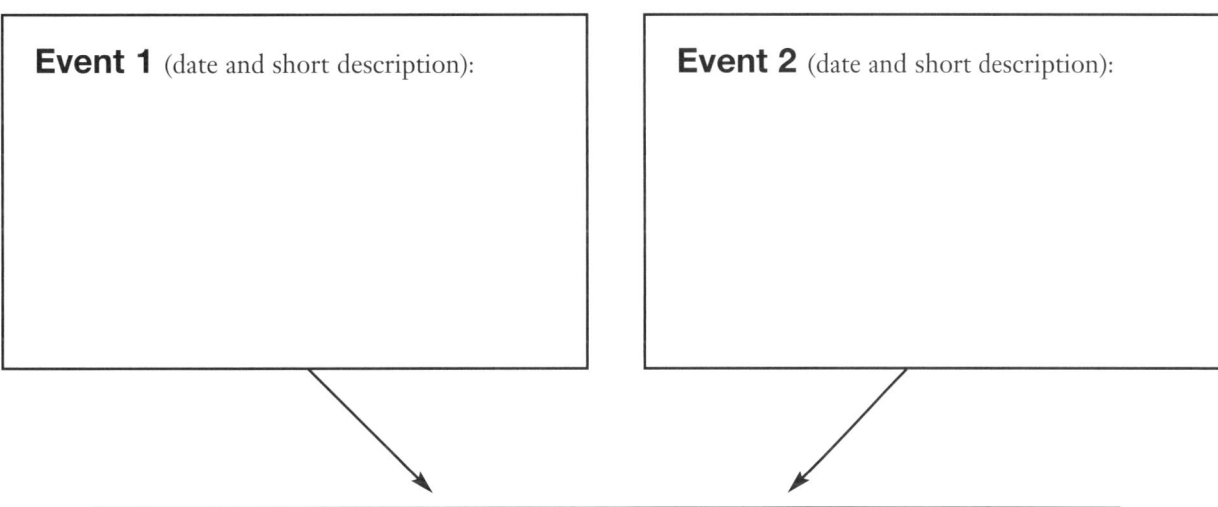

Event 1 (date and short description):

Event 2 (date and short description):

How the two events are related:

BECOMING FOOD SCIENTISTS : INTERACTING PARTS : FOOD PROCESSING : ENVIRONMENTAL EFFECTS : **WASTE** : MAKING CHOICES

Name	Date

Food-Related-Waste Data

Your assignment is to collect data about your own lifestyle. For one day, you will record everything you eat. In the "Food I Ate" column, record the name of the food (for example, "slice of pizza"). In the "Packaging" column, write down all of the packaging that was used to wrap the food. For example, the slice of pizza may have been placed on a paper plate, wrapped in foil, and placed inside a paper bag. Leave the other two columns blank for today. We will come back to them and the other tables on the next page in the next lesson. It is important to keep an accurate record of this data.

Food-Related-Waste Log

Food I Ate	Packaging	Factory Points	Transportation Points
Example: Slice of pizza	Paper plate, foil, paper bag, paper napkin		

(continued on next page)

Farm to Table & Beyond
©2008 Teachers College Columbia University

Name	Date

Food-Related-Waste Data

My Data

Metal	Plastic	Glass	Paper	Factory Points	Transportation Points

My Group's Data

In the columns below, record the names of each group member. Use one column for each group member.

Packaging Material	Name:	Name:	Name:	Name:	Name:	Group Totals
Metal						
Plastic						
Glass						
Paper						
Factory Points						
Transportation Points						

Waste Inventory

AIM

To analyze the amount of waste individuals generate and to develop a method for surveying school-cafeteria waste.

SCIENTIFIC PROCESSES

- observe, analyze, predict, develop research questions

OBJECTIVES

Students will be able to:

- use data to tally the amount of food-related waste the class generates;

- predict how much food-related waste they might generate over time;

- discuss actions they can take to reduce food-generated waste;

- develop a research question and method for collecting data about school-cafeteria waste.

OVERVIEW

In this lesson, students explore different ways of analyzing waste. As a class, they read about research scientists studying waste. Then students analyze their personal food-waste data, which they collected as homework in Lesson 25. They tally their individual results, then they work in small groups to compile the group's data. This lesson's examination of personal data is a practical way for students to grasp the amount of packaging that they use. By looking at the class data, students gain an understanding of their peers' food practices. This provides students with an opportunity to compare and contrast their own food practices with those of their classmates. Estimating how much waste the class might generate over a week, a month, and a year further illustrates the magnitude of the problem of food-related waste. Finally, students apply what they have learned to develop a class research question and methodology for conducting an observational study to determine how much waste is generated in their school cafeteria.

MATERIALS

For the teacher:
- *Displaying Food-Related-Waste Data* lesson resource
- *Investigating-Waste Project* lesson resource
- Chart paper
- Markers

For each student:
- *Studying Waste* student reading
- *Food-Waste-Data Analysis* student reading
- *Food-Related-Waste Data* activity sheet (pp. 384–385)
- *Developing Questions about Waste* activity sheet
- *Data Collection and Analysis* activity sheet
- *Landfill Observations* activity sheet (p. 370)
- LiFE Log
- (Optional) 4 colored pencils, each a different color

PROCEDURE

Before You Begin:

- Remind students to bring in their homework from Lesson 25.

- Review the *Displaying Food-Related-Waste Data* and *Investigating-Waste Project* lesson resources.

- Review the *Studying Waste* and *Food-Waste-Data Analysis* student readings and the *Developing Questions about Waste* and *Data Collection and Analysis* activity sheets. Make copies for each student.

- If you have not already done so, post the Module Question and Unit 5 Question at the front of the classroom.

MODULE QUESTION

What is the system that gets food from farm to table, and how does this system affect the environment?

UNIT QUESTION

How can we reduce the food-related waste that we produce?

 QUESTIONING

1. Review Module and Unit Questions

Remind students of the Module and Unit questions. Explain that in this lesson they will analyze data and estimate how much food-related waste they produce, both as individuals and as a class. Announce that students will then use this knowledge to help them develop a study to learn more about the waste generated in their school cafeteria. If necessary, remind students to complete their three-day food logs from Lesson 24.

 SEARCHING

2. Study Waste

Distribute the *Studying Waste* student reading to each student. This reading illustrates some ways that scientists study waste. Have students take turns reading aloud. Remind them that they will be researchers studying waste, too. Encourage them to pay particular attention to the methods the scientists use. Stress that there are different ways of looking at the same problem. Students can refer back to this reading as they work on their own data display.

Give students a few minutes to review the reading. *What are some ways to analyze waste?* (By weight, by use, by material.) *Do people overestimate or underestimate the amount of food they waste? Do you think students at our school will be accurate in their estimation of how much food they waste in the cafeteria? How can we find out?*

Before moving to the next task, have students take turns making and recording their observations of the landfill.

 EXPERIMENTING

3. Analyze Food-Related-Waste Data

Have students take out their homework. If you are using colored pencils, distribute them to students. If students have not completed the *Food-Related-Waste Data* activity sheet, give them a few minutes to try to remember everything they ate the day before. Point out that this data will not be as accurate as if students had completed the log at the time they ate the food. *Why won't the data be as accurate if you record it from memory? Why is it important for data to be accurate?*

Go over the ***Food-Waste-Data Analysis*** student reading with the class. Invite one student to share a log entry with the class. Record it on chart paper. Make sure it has each of the different categories of packaging listed. If not, add some items so each category is represented. Next, make a key. Use a different colored marker to represent each of the packaging categories: metal, plastic, glass, and paper. Tell students to make a key on their log. Have them write "Metal" and then choose a color to represent metal. Next, write "Plastic" and choose a color to represent plastic, and so forth. Once students have completed their keys, demonstrate how to use the different colors to help count and categorize the packaging material. Next, have students complete the "Factory Points" and "Transportation Points" sections of their activity sheets.

When they have finished, divide them into small groups. Have each group calculate totals for metal, plastic, glass, and paper, using the group members' data. Then, have each group use the factory and transportation points from each group member's My Data table to calculate the total factory and transportation points for their group. As students work with these data, encourage them to compare and contrast their individual results.

Draw a table on the board, using the ***Displaying Food-Related-Waste Data*** lesson resource as a guide. Invite a volunteer from each group to record the group's data in the appropriate column. Analyze the results with the class. *Which packaging material did our class use most? Now compare it to your personal data — which material did you use most? As a class, how did we do with transportation points? What factors might influence this number?* (We live in a city and the food comes from farms outside the city. It's winter and the farmers' market is closed.) *Do you see a relationship between the amount of snack food, like cookies and chips, that you ate and the amount of packaging?*

After students have discussed the data, ask them to think of ways that they can display the data. Draw the line plots shown on the ***Displaying Food-Related-Waste Data*** lesson resource on the board. Use the information on the lesson resource to demonstrate to students how the class can input their data on a line plot. Have students use their own data to fill in the line plots. Discuss these results. Encourage students to ask each other questions as they compare and contrast their own data with their classmates'. If you have time, extend the discussion, using the sample graph as another way to analyze the data.

4. Estimate Long-Term Waste

Tell students that they are going use their class data to estimate how much food-related waste the class might generate in a week, a month, and a year. *Any ideas how we can use the data we've just tallied to estimate how much waste we might generate over a week, a month, or a year?* If necessary, guide students to the understanding that they should multiply their totals by 7 (for a week), 30 (for a month), and 365 (for a year) to get an idea of how much food-related waste they would generate over time.

Briefly discuss students' thoughts about the data they analyzed from their ***Food-Related-Waste Data*** activity sheets. *Do you think you created a small amount of waste, a medium amount, or a large amount? Was it more or less than you might have imagined? What surprised you the most?* Remind students that they will conduct a research project about cafeteria waste. The next task for the class is to develop a research question.

5. Develop Research Question

Explain to students that they are going to design an investigation to collect data on waste generated by their school food system. Refer to the ***Investigating-Waste Project*** lesson resource for this discussion. Remind students that the first step is to develop a research question.

Explain that the purpose of the question is to help them learn more about the waste generated by the school food system.

Have students complete the *Developing Questions about Waste* activity sheet. Then, engage them in a whole-class discussion. *What would you like to know about cafeteria food waste? What types of waste do we throw away in the cafeteria? Do younger children throw away different waste than older children? What will we measure? What data will we record? How long will we need to complete our investigation?* Invite students to share their ideas. Record all possible questions on the board. Have the class agree upon one research question. Make sure that the question states clearly the time span and scope of the project so that students are clear about what information they are gathering. Help students understand that, in order for them to avoid frustration, their questions must be answerable given the resources they have available.

Students will learn a great deal by studying how much trash is generated and how much food is thrown away. Remind students what they learned about the three R's. *How can we reduce the amount of food we throw away each day?* (Not take as much.) *How can we recycle food waste?* (Make compost.) *What natural resources are wasted when food is thrown away?* (Wasted food represents the resources that were used to grow the food, preserve it, process it, package it, and transport it.) Make sure students are aware of the result of wasting food.

6. Develop Questionnaire

Once the class has selected its research question, help students design a questionnaire or form they can use to collect data. Explain that they may use a chart like the one used to collect data on their own food-related waste. They may also create a check sheet that they can use to observe other classes during other lunch periods. Refer to the *Investigating-Waste Project* lesson resource for samples. Distribute the *Data Collection and Analysis* activity sheet and

review it with students through "Recording Data." Tell students they will use this activity sheet again in the next lesson when they organize and analyze the data they collect.

Students may wish to interview the food-service staff if their question involves how far the foods are transported or to what extent they are processed. Be sure the data-collection method your students use is appropriate to answer their research question. Encourage students to be creative. The purpose of this activity is for your class to learn more about the waste generated through the school lunch program so they can come up with a plan or campaign to reduce cafeteria waste. Once students have developed their data-collection tools, have them develop hypotheses about what the answer to their research question will be.

APPLYING TO LIFE

7. LiFE Logs

Have students write at least two paragraphs to respond to one of the following statements in their LiFE Logs: (1) "I have changed my thinking about food-related waste because…" (2) "I have not changed my thinking about food-related waste because…"

8. Homework

Remind students to complete their three-day food log, if they have not already done so.

Displaying Food-Related-Waste Data

This chart provides information about group totals the class gathers from its *Food-Related-Waste Data* activity sheets. It doesn't tell you about individual data, whether members of the group live in homes with gardens and grow their own food, have access to locally grown food, and so forth. It tells you what food class members bought and ate, but not why.

	Group A	**Group B**	**Group C**	**Group D**	**Total**
Metal					
Plastic					
Glass					
Paper					
Factory Points					
Transportation Points					

Another way to display the information students have collected is to use a line plot to quickly organize the numerical data. A line plot shows the range of the data — from the lowest to the highest value. Draw two line plots on the board, like the ones below.

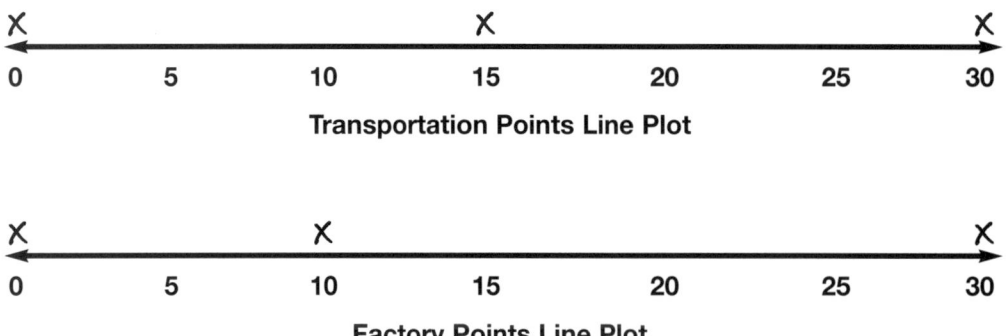

Transportation Points Line Plot

Factory Points Line Plot

Put an X at "0" on both line plots. Explain that the marks represent Anna, a person who grows all of her own food. She has zero transportation points. Anna prepares almost all of her food at home, so she has very few factory points. Place an X at 15 on the transportation line. This represents Evan, someone who eats food that comes from local farms. Place an X at 10 on the factory line. Evan also eats mostly unprocessed foods or minimally processed foods. Next, place an X on each line plot at 30. These marks represent Nick, who eats highly processed food. Remove these Xs and put the class data on the line plots.

If you place the data into a graph and compare each student's factory points with their transportation points, what do you see? You may find students have more questions. You may find that some students have many transportation points compared to their factory points. Challenge students to describe what that suggests. Students may find that they want to go back to collect more data to try to answer these questions.

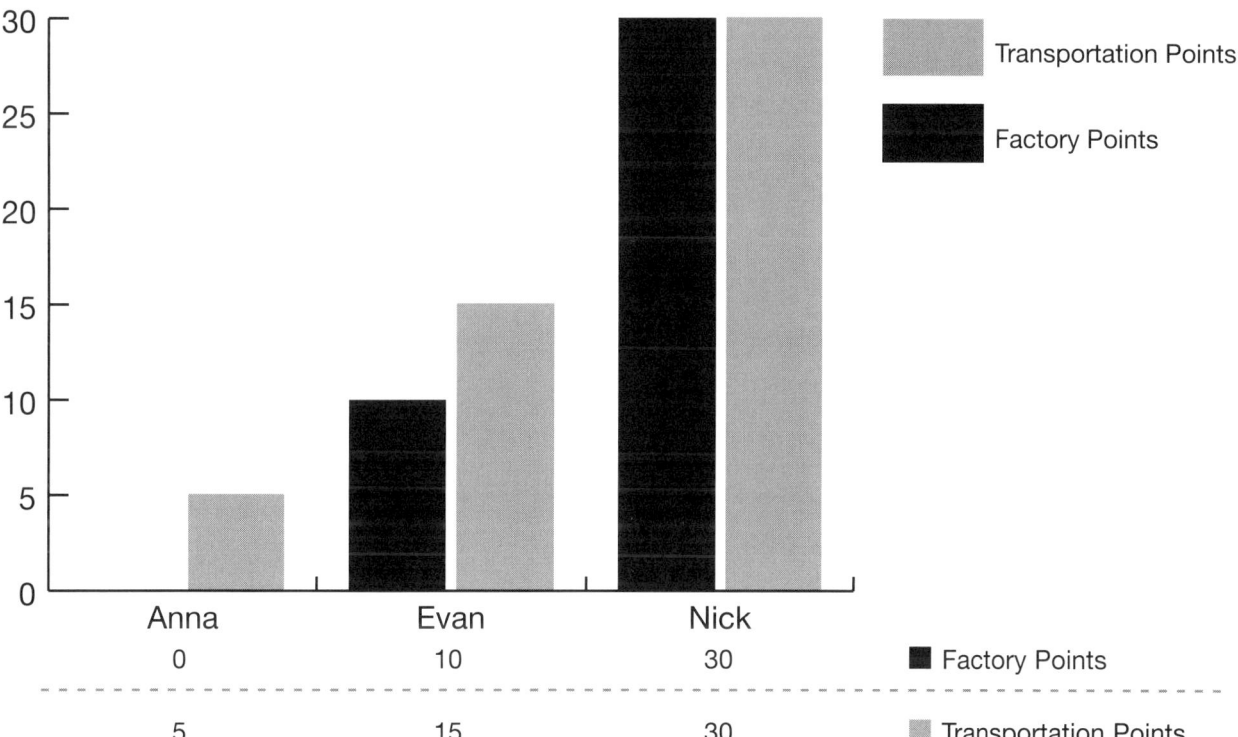

Investigating-Waste Project

With this project, students study human behavior by forming a research question, creating a questionnaire to gather data, and collecting data through observation. You may need to help your class think of a testable question. Have students review their notes. Prompt them to think about what they have observed and the questions they still have. Use the sample questions below to stimulate student thinking.

Sample Questions

1. *How much of our cafeteria waste comes from paper products like cardboard trays, plates, milk cartons, and other food packages?*

2. *How much solid waste is thrown away each day in our cafeteria?*

3. *How many pounds of food are thrown away each day in our cafeteria?*

4. *Who throws away more food in our cafeteria, the younger students (Grades K–3) or the older students (Grades 4–6)?*

5. *How much of the food we are served in the cafeteria comes from local farms?*

6. *Do classes who have studied* Farm to Table & Beyond *generate more or less waste than classes who have not studied* Farm to Table & Beyond?

7. *How much of the food we eat in the cafeteria is processed in a factory?*

8. *How much of the food waste that is generated each day could be composted?*

9. *How many of the recyclable materials are put in the recycling bin? How many are put into the garbage instead of being recycled?*

After the class has its testable question, remind students that the next step is their investigation design. Have them record where they will conduct their study. Then, discuss with them how they will collect the data. Will they stand in the cafeteria and record what they observe? Or will they observe indirectly by looking through the trash cans and recycling bins? Next, ask students to think about what they can measure that will help them answer their question. Encourage students to discuss the methods they will use to collect their data. Use these examples to guide your work.

Example #1

Question: How much of our cafeteria waste comes from paper products?

Method: We will make observations in the cafeteria. We will stand near the trash bins and watch what students throw away. We will record what we observe on a chart.

Date_____ Lunch period observed_____

Observer (your name) _____

Give each student a number. Put one check in each box to represent the material that was thrown away. If the person threw away more than one of each item, put one check in for each item thrown away. Count how many "other paper materials" they throw away and put a check for each item in the "Other Paper Materials" box.

Student #	Tray	Plate	Bowl	Milk Carton	Other Paper Materials
1					
2					
3					
4					
5					
6					
7					
8					
9					
10					
etc.					

Example #2

Question: How many pounds of food are thrown away each day in our cafeteria?

Method: We will weigh an empty garbage can or pail and record the weight. At the end of a lunch period, when students have thrown away their food waste, we will weigh the garbage can again. Then we will subtract the weight of the empty garbage can. The remaining amount will be the amount of food waste that was thrown away.

Date_____ **Lunch period observed**_____

Observer (your name)_____

Describe the container that you will use to collect the food waste. For example, empty plastic 10-gallon garbage can or 2-gallon plastic pail.

_____

How much does the garbage can or pail weigh when it is empty?_____

How much does the garbage can or pail weigh with the food waste in it? _____

Subtract the weight of the empty garbage can or pail.

What is the remaining weight?_____

How much food waste was thrown out during your investigation? _____

$$\underline{\hspace{4cm}} \quad - \quad \underline{\hspace{4cm}} \quad = \quad \underline{\hspace{5cm}}$$

| Weight of garbage can or pail with food waste | Weight of empty garbage can or pail | Weight of food waste |

Example #3

Observation: At our school, we recycle paper, plastic, cans, glass. (If there are other items that your school recycles, add them to the list. Delete any item your school doesn't recycle.)

Question: How many of the recyclable materials are put in the recycling bin? How many are put into the garbage instead of being recycled?

Method: We will stand by the waste and recycling containers in the cafeteria. We will assign each student a number. Then we will count the number of recyclable items each student puts into the solid-waste container and the number of recyclable items each student puts into the recycling bin.

Date_____ **Lunch period observed**_____

Observer (your name)_____

Put a check in the "trash" column for each recyclable item that the student throws away. Then, put a check in the "recycling bin" column for each item the student recycles.

Student #	Trash	Recycling Bin
1		
2		
3		
4		
5		
6		
7		
8		
9		
10		
etc.		

Name   Date

Studying Waste

Imagine going to work every day and studying what people throw away. Why would anyone want to do that? It's garbage. It's yucky. It's stuff that people don't want anymore. While it is true that waste is what we throw away and it can get pretty messy, to a scientist, all that stuff is loaded with information. They can learn a lot from garbage, and so can you!

Imagine you are an archaeologist, but instead of studying ancient cultures, you are digging through modern garbage. You are gathering data about all the food that is wasted between the time it is grown on a farm until it leaves your plate. Researchers will use this data to estimate the value, in dollars, of the food that is wasted. They will also analyze the data to learn more about the environmental effect of food waste.

Dr. Tim Jones and students at the University of Arizona didn't have to imagine…they were some of the scientists who collected this food-waste data. They analyzed the garbage of 200 families. They interviewed people, looked at grocery-store receipts, and sorted and weighed food waste. They wanted to find out how much edible and nonedible food is thrown out each day.

Here's what they found. On average, an American family throws out 1.28 pounds of food a day. This is food that goes into the trash can. It isn't composted, it isn't put in a garbage disposal, and it isn't fed to the family pet. It's simply thrown away. That's about 470 pounds per family per year. Dr. Jones estimates that each year a family of four throws away close to $590 worth of fruit, vegetables, meat, and grains.

Dr. Jones compared his data to data from the 1980s to see if people throw away more or less than they did about 20 years ago. He discovered that people throw away about three times as much food today as they did in the 1980s.

The research team also studied the amount of crops lost at farms. They looked at farms that grow citrus (oranges, lemons), apples, and fresh vegetables. Here they found that one reason crops were lost was due to natural causes, like weather conditions. Sometimes the crops are left in the field and plowed into the ground. Sometimes portions of crops are wasted in processing. There is a lot of waste when products like prepared salad mixes, presliced carrots, and broccoli flowers are prepared and packaged. Some of the parts of the plant are cut off, not used, and thrown away.

What does all of this mean? According to Dr. Jones's study, Americans waste more than

(continued on next page)

Name   Date

Studying Waste

$43 billion worth of food each year. That's a lot of waste, but it doesn't stop with food. In 2006, Americans generated about 251 million tons of solid waste.

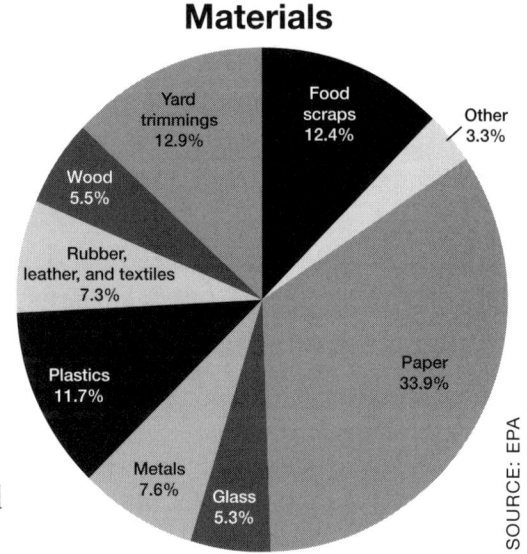

Materials

SOURCE: EPA

Researchers at the U.S. Environmental Protection Agency (EPA) analyze municipal solid waste, or garbage, that comes from homes, businesses, and schools. One method they use is to study it according to the kind of material it is made of, such as paper, food scraps, plastics, metals, leather, textiles, wood, and glass. Researchers found that about 34 percent of the solid waste is paper and about 12 percent is plastic.

Another method used to study solid waste is to put it into categories according to product. Look at the pie chart. What are some of the categories? **Nondurable goods** are products not meant to last, like newspapers, paper plates, and magazines. **Durable goods** are products meant to last a long time, like refrigerators, cars, and furniture. According to the 2006 data, about 32 percent of the solid waste was containers and packaging and 25.5 percent was nondurable goods.

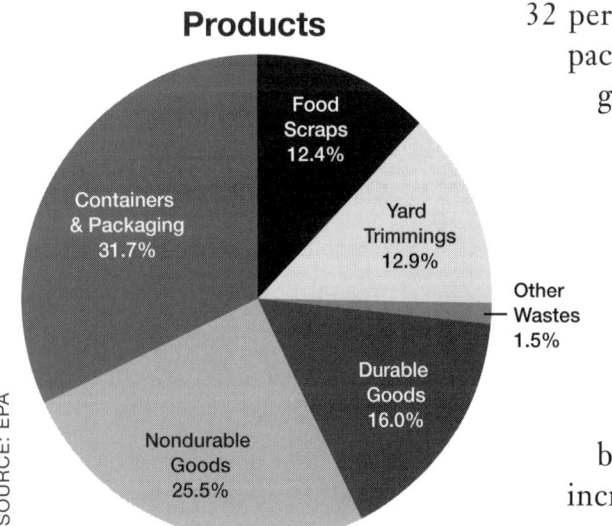

Products

SOURCE: EPA

EPA researchers have been studying waste for about 30 years. They have found that the waste produced in the United States has more than doubled. In 1960, Americans produced 88 million tons of waste. By 2003, that number was about 236 million tons. The population has increased, but that's not the whole reason for the increase. Our lifestyle has changed. People buy more disposables and convenience items, like fast foods, and single-serving food items. More packaging and more nondurable products means more garbage. What can we do to change this? Researchers say it's not hard. We can plan menus, shop wisely, compost, and recycle. What do you think?

Name	Date

Food-Waste-Data Analysis

Use this information to guide you as you complete the Food-Related-Waste Log and the My Data table.

Packaging

Look at the "Packaging" column on the Food-Related-Waste Log. How many items made of metal do you see? Record the total number of metal items in the "Metal" column of the My Data table. How many plastic items do you see? Record that number in the "Plastic" column. How many glass items do you see? Record that number in the "Glass" column. How many paper items do you see? Record that number in the "Paper" column.

Factory Points

Decide how much you think this food was processed in a factory and assign 0, 1, 2, or 3 points for each food. These are the factory points.

Number of Points	Characteristics
0 points	Foods that have not been processed or changed at all. These foods are still in their whole form. Some examples: eggs, fresh fruits and vegetables, and nuts inside their shell.
1 point	Minimally processed foods. Some examples: vegetables cut into pieces, applesauce, peanut butter, canned or frozen fruits or vegetables, rolled oats, plain yogurt, pasteurized milk, white rice, 100-percent-fruit juices, or whole-grain breads.
2 points	Moderately processed foods. Some examples: fruit yogurt, canned soups, pasta, pasta sauce, cheese, or tofu.
3 points	Highly processed foods. These no longer look like food fresh from the farm. Some examples: packaged cakes and cookies, chips, most breakfast cereals, candy, frozen meals, and hot dogs. Highly processed foods are mixed foods and they do not need much preparation, if any, at home.

(continued on next page)

BECOMING FOOD SCIENTISTS : INTERACTING PARTS : FOOD PROCESSING : ENVIRONMENTAL EFFECTS : **WASTE** : MAKING CHOICES

Name   Date

Food-Waste-Data Analysis

Transportation Points

Think about how far the food was transported. Next, assign 0, 1, 2, or 3 points depending on how far away it came from. "Transportation Points" may be harder to assign than "Factory Points." If you are not certain, assign the food 3 points.

Number of Points	Characteristics
0 points	Foods that did not need to be transported at all. Some examples are foods grown in a community garden near your home, or foods grown in your backyard.
1 point	Foods that were transported a short distance. Some examples are foods that you bought from a farmers' market, from a community-supported agriculture farm (CSA), or foods from a supermarket that were grown and processed in your state or region.
2 points	Foods that were transported a medium distance. Some examples are foods from a place a few states away.
3 points	Foods that were transported across long distances. Some examples are fresh foods that came from far across the country or from another country. Also, most highly processed foods would be 3-point foods because all of the added ingredients are transported to the factory to make the processed food.

Name   Date

Developing Questions about Waste

Use this activity sheet to help you develop a testable question for the class to investigate your school cafeteria's waste.

1. Think about what you already know about waste in your school cafeteria. For example, are different kinds of waste generated in your school cafeteria? What are they? What happens to the waste? Is any recycled? Record your observations.

_____

_____

_____

_____

Looking at your notes, develop a research question.

2. Based on what you already know, what are you curious about? What do you want to learn?

Research question: _____

_____

3. What do you think your investigation will show? Write a list of some hypotheses you have. For example, if your research question is about whether or not different groups of students throw away different amounts of waste, your hypothesis might be: More younger children throw away recyclable waste than older children. Or, students who have not studied *Farm to Table & Beyond* recycle less material than students who have studied *Farm to Table & Beyond*.

If your question is about differences in the kinds of food that are thrown away, your hypothesis might be: More students throw away vegetables than fruit. Or, more students throw away processed food than fresh fruit.

(continued on next page)

Name

Date

Developing Questions about Waste

Think about your question. What is your hypothesis?

_____

_____

_____

_____

4. To test your hypothesis, what do you need to do? Think about all the steps that you will need to take and write them down.

_____

_____

_____

_____

5. Ask yourself: Where will the class collect the data? What does the class need to measure? What data will we record?

_____

_____

_____

_____

(continued on next page)

Name	Date

Developing Questions about Waste

6. What materials does the class need in order to collect these data? Do we have all the materials and equipment we need?

_____

_____

_____

_____

_____

7. How long will it take the class to collect the data? Is there enough time to complete the investigation and analyze the data?

_____

_____

_____

BECOMING FOOD SCIENTISTS : INTERACTING PARTS : FOOD PROCESSING : ENVIRONMENTAL EFFECTS : **WASTE** : MAKING CHOICES

Name   Date

Data Collection and Analysis

So far, you have a research question and a data-collection plan for your food-related waste project. The data you collect will give you information about your project. How will you record this data? When it is recorded, how will you find meaning in your data? Begin by choosing a collection method. Two ways you can collect data are by measuring and by counting.

Measuring

Think about different kinds of data that you can collect by measuring. You can measure how much time it takes students to eat lunch. You can measure the size of your cafeteria. You can also measure the weight of the waste that is thrown away.

Counting

What data can you collect by counting? You can find out how many milk cartons are put into recycling and how many are put into the trash. You can find out how many plastic forks or plastic spoons are thrown away. You can compare the number of students who have studied *Farm to Table & Beyond* to the number who have not yet studied it.

Recording Data

After you decide how to collect your data, you will decide how to record it. If you are going to count the number of milk cartons that are thrown away, one way to do that is to wait until everyone has finished lunch. Then you carefully pour out the contents of the garbage can onto a large plastic sheet or garbage bag. Next, you count each milk carton and record the number. Then you do the same thing with the recycling bin. You record two numbers: the number of milk cartons in the garbage can and the number of milk cartons in the recycling bin.

Can you think of another way to count the milk cartons? What if you were standing in the cafeteria and counting the milk cartons as students placed them in either the garbage can or the recycling bin? You might make two columns and use a check mark to record each milk carton that was put in either the garbage can or the recycling bin. Then you record the total for each column.

	Garbage Can	Recycling Bin
Milk Cartons	✓✓✓✓✓✓✓✓✓	✓✓✓✓✓✓
Total Number of Milk Cartons	12	8

(continued on next page)

Name   Date

Data Collection and Analysis

What are some other ways that you can keep count? What if you wanted to compare the number of students in kindergarten through third grade who recycle to the number of students in fourth through sixth grade who recycle? How would you record that data? Make a sketch below.

If you wanted to weigh the amount of garbage that is thrown away each day in the cafeteria, how would you collect and record that data? What do you need to know to find out how much the garbage weighs? Is it important to find out how much the empty garbage can weighs? Why?

_____

_____

_____

Organizing Your Data

Now that you have some raw data, it is time to interpret it. What if you want to know if one grade at school recycles more milk cartons than another? Each day for a week, you and your team members stand near the recycling bin. You record the number of students in each class and you put a check mark next to each student who recycles a milk carton. At the end of the week, you have lots of check marks and boxes. What do you do with this data?

You could show the range of data with a line plot. You could put a check for each milk carton that is recycled above each grade level.

(continued on next page)

Name	Date

Data Collection and Analysis

When you look at this line plot, it looks like there are more milk cartons recycled in third grade than in first grade. What else can you say about this data?

What if you wanted to collect data about the different kinds of material that your school recycles? How could you collect that data?

Material												
Plastic spoon	✓	✓	✓	✓	✓	✓						
Plastic fork	✓	✓	✓	✓								
Plastic knife	✓	✓	✓									
Paper plate	✓	✓	✓	✓	✓	✓	✓	✓				
Milk carton	✓	✓	✓	✓	✓	✓	✓	✓	✓			
Cardboard tray	✓	✓	✓	✓	✓	✓	✓	✓	✓	✓	✓	✓

Once you have collected the data, you can organize it in different ways.

Cardboard tray   ✓ ✓ ✓ ✓ ✓ ✓ ✓ ✓ ✓ ✓ ✓ ✓

Milk carton   ✓ ✓ ✓ ✓ ✓ ✓ ✓ ✓ ✓

Paper plate   ✓ ✓ ✓ ✓ ✓ ✓ ✓ ✓

Plastic spoon   ✓ ✓ ✓ ✓ ✓ ✓

Plastic fork   ✓ ✓ ✓ ✓

Plastic knife   ✓ ✓ ✓

(continued on next page)

Name	Date

Data Collection and Analysis

Organizing the data from most to least, it's easy to see that more milk cartons were recycled than plastic knives. When you look at this information, you may have more questions. For example, did every student who had a milk carton also have a plastic knife? What if only three students took plastic knives and they all recycled them? You might want to rethink what data you collect. This is one example of why it is very important to think about the data you want to collect before you begin your investigation. Think it through and sketch out some ideas. When you are ready, discuss it with your teacher.

Analyzing the Data

Once you have gathered the data, it's time to interpret it. To interpret it, you look for patterns or relationships that might tell you something. You think about this new information and try to connect it to what you already know. You can ask yourself some questions that might help you understand the data. For example, is there a relationship between the students who have studied recycling and the amount of food-related waste that they recycle or compost compared to the students who have not studied recycling? What did we observe? How can we summarize the data? Are there any patterns? Do more sixth graders recycle than first graders? What can you infer from your data? What generalizations can you make from the data you collected? Does this evidence support your hypothesis? What new questions do you have?

BECOMING FOOD SCIENTISTS : INTERACTING PARTS : FOOD PROCESSING : ENVIRONMENTAL EFFECTS : **WASTE** : MAKING CHOICES

Analyzing Cafeteria Waste

AIM

To collect and analyze data about food-related waste in the school cafeteria.

SCIENTIFIC PROCESSES

- question, gather data, build theories, apply

OBJECTIVES

Students will be able to:

- gather data to answer their research question about waste generated in the school cafeteria;

- organize the data they collect in a way that will allow for visual analysis;

- analyze and draw conclusions from their data;

- summarize what they learned about food-related waste.

OVERVIEW

In the previous lesson, students developed a research question and designed an observation study to investigate waste in their school's cafeteria. In the beginning of this lesson, students make their final landfill observations. Then they carry out the study they designed in the previous class. First, the class reviews the research process, with a focus on the **Data Collection and Analysis** activity sheet from Lesson 26. Then students conduct their study and collect data. Next, they organize and analyze their data. They look for patterns and make generalizations about the cafeteria waste. Your class's research question and data-collection methods will determine how much time the lesson takes. This activity provides students with an opportunity to use a real-world example — their own school cafeteria — to reflect on the Unit Question. For homework, students continue to examine personal data. They look at the data in their food logs and analyze their own eating habits.

MATERIALS

For the class:
- Cafeteria research-project materials developed in Lesson 26
- Markers
- Chart paper
- Rulers, calculators, and other materials to tabulate and display survey results
- Graph paper

For each student:
- *Food-Log Analysis* activity sheet
- *Data Collection and Analysis* activity sheet (pp. 402–405)

- Data collection form or questionnaire from Lesson 26
- *Landfill Observations* activity sheet (p. 370)
- *Studying Waste* student reading (pp. 395–396)
- Personal Three-Day Food Log from Lesson 24
- LiFE Log

PROCEDURE

Before You Begin:

- Decide when and how the data-collection process will take place. Depending on the scope of the project, this may take more than one day. Consider having students work in groups. This way they can take turns collecting the data.

- Review the *Food-Log Analysis* activity sheet.

- Remind students to bring in their *Landfill Observations* and *Data Collection and Analysis* activity sheets.

- If students have developed data-collection forms, make copies for student use.

- Remind students to bring their completed three-day food logs to class.

- If you have not already done so, post the Module Question and Unit 5 Question at the front of the classroom.

MODULE QUESTION

What is the system that gets food from farm to table, and how does this system affect the environment?

UNIT QUESTION

How can we reduce the food-related waste that we produce?

 **QUESTIONING**

1. Review Module and Unit Questions

Remind students of the Module and Unit questions. Explain that this lesson is about doing the research project they developed in the previous lesson.

 **THEORIZING**

2. Final Landfill Observations

Have students make their final landfill observations. Remind them to look for any liquid that is leaching out. *Where is the liquid coming from?* (Water that was simulating rain is leaching out.) *Would you drink this? What would happen to our groundwater if the garbage included toxic materials such as motor oil, batteries, lead paint, or household cleansers?* (The groundwater would become contaminated. We couldn't drink the water.) Have students look at their predictions. *Based on your observations, do you want to change your prediction?* Invite students to share what they have learned about landfills.

 **EXPERIMENTING**

3. Prepare Forms

Check to see if students need more time to prepare their data-collection forms. Make sure the class has a testable research question and that students understand what information they are gathering. Be sure they have a form for collecting their data.

4. Review Research Process

Review the entire research process, including data collection, analysis, and display. Have students take out their *Data Collection and Analysis* activity sheet. Review the sections on organizing and analyzing data. Check for student understanding.

Remind students that they have already done the initial steps of the research process. They have developed a research question, designed a method to collect data, and developed a hypothesis about what the answer to the research question might be. Explain to students that in the next steps of the research process they will collect, organize, analyze, and display the data.

5. Collect Data

Tell students that they are going to gather their data. Emphasize that the whole class is a research team. Stress the importance of working together to collect the best possible data. Allow students ample time in the cafeteria to collect their data.

6. Organize Data

Once the data are collected, explain that the next step is to organize the information so students can better interpret their results. Remind students that they practiced organizing data when they tallied the class results from the **Food-Related-Waste Data** log. As a class, decide the best way to display the data. Encourage students to sketch quick graphs to get a sense of the data. *What can we say about the cafeteria-waste data?*

 **THEORIZING**

7. Analyze Data

After students have displayed their data, hold a class discussion about the findings. Invite students to share what they learned about waste in the cafeteria during their research. Accept all answers and record them on chart paper. Save the chart paper for the next lesson. Emphasize that findings should be listed in a clear and concise way. Remind students that they will be drawing their conclusions based on their findings.

8. Draw Conclusions

Ask students to make some conclusions based on their findings. Make a list of conclusions next to the list of findings. Here are some sample findings and conclusions from students we have worked with: (1) students throw away 1,200 eating utensils each day. This adds up to lots of space in a land-fill, especially as these are eating utensils that could be washed and used again; (2) only 30 percent of recyclables make it into the recycling bin. This means the other 70 percent is going to end up in a landfill and take up space; (3) apples are the only food item that comes from a local farmer. Local farms grow more than apples. If we bought more local food we would reduce the amount of fossil fuel needed to transport food.

Discuss with students what they learned about food-related waste from this research project. *What was the most significant finding? Were you surprised by any of your results?* Remind students that research generates new questions along with answers. Encourage students to share any new questions that have come up for them as a result of their cafeteria research.

 **APPLYING TO LIFE**

9. LiFE Logs

For homework, have students use the activity sheet to help them analyze their personal food habits. Tell students they will use the data to help them change one food-related behavior. Point out that changing a behavior is not easy. Stress that it is important to choose one small specific behavior in order to be successful. Choosing something too vague to put into practice or too vast, such as never eating candy again, is rarely successful.

Tell students to record these three types of changes in their LiFE Logs: (1) Substitute one food for another, such as eating an apple for a snack rather than chips. (2) Add a certain food. For example, eating more fruit or vegetables. (3) Reduce the amount of a certain food. For example, not eating a candy bar after school. Help students understand these types of changes. Tell them that they are to review their personal food habits and then choose one type of behavior change to make in their own lives. Tell students they will report on their food habits and the progress they make with their behavior change in Unit 6.

Name   Date

Food-Log Analysis

Use this activity sheet to guide you through your analysis of your three-day food log.

1. Find the six foods you ate most often.

_____   _____   _____

_____   _____   _____

2. Were any of the foods you ate wrapped in individual packages? _____
   If yes, which ones?

_____

3. How many times did you eat fast food? _____

4. What beverages did you drink most often? _____

5. Think about what you have learned about the farm-to-table system. Write down three foods you would like to eat less often. Explain why.

_____   _____   _____

_____

6. Write down three foods you would like to eat more often. Explain why.

_____   _____   _____

_____

7. What change will you try to make to your food choices?

_____

_____

Taking Action

AIM

To develop, implement, and evaluate an action plan to reduce food-related cafeteria waste.

SCIENTIFIC PROCESSES

- question, design, implement, evaluate

OBJECTIVES

Students will be able to:

- develop goals for their action plan;

- design the methods to implement their plan;

- implement their action plan through gathering, organizing, and analyzing data on their implementation;

- evaluate the implementation of their plan.

OVERVIEW

In this lesson, students develop a plan to reduce waste based on their findings. First they identify goals. Using these goals, they develop an action plan to reduce cafeteria waste that involves the entire school community. Students develop the methods to implement their plan. Once the plan is in place, they collect data and compare this new data to their initial data to see if their plan has had an impact on the amount of food-related waste generated in the cafeteria. Finally, students evaluate their plans and revise them, if necessary. This project can range from simple actions students can do on a daily basis, to more complex plans involving major changes to the lunch program. It is up to you and your students to decide how to design, implement, and evaluate this project.

MATERIALS

For the teacher:
- *Sample Action Plans* lesson resource
- Chart paper
- Markers

For the class:
- Chart-paper list of food-related waste findings and conclusions from Lesson 27
- Graph paper
- Chart paper
- Markers

For each student:
- *In My Opinion* activity sheet
- (Optional) *Data Collection and Analysis* activity sheet (pp. 402–405)
- LiFE Log

PROCEDURE

Before You Begin:

- Take out the chart-paper list of food-related waste findings from the class research project. Post them at the front of the classroom.

- Review the *Sample Action Plans* lesson resource.

- Review the *In My Opinion* activity sheet and make copies for each student.

- If you have not already done so, post the Module Question and Unit 5 Question at the front of the classroom.

MODULE QUESTION

What is the system that gets food from farm to table, and how does this system affect the environment?

UNIT QUESTION

How can we reduce the food-related waste that we produce?

1. Review Module and Unit Questions

Remind students of the Module and Unit questions. Explain that this lesson is about developing an action plan to reduce food-related waste in your school's cafeteria. Briefly review with students the findings and conclusions that were made from analyzing their data in the previous lesson.

2. Develop Goals

Explain to students that they are going to use their data in order to come up with some goals for reducing food-related waste in their cafeteria.

Emphasize that goals will have the greatest chance of success if they are clear, targeted toward a specific behavior or action, big enough to make a difference, small enough to be achievable, and involve an action or behavior that is within their control. Some sample goals include: (1) decrease the amount of food that gets thrown in the garbage; (2) decrease the amount of solid waste from trays, milk cartons, and eating utensils; (3) increase the number of students putting recyclable items in the recycling bins instead of the garbage cans.

3. Develop Plan

Once students have established their goals, they can develop and implement a plan to meet them. Students should think hard about the kinds of actions that will be required to meet their goals. Make sure they develop a specific plan that will help them meet their goals. Encourage students to be creative. Remind them that it's important to remember that they may not see a significant change overnight. It will take time to build up support for the plan. You may want to encourage them to start small. See the *Sample Action Plans* lesson resource for examples.

4. Implement Plan

If students have not already reported their earlier findings to the whole school, encourage them to do so. Remind them that it is important for everyone to understand why your class is trying to make a change. Encourage them to think about how they can sell their idea to their peers. When students are ready to implement their plan, consider making an announcement beforehand and rallying the student body to help reduce food-related waste. It is important to publicize this project. Challenge students to think about how to get their message across. Encourage them to think about both the content of the message and the medium for delivering it. Once the word is out, implement the plan.

5. Gather Data

Review data-collection methods with students. Remind students that one of the most important steps after implementing any type of action plan is to collect data about the plan to evaluate its effectiveness. Invite students to share ideas for how they might collect data about their action plan. Data-collection techniques can range from very simple to complex. Encourage students to develop their own ideas about how to collect data about their action plans and record all ideas on the board or on chart paper. Remind students to think about what they want to know. If students do not think of it themselves, suggest that they use the same data-collection methods they used for their initial data. This way they can compare their two sets of data to determine what changes have been made as a result of their plan. Have students decide on the best method to collect data and work together to create a specific time frame during which they will do the data collection.

6. Organize Data

Once students have their data, remind them that the next step is to organize it. You may wish to have them review the *Data Collection and Analysis* activity sheet. Remind students that they organized data when they did their first food-related-waste research project. Have students work in small groups. Invite volunteers to share their methods for organizing the data. *What can we say about this recent cafeteria-waste data? How does it compare to our earlier data?*

7. Draw Conclusions

Once students have organized their data, hold a class discussion about the findings.

Ask students to think about whether their action plan is really effective at reducing waste in the school's cafeteria. Invite them to share their observations about the plan's effectiveness. Remind them to use evidence to support their conclusions. Pose the following questions for discussion. *Is your goal a concrete action that people can do? Is the plan mostly in your control? If your plan continues to be implemented for years into the future, do you think it will significantly reduce the waste generated at school? What can you do to make your action plan a permanent part of our school policy?*

8. Consider Next Steps

Ask students to consider how effective they think their plan is now in order to consider what steps they might take next to reduce waste even further. Help students envision what the next level of action might be. For example, if the class's plan is about reducing the amount of highly processed food the cafeteria serves, urge the students to expand this line of thinking and try to have the school purchase more foods grown by local farmers. Ask students to speculate on how such a plan might work. If students feel that their plan is not at all effective, ask them to think about the reasons.

9. LiFE Logs

Remind students that this is the final lesson in Unit 5. As a wrap-up to the unit, ask students to write a newspaper op-ed piece, at least five paragraphs long, that describes their action plan and why it should be implemented in schools. Distribute the *In My Opinion* activity sheet. Review it with students. Point out that there are five boxes. The information in each box represents the content for one paragraph. Tell them to use it to organize their thoughts. Remind students to cite evidence from their readings and research to demonstrate what they have learned about food-related waste through their food-systems studies.

Sample Action Plans

Your students may find it challenging to develop an action plan that is reasonable in scope yet yields the results they desire. If developing and implementing an action plan is something they have never done before, encourage them to start small. Remind them that they can build on their small successes.

Here are some sample plans to spur on your students' thinking.

- Students who bring their lunches from home can use airtight containers instead of wrapping their lunches in plastic wrap, aluminum foil, or waxed paper. How much disposable wrap is removed from the solid-waste stream by using such containers?

- Ask each student to set a personal goal of recycling at least one food container or package every day at lunch. How many disposable food containers are removed from the solid-waste stream each day?

- Start a campaign to encourage students to take only the food that they know they will eat. How much food waste is removed from the solid-waste stream?

- Start a campaign to encourage students to use reuseable drink bottles rather than disposable juice boxes, plastic water bottles, or milk cartons. How much solid waste does this remove from the waste stream?

- If your school does not have bins set up for recycling, solid waste, and food scraps to compost, ask your school custodians to work with you to help students separate their waste. Make signs to place near each container. Did we increase the amount of food that is composted and increase the amount of waste that is recycled?

- If your school does not have a composting program, there are numerous resources available on the Internet. To get started, check out Cornell University's "Composting in School" (*http://compost.css.cornell.edu/schools.html*) and composting resources in the National Gardening Association's *Gardening with Kids* online store (*www.garden.org*).

- If the youngest students at your school do not recycle, plan a recycling presentation to teach them about the three R's. You could put on a puppet show, design posters, or tell them stories about the importance of recycling. Did primary students increase the amount of material they recycled after we gave our presentation?

- Each student brings in a reusable fork and spoon from home. After lunch, students wash them and they are stored in a ziplock plastic bag. Every day, students bring their forks and spoons to the cafeteria. How many disposable eating utensils are removed from the solid-waste stream by using this plan?

Name   Date

In My Opinion

Use this graphic organizer to help you plan your five-paragraph essay.

Main Idea:

Evidence to support my idea:

Evidence to support my idea:

Evidence to support my idea:

Conclusion:

Farm to Table & Beyond
©2008 Teachers College Columbia University

Making Choices

Farm to Table & Beyond Expo

AIM

To use what we have learned about our farm-to-table system and how it affects the environment.

SCIENTIFIC PROCESSES

- **synthesize information, communicate knowledge, apply**

OBJECTIVES

Students will be able to:

- **discuss our farm-to-table system and how it affects the environment;**

- **communicate about our farm-to-table system clearly and effectively by oral and written means;**

- **apply what they learned during this module to design an exhibit or poster.**

OVERVIEW

Throughout this module, students have been investigating the farm-to-table system and how it affects the environment. In this lesson, students work in groups to design exhibits and posters to teach others what they have learned. Putting on an expo provides students with an opportunity to pull together their understanding of the global food system. The expo can be as big or as small as you and your class choose to make it. In developing their exhibits — which can be hands-on experiments, demonstrations, or posters — students gain experience with reading, writing, researching, planning, the visual arts, critical thinking, and making presentations.

MATERIALS

For the teacher:
- Planning a *Farm to Table & Beyond* Expo (p. 27)
- ***Sample Expo Projects*** lesson resource

For each group:
- Chart paper
- Markers, crayons and/or colored pencils
- Magazines (for photos to cut out)
- Scissors
- Glue
- (Optional) 1 exhibit board

For each student:
- ***Farm-to-Table Expo Project*** activity sheet
- File folder
- LiFE Log

PROCEDURE

Before You Begin:

- Review Planning a *Farm to Table & Beyond* Expo and the **Sample Expo Projects** lesson resource.

- Review the **Farm-to-Table Expo Project** activity sheet and make a copy for each student.

- Gather materials.

- If you have not already done so, post the Module Question and Unit 6 Question at the front of the classroom.

MODULE QUESTION

What is the system that gets food from farm to table, and how does this system affect the environment?

UNIT QUESTION

How can we use the science we learned to make ecologically sound food-system choices?

 **QUESTIONING**

1. Introduce the Unit 6 Question

Review with students the Module Question and the first five Unit questions.

Engage students in a discussion of what they have learned about the environmental effects of the global food system. Introduce the Unit 6 Question. Explain that in this lesson students pull together what they have learned about the farm-to-table system and share it with others. They will work in groups to plan and present an expo.

 **APPLYING TO LIFE**

2. Choose Theme

Distribute the **Farm-to-Table Expo Project** activity sheet and a file folder to each student. Tell students to use the folders to store their expo materials as they develop them. Have students think about what they want to tell others about the farm-to-table system and how it affects the environment. *What have you learned that you would like to share?* Remind students of the Unit 6 Question. *What science have we learned that we can share with others so they can make ecologically sound food choices? Who do you want to invite to the expo? What information would this audience like to hear?*

3. Plan Projects

Brainstorm different ideas for expo projects as a whole group. Refer to the **Sample Expo Projects** lesson resource for suggestions. Record the ideas on the board or on chart paper.

Give students presentation tips. Remind them that communication includes four things: (1) the message; (2) the person or persons delivering the message; (3) how the message is being delivered; and (4) how the audience receives the message. Engage students in a discussion of different ways they can share their information with their audience. *What are some ways we can communicate information?* (Write, draw, give a dramatic presentation.) *What can you make if you want people to take information away?* (A brochure or a handout.) *What can you do if you want to get someone's attention as they walk through the expo?* (Make a colorful poster, put on a skit, have music.) Challenge students to think of creative ways to share their information.

Divide the class into small groups. Review the **Farm-to-Table Expo Project** activity sheet with the class. Once each group has selected its project, have teams identify tasks and make assignments. Have each team elect a leader to guide the project and keep track of the details and deadlines. Have team leaders record this information on a chart.

4. Organize Project Work

Meet with team leaders to discuss project deadlines. Refer to the project checklist in Planning a *Farm to Table & Beyond* Expo.

Remind students to organize their information so it makes sense. Tell them to keep asking themselves what their audience really needs to know to understand your message. *What is the essential information you need to include?*

5. Preview Projects

Invite each group to display its project. Ask each team leader to discuss the team's project design and why the team chose this project. Ask the groups to critique one another's projects. Remind students to offer only constructive criticism. Encourage them to practice several times before the expo.

6. Present the Expo

Remind students to introduce themselves to the audience. Make sure you leave enough time for students to set up their displays. Tell them to make sure they have everything they need to give their presentation. Be sure to have extra tape, scissors, and glue for last-minute repairs.

After the expo, congratulate students on their fine work.

7. Homework

Ask students to think about all that they have learned about the farm-to-table system. Tell them to write two paragraphs in their LiFE Logs that answer the Unit 6 Question, *How can we use the science we learned to make ecologically sound food choices?*

Sample Expo Projects

Throughout this module, students have investigated the farm-to-table system and how it affects the environment. Use the expo as a way to pull together all that students have learned. The suggestions below are just a sampling of projects drawn from the module for you and your class to consider.

Where Did This Pancake Come From?

Develop a poster referring to the *Pancake Ingredient Chart* (p. 210). Have the *Pancakes* student recipe (p. 217) as a handout for visitors.

How Does a Landfill Work?

Refer to *Investigating Landfills* (pp. 367–368) to set up a demonstration. Include a poster that explains how a landfill works.

From Farm to Table

Refer to *Exploring Our Food System* (pp. 80–81). Make an educational poster to teach others about our food system.

How Did This Food Get to My Plate?

Refer to *Exploring Our Food System* (pp. 80–81) and the *Food-Systems Flowchart* (p. 171). Make an illustrated food-system flowchart.

Food Change

Refer to *Food-Change Experiment* (p. 85), *Food-Change Observations* (pp. 95–99), and *Thinking about Food Change* (p. 256). Develop an educational poster that describes how food changes.

Where Does Packaging Come From?

Students present their product research from *Sand to Glass, Plant to Paper, Petroleum to Plastic,* and *Ore to Aluminum Can* (pp. 130–133).

Packaging Analysis

Refer to *Investigating Cereal Packaging* (p. 87) and *Packaging Analysis* (pp. 92–93). Explain why we use packaging. Offer alternatives. Discuss ways to help reduce waste from packaging. Make a brochure or a handout.

Bread-Mold Demonstration

Refer to *Do Preservatives Affect Mold Growth?* (p. 238) and *Mold-Growth Predictions* (pp. 243–246). Include a poster that explains how the use of preservatives affects mold growth.

(continued on next page)

Transportation Systems

Refer to *Transportation System Guiding Questions* (pp. 163–165). Make educational posters that describe the different modes of transportation. Include the advantages and disadvantages of using the various kinds of transportation.

X-tremely Healthy Snacks Project

Refer to *Snack Assessment* (p. 356) and *Snack-System Design* (p. 357). Make a poster that describes the snack-system design. Explain how this system will have a low impact on the environment.

All about Fossil Fuels

Refer to *Fossil Fuels* (pp. 303–306). Develop a poster that describes what fossil fuels are made from and common ways that we use them. Include illustrations of products made from fossil fuels.

Earth Systems

Refer to *Earth-Systems Worksheet* (pp. 294–295). Make educational posters describing Earth's systems.

Which Wrap?

Refer to *Which Wrap?* (pp. 123–129). Set up a demonstration or invite expo attendees to test the different wraps (p. 123). Develop a poster that discusses the environmental impacts of the wraps that students developed for the snack company (pp. 124–129).

Name   Date

Farm-to-Table Expo Project

Here are some tips to help guide your work. This is a big project, so think about dividing the work into six separate stages.

Stage One: Gather Research Materials

Collect all the posters, drawings, flow charts, diagrams, essays, and graphic organizers that you have created while studying *Farm to Table & Beyond*. Put them in your expo folder. If any of your materials are too large to fit in the folder, ask your teacher for a safe place to store them.

Stage Two: Plan Your Exhibit

Think about what you have learned. How can you use this information to help you make choices in the future? What have you learned that might affect your choices when you buy a snack? Will you think about packaging? Will you consider how far the food had to travel before it arrived at your local store? Will you think about fossil fuels and climate change? Choose one topic that you want to share with other students your age or younger.

Stage Three: Assign Roles

You are working with others in your group. Once you know what your project will be, think about all the tasks that you need to finish before your project is complete. Make a list. For example, does someone need to draw pictures or find photographs? Who will do this task? Who will be the presenter? Will the presenter write the script? Who will make the display? Who will collect the materials that you need?

Stage Four: Create Your Exhibit

Are you doing a demonstration or making a poster? Whatever your project is, you need to make a list of materials. As a group, think through all the supplies that you will need. Check with your teacher to find out which supplies are already in the classroom.

Stage Five: Write Your Script

Plan to give a presentation that lasts about two minutes. Write a script that lists all of the important facts that you want to share. Review the script as a group. Discuss the script. Make changes if necessary.

Stage Six: Present

Practice the presentation. Set up your display. Have fun while you teach others!

Bringing It All Together

AIM

To reflect on and synthesize what we have learned about the farm-to-table system.

SCIENTIFIC PROCESSES

- synthesize information, communicate knowledge

OBJECTIVES

Students will be able to:

- articulate their reason for their behavior change;

- express in writing and drawing an answer to the Module Question;

- evaluate changes in their understanding of the farm-to-table system and how it affects the environment;

- create a list of food-choice guidelines based on what they have learned;

- describe why each of their guidelines is important.

OVERVIEW

This final lesson in the module is an opportunity for you and your students to contemplate, synthesize, and establish what changes have occurred in students' answers to the Module Question, *What is the system that gets food from farm to table, and how does this system affect the environment?* To do this, students revisit the food-system posters they created in Lesson 4 that demonstrated all the steps involved in getting food from the farm to the table. Students also write a new answer to the Module Question and compare their current answer to what they wrote in Lesson 4. Finally, they create food-choice guidelines based on their understandings of our food system and how it generates waste in the environment. Students discuss the data they gathered from their food logs and share their progress on their behavior change.

MATERIALS

For the teacher:
- *Food-Choice Guidelines* lesson resource

For the class:
- Chart paper
- Markers

For each student:
- Completed *Food-Related-Waste Data* activity sheet (pp. 384–385)
- *Food-Waste Data Analysis* activity sheet (pp. 397–398)
- Completed Personal Three-Day Food Log
- Food-system diagrams drawn in Lesson 4
- 1 sheet of drawing paper
- Markers or crayons
- LiFE Log

PROCEDURE

Before You Begin:

- Remind students to bring in their completed *Food-Related-Waste Data* and *Food-Waste Data Analysis* activity sheets and their three-day food logs.

- Ask students to review the food-system diagrams they drew at the beginning of the module as a pre-assessment of their understanding of the farm-to-table system and how this system affects the environment.

- Make sure students have their LiFE Logs with the answer to the Module Question that they wrote as a pre-assessment.

- Review the *Food-Choice Guidelines* lesson resource.

- Be sure you have the Module Question and all six Unit questions posted at the front of the classroom.

MODULE QUESTION

What is the system that gets food from farm to table, and how does this system affect the environment?

UNIT QUESTION

How can we use the science we learned to make ecologically sound food-system choices?

1. Review the Module Question and All Unit Questions

Review the homework from Lesson 29. Invite students to share their answers to the Unit 6 Question. Explain to students that this is the final lesson in the module, and it is an opportunity to reflect and synthesize what they have learned. They will think about the answer to the Module Question and how the Unit questions helped them expand their answer to the Module Question.

2. Share Progress in Behavior Change

Review the food-log analysis and behavior-change homework. Do a quick tally on the board to see if there are patterns in students' food habits. Invite students to discuss what they learned about their own food habits. *What did you learn? Did anything surprise you?*

Ask several volunteers to describe the three different kinds of behavior change they learned about. Write each type of change on the board. Ask students which change they selected. Tally the results. Invite students to share their progress on their behavior-change goal. Ask students if modifying their behavior has generally been easy or hard for them to do. If students have been successful at changing their behavior, ask them to share with the class what they did differently in their daily routine to help them stick with their behavior change.

How does making a change in your behavior affect the environment? What if you substitute an apple for a processed food like candy? How will this affect the environment? What if you eat fewer chips or drink less soda? Encourage students to think about what they have learned about the farm-to-table system and how it affects the environment as they ponder these questions.

3. Create a Poster

Ask students to draw a poster of all the steps that food goes through in the farm-to-table system. Remind students that they did this same activity when they first started this module. Encourage students to include new information they have gained in studying this module in their pictures. Remind students to use words, arrows, and drawings in their representation of the farm-to-table system.

4. Share Posters

After students have finished drawing, invite them to share their work with the class. Encourage students to be supportive of each other throughout the questioning and discussion of what was learned. Pose the following questions for discussion: *What did you add to the drawing today that you did not include in the original? Did you leave anything out that you had in your original drawing? Compare your two drawings — how are they similar? Did you include ways that the farm-to-table system affects the environment in both drawings? Has your thinking changed? How has it changed?* Bring the discussion to a close. You may wish to use these posters to replace the preassessment drawings and keep them posted in the classroom.

5. Develop Food-Choice Guidelines

Explain to students that food-choice guidelines are rules that can be used when they are trying to decide what to eat. Hang up a sheet of chart paper. Refer to the ***Food-Choice Guidelines*** lesson resource. Make two columns, one with the heading "Food-Choice Guideline" and the other headed "Why It Is Important." Work with students to generate a list of guidelines, including the samples listed on the lesson resource. Encourage students to think of good reasons that their guidelines are important. Challenge them to use what they have learned in this module to defend their reasoning. Keep the guidelines hanging in the classroom and revisit them from time to time, discussing with students how they are, or are not, using the guidelines to make their personal food choices.

6. LiFE Logs

Have students write an answer to the Module Question, *What is the system that gets food from farm to table, and how does this system affect the environment?* They can use their drawing as a guide. Encourage students to write as complete an answer to the Module Question as they can. Remind them to look at and think about the Unit questions and to incorporate what they learned in each unit into their answer. Refer to the Assessment section of the introduction for more information on assessing your students' learning.

7. Reflect on What They Learned

After students complete their answer to the Module Question, have them compare what they wrote now to what they wrote at the beginning of the module. Ask students to write a short paragraph that describes how their answer has changed. Have some students read their current answers to the Module Question. Ask students to share what they have learned. Encourage students to continue to think about and apply what they have learned by following the food-choice guidelines just developed and talking about food-production issues with family and friends.

Food-Choice Guidelines

This is a sample of the types of food-choice guidelines that students may develop. They may have other ideas not represented here. This is fine. If students don't think of some of the ideas in these sample guidelines, present these ideas and ask students if they want to add them to their list.

Food-Choice Guideline	Why It Is Important
Eat foods grown or raised on local farms whenever we can.	When we eat local food we shorten the distance between farm and table. This reduces the amount of transportation that is needed.
Eat whole foods like fruits and vegetables instead of processed, packaged foods.	Natural resources and fossil fuels are used to process and package foods. Eating foods that are not packaged or processed means that fewer natural resources, including the fossil fuels that power factories, are used.
Preserve the harvest's surplus by canning, drying, and freezing foods.	Preserving helps food last longer. With surplus food preserved, we do not need to bring in food from other places during the winter.
Eat more food that is grown without the use of synthetic chemicals and pesticides.	Food grown this way minimizes the contamination of soil and water.
Don't waste food.	Many natural resources are used to produce the food we eat. When we throw food away, we not only turn food into garbage, we waste all the resources used to grow the food.
Compost food scraps whenever possible.	This helps return nutrients to the soil and reduces the amount of waste that goes to landfills.
Reduce the amount of waste that goes to landfills. Reduce, reuse, and recycle.	Landfills can leak, which can contaminate the surrounding soil and water.

Bibliography

American Association for the Advancement of Science. (1990). *Science for all Americans.* New York: Oxford University Press.

American Association for the Advancement of Science. (1993). *Benchmarks for science literacy.* New York: Oxford University Press.

Ausubel, D.P. (1968). *Educational psychology: A cognitive view.* New York: Holt, Rinehart, & Winston.

Bad Greenhouse. (2008). Retrieved June 9, 2008, from *www.ems.psu.edu/~fraser/Bad/BadGreenhouse.html.*

Ball, D.L., & Cohen, D. (1996). Reform by the book: What is — or might be — the role of curriculum materials in teacher learning and instructional reform? *Educational Researcher, 25 (9),* 6–8, 14.

Ball, D.L., & Feiman-Nemser, S. (1988). Using textbooks and teachers' guides: A dilemma for beginning teachers and teacher educators. *Curriculum Inquiry, 18 (4),* 401–23.

Calabrese Barton, A., Koch, P.D., Contento, I.R., & Hagiwara, S. (2005). From global sustainability to inclusive education: Understanding rural children's ideas about the food system. *International Journal of Science Education, 27 (10),* 1163–86.

Driver, R., Squires, A., Rushworth, P., & Wood-Robinson, V. (1994). *Making sense of secondary science: Research into children's ideas.* London/New York: Routledge.

Gautier, C., Deutsch, K., & Rebich, S. (2005). Misconceptions about the greenhouse effect. *Journal of Geoscience Education, 54 (3),* 386–95.

Grossman, P., & Thompson, C. (2004). *Curriculum materials: Scaffolds for new teacher learning?* Seattle, WA: Center for the Study of Teaching and Policy and Center on English Learning & Achievement (CELA).

Kempton, W. (1997). How the public views climate change. *Environment, 39 (9),* 12–21.

Kesidou, S., & Roseman, J.E. (2002). How well do middle school science programs measure up? Findings from Project 2061's curriculum review. *Journal of Research in Science Teaching, 42 (7),* 767–90. (See also: *www.project2061.org/publications/textbook/mgsci/report/crit-used.htm.*)

Koulaidis, V., & Christidou, V. (1999). Models of students' thinking concerning the greenhouse effect and teaching implications. *Science Education, 83 (5),* 559–76.

Kyle W. & Shymansky, J. (1989). Enhancing learning through conceptual change teaching. *Research Matters — to the Science Teacher, 2.* Retrieved October 25, 2005, from *www.educ. sfu.ca/narstsite/publications/research/concept.htm.*

McWilliams, M. (1998). *Food fundamentals, 7th edition.* Redondo Beach, CA: Plycon Press.

McWilliams, M. (2001). *Foods: Experimental perspectives.* Upper Saddle River, NJ: Prentice Hall.

Meadows, G., & Wiesenmayer, R. (1999). Identifying and addressing students' alternative conceptions of the causes of global warming: The need of cognitive conflict. *Journal of Science Education and Technology, 8 (3),* 235–39.

Murray, S. (2007). *Moveable Feasts: From ancient Rome to the 21st century, the incredible journeys of the food we eat.* New York: St. Martin's Press.

National Research Council. (1996). *National science education standards.* Washington, DC: National Academies Press.

Nelson, B.D., Aron, R.H., & Francek, M.A. (1992). Clarification of selected misconceptions in physical geography. *Journal of Geography, 91 (2),* 76–80.

National Science Teachers Association. (2007). *Resources for Environmental Literacy.* Arlington, VA: NSTA.

Renner, J.W., & Marek, E.A. (1990). An educational theory base for science teaching. *Journal of Research in Science Teaching, 27 (3),* 241–46.

Rule, A.C. (2005). Elementary students' ideas concerning fossil fuel energy. *Journal of Geoscience Education, 53 (3),* 305–18.

Stern, L., & Roseman, J. (2004). Can middle-school science textbooks help students learn important ideas? Findings from Project 2061's curriculum evaluation study: Life science. *Journal of Research in Science Teaching, 41 (6),* 538–68.

Tsurusaki, B.K., & Anderson, C.W. (2007). Students' understanding of connections between human engineered and natural environmental systems: Similarities and differences across grade level and context. Paper presented at the National Association for Research in Science Teaching's annual conference, New Orleans, LA.

Volk, T. (1993). Educating for responsible environmental behavior in environmental education, teacher resource handbook: A practical guide for K-12 environmental education (Richard Wilke, ed). Millwood, NY: NSTA & Kraus International Publishers.

Sources

Lesson 8

Africa: The Cradle of Civilization, "Rise to Glory: Trading Empires: Indian Ocean and Red Sea," *http://library.thinkquest.org/C002739/AfricaSite/LMTrade RedIndian.htm* (accessed October 30, 2007).

BNSF Railway, "BNSF Facts," *www.bnsf.com/media/ bnsffacts.html* (accessed December 8, 2007).

Boeing, "747 Fun Facts," *www.boeing.com/commercial/ 747family/pf/pf_facts.html* (accessed December 1, 2007).

Boston University, "Boston University–University of Naples 'L'orientale' Team Find Artifacts of Ancient Egyptian Sea Vessel," news release, April 20, 2005, *www.bu.edu/phpbin/news/releases/display.php?id=913* (accessed October 10, 2007).

Energy Information Administration, "Energy Kid's Page, Transportation Timeline," *www.eia.doe.gov/kids/history/ timelines/transportation.html* (accessed May 15, 2008).

Great Lakes Information Network, "Zebra Mussels in the Great Lakes Region," *www.great-lakes.net/envt/ flora-fauna/invasive/zebra.html* (accessed November 10, 2007).

Growing a Nation: The Story of American Agriculture, "A History of American Agriculture," *www.agclassroom. org/gan/timeline/* (accessed August 12, 2007).

History World, "History of Domestication of Animals," *www.historyworld.net/wrldhis/PlainTextHistories.asp? historyid=ab57* (accessed August 10, 2007).

"Largest ships in the world steam ahead to the Port of Charleston," *Charleston Regional Business Journal*, June 19, 2000, *www.charlestonbusiness.com/pub/3_13/news/ 1052-1.html* (accessed November 15, 2007).

Lovett, Richard A., "World's Oldest Sea Vessels Discovered in Egypt," National Geographic News [Internet], March 7, 2006, *http://news.nationalgeographic. com/news/2006/03/0307_060307_egypt_ships.html* (accessed August 12, 2007).

Minnesota Department of Transportation, "50 Fun Facts," *www.dot.state.mn.us/interstate50/50facts.html* (accessed October 31, 2007).

National Museum of American History, "America on the Move," *http://americanhistory.si.edu/onthemove/ exhibition/* (accessed August 12, 2007).

Port Canaveral: Recreation, "Endangered Species Watch," *www.portcanaveral.org/recreation/endangered.php* (accessed December 2, 2007).

The Port of Los Angeles, *www.portoflosangeles.org* (accessed December 2, 2007).

Port of Seattle, "Marketing Across the Ocean: What Do You Need to Do?," *www.portseattle.org/downloads/ community/education/casestudy/marg_s.doc* (accessed November 20, 2007).

The St. Lawrence Seaway: Gateway to North America, "Commercial Shipping," *www.greatlakes-seaway.com/en/ commercial/index.html* (accessed November 10, 2007).

United States Centennial of Flight Commission, "A History of Commercial Air Freight," *www.centennialofflight.gov/essay/Commercial_Aviation/Air Freight/Tran10.htm* (accessed August 9, 2007).

United States Department of Agriculture. "Aquatic Species, species profile: Zebra mussel," National Agricultural Library, *www.invasivespeciesinfo.gov/aquatics/ zebramussel.shtml* (accessed December 2, 2007).

Lesson 11

Haard, Norman F., S.A. Odunfa, Cherl-Ho Lee, Dr. R. Quintero-Ramirez, Dr. Argelia Lorence-Quinones, & Dr. Carmen Wacher-Radarte, "Botanical Structure of Cereals," *Fermented Cereals: A Global Perspective.* FAO Agricultural Services Bulletin No. 138 (1999), *www.fao. org/docrep/x2184e/x2184e00.htm#con* (accessed June 25, 2008).

Iowa Corn, *www.iowacorn.org* (accessed November 10, 2007).

National Corn Growers Association, "Frequently Asked Questions," *www.ncga.com/education/main/FAQ.html* (accessed November 10, 2007).

National Grain and Feed Association, *www.ngfa.org/ trygrains_corn.asp* (accessed November 10, 2007).

North American Millers' Association, *www.namamillers. org/prd_c_mill.html* (accessed November 10, 2007).

United States Department of Agriculture/NASS South Dakota Field Office, "South Dakota's Rank in United States Agriculture," March 2008, *www.livestock.doa.sd.gov/ index_docs/SD_AG_rank_07.pdf* (accessed May 3, 2008).

Lesson 13

Intute: social sciences timeline, "History of Food," *www.intute.ac.uk/socialsciences/timeline_History_of_food.html* (accessed December 10, 2007).

Lesson 16

Just Good Eats, "Kosher Garlic Dill Pickles," *www.justgoodeats.com/recipes/show.php?record_ID=158* (accessed January 31, 2005).

Lesson 17

Algalita Marine Research Foundation, *www.algalita.org/* (accessed April 10, 2008).

Ebbesmeyer, Curtis C., "Using Flotsam to Study Ocean Currents," NASA, *http://oceanmotion.org/html/gatheringdata/flotsam.htm* (accessed April 10, 2008).

My NASA Data, "Glossary: gyre," *http://mynasadata.larc.nasa.gov/glossary.php?&word=gyre* (accessed April 10, 2008).

NOAA, "National Weather Service: JetStream — Online School for Weather," *www.srh.noaa.gov/jetstream/atmos/layers.htm* (accessed March 15, 2008).

Sherman, Maurina S., "Going to Extremes," *Science & Technology Review*, Lawrence Livermore National Laboratory, July 12, 2004, *www.llnl.gov/str/JulAug04/Fried.html* (accessed March 15, 2008).

World Book @ NASA, "Earth," *www.nasa.gov/worldbook/earth_worldbook.html* (accessed March 15, 2008).

Lesson 18

United States Department of Energy, Office of Fossil Energy, "Common Products Made from Oil & Natural Gas: Educational Poster" (pdf), *www.fossil.energy.gov/education/index.html* (accessed April 30, 2008).

Lesson 19

Department of Materials Science and Engineering, University of Illinois Urbana-Champaign, "Energy Historical Timeline 1 Million B.C.–Present," *http://matse1.mse.uiuc.edu/energy/time.html* (accessed June 28, 2008).

Energy Information Administration, "Energy Kid's Page, Energy Timelines," *www.eia.doe.gov/kids/history/timelines/index.html* (accessed May 15, 2008).

Lesson 20

Arbor Day Foundation, "Zone Changes," *www.nationaltreetrust.org/media/mapchanges.cfm* (accessed March 22, 2008).

Winter, Maiken, "Birds and Global Climate Change," *BirdScope*, Cornell Lab of Ornithology, Summer 2007, Volume 21, Number 3, *www.birds.cornell.edu/Publications/Birdscope/Summer2007/index.html* (accessed March 20, 2008).

Lesson 21

Nevala, Amy E., "What Does It Take to Break a Whale?," *Oceanus*, 20 June 2007, Woods Hole Oceanographic Institution, *www.whoi.edu/oceanus/viewArticle.do?id=28654* (accessed May 20, 2008).

NOAA, "NOAA Takes Steps to Reduce Ship Collisions with Endangered North Atlantic Right Whales," *ScienceDaily*, July 18, 2005, *www.sciencedaily.com/releases/2005/07/050710172743.htm* (accessed June 28, 2008).

NOAA, "Northeast US Right Whale Sighting Advisory System (SAS)," *http://rwhalesightings.nefsc.noaa.gov/* (accessed May 10, 2008).

Right Whale Listening Network, "Bioacoustics Research Program," Cornell Lab of Ornithology. *http://listenforwhales.org/* (accessed May 20, 2008).

Swink, Simone, "Whales Win Right-of-Way in Atlantic Shipping Lanes," *National Geographic News*, March 5, 2003, *http://news.nationalgeographic.com/news/2003/03/0305_030305_tvrightwhales.html* (accessed April 4, 2008).

University of Rhode Island, "Rhode Island Researcher Develops Sonar Imaging Device to Protect Whales," *ScienceDaily*, July 7, 2000, *http://www.sciencedaily.com/releases/2000/07/000703090140.htm* (accessed June 28, 2008).

Woods Hole Oceanographic Institution, "Right Whales," *www.whoi.edu/page.do?pid=12639* (accessed May 10, 2008).

Lesson 22

Gigli, Rossella, "Italy: Increasing Interest in Apple Paper," *Fresh Plaza*, December 17, 2007, *www.freshplaza.com/news_detail.asp?id=13262* (accessed March 15, 2008).

Lesson 23

Blouin, Melissa, "Researchers Solve Pickles Pickle," University of Arkansas Daily Headlines, August 21, 2000, *http://dailyheadlines.uark.edu/1242.htm* (accessed March 3, 2008).

Lesson 24

The California Integrated Waste Management Board, "Lesson 2: Away to the Landfill,"*Closing the Loop*, Unit 1: Managing and Conserving Natural Resources, *www.ciwmb.ca.gov/Schools/curriculum/CTL/46Module/Unit1/Lesson2.pdf* (accessed April 10, 2008).

University of Wisconsin-Madison, "Bottle Basics," *Bottle Biology*, *www.bottlebiology.org/basics/index.html* (accessed April 10, 2008).

Lesson 25

Rotten Truth (about Garbage), 1998, *www.astc.org/exhibitions/rotten/rthome.htm* (accessed May 15, 2008).

United States Environmental Protection Agency, Municipal Solid Waste (MSW), "Basic Information," *www.epa.gov/epaoswer/non-hw/muncpl/facts.htm* (accessed June 25, 2008).

United States Environmental Protection Agency, Municipal Solid Waste (MSW), "Milestones in Garbage," *www.epa.gov/epaoswer/non-hw/muncpl/timeline_alt.htm* (accessed March 30, 2008).

Lesson 26

Harrison, Jeff, "Study: Nation Wastes Nearly Half Its Food," University of Arizona, *UA News*, November 18, 2004, *http://uanews.org/node/10448* (accessed March 31, 2008).

Glossary

Abiotic — Nonliving. The abiotic elements found in an ecosystem include air, rocks, sunshine, and water.

Acetic acid — A clear, colorless organic acid. It is the chief ingredient that makes vinegar sour.

Acid — A compound that tastes sour, turns litmus red, and reacts with certain metals to produce hydrogen gas. Examples: vinegar and lemon juice.

Air cavity — In a seed, the space between the seed coat and the endosperm.

Air-circulation pattern — See **wind.**

Amphora — A large clay jar used for storing and transporting food in ancient Greece.

Annual — A plant that germinates, flowers, and dies within one year. Many grains are considered annuals.

Archaeologist — A scientist who learns about early cultures by studying the artifacts left behind by them.

Atmosphere — A mixture of gases surrounding the Earth, including nitrogen, oxygen, carbon dioxide, and water vapor.

Atom — The smallest particle of mass that contains the chemical properties of an element.

Awn — The bristle-like structure on the tip of a grass flower.

Baffles — Dividers used inside a tanker that prevent the liquid it carries from sloshing around inside.

Base — A compound that tastes bitter, feels slippery, and turns litmus blue. Examples: baking soda and ammonia.

Baseline data — Data taken at the beginning of an experiment, or before it begins. Baseline data may be compared to data collected at the end of an experiment to show changes that have occurred.

Beard — In grasses, a common term for the **awn.**

Big rig — A tractor-trailer truck used to haul **goods** along the highway system across the country. Also called an 18-wheeler or a semi truck. The goods are packed into the trailer and the driver sits in the tractor.

Biosphere — The subsystem on Earth within which life occurs.

Biotic — Living. The biotic elements found in an ecosystem include plants, animals, and microbes.

Blanch — To submerge a food in boiling water, then plunge it into ice water to stop it from cooking. Foods are boiled for a few minutes to reduce enzyme activity, preserve color, and prepare food for freezing.

Boxcar — A type of car on a freight train that is designed to protect freight from weather and rough travel conditions.

Bran — The protective outer layer of a grain kernel that contains the most fiber. The bran also contains starch, protein, vitamins, and minerals.

Brine — A salt-and-water solution used for pickling.

Canning — A method of food preservation that encloses food in an airtight can or jar. Canning prevents reactions of enzymes, interactions with oxygen, and moisture loss in food.

Carbon footprint — The amount of greenhouse-gas emissions caused by human activities. The footprint is measured in units of carbon dioxide over a period of time, often for one year.

Cellular respiration — The process by which a cell uses sugars and oxygen to produce usable energy. The byproducts are carbon dioxide and water.

Cereal — An edible grass, called grain, such as wheat, oats, corn, rye, and rice. Grains serve as a large component of diets worldwide.

Chemical preservation — Using chemical preservatives to stop growth of microorganisms in food.

Constraints — Limitations that arise when finding a solution to a problem.

Control group — In an experiment, the group that has the usual or normal conditions to compare with the **experimental group.** For example, in an experiment to determine what would happen to a plant that does not get sunlight, the control group would be plants that are exposed to sunlight. The **experimental group** would be plants that are not exposed to sunlight.

Covered hopper — A type of car on a freight train in which freight is loaded through hatches on top of the car and emptied through chutes in the bottom. Covered hoppers usually carry grain, salt, sugar, soybeans, or flour.

Crude oil — Another word for petroleum. Crude oil is formed from marine plants and animals that sink to the bottom of the ocean, become covered with sediment, and are placed under great pressure and heat for millions of years.

Cryosphere — The regions of Earth where the temperature is 32°F or below and water exists in solid form such as ice, snow cover, or frozen ground.

Deforestation — The process of changing forested lands to non-forest use.

Dehydrator — An electrical appliance used to remove moisture from foods as a method of food preservation. The machine has several trays and dries food using electric fans or convection air currents.

Dent corn — A type of corn that is inedible when fresh, but is made into animal feed or processed to make cornstarch, corn syrup, or other products. It is grown more than any other crop in the United States.

Dormant — A phase for plants during which they are not actively growing, but are protected from the environment.

Drying — A method of food preservation that removes moisture from the food in order to **inhibit** microorganism growth. Food may be dried using the sun, an oven, or a special **dehydrator.**

Durable goods — **Goods,** such as household appliances, that are not consumed with use and last for a period of time, usually at least three years.

Ecosystem — A **system** of all living and nonliving parts that interact in an environment.

Embryo — The earliest stage of development of an animal or plant.

Endosperm — The starchy portion of a grain kernel. The endosperm provides nutrition to the developing seed. Composed mostly of starch, the endosperm also contains small amounts of oil and protein.

Environmental Protection Agency (EPA) — The governmental agency formed to protect human health and the environment. The EPA is responsible for developing and enforcing environmental regulations, performing research, and educating the public on environmental issues.

Exosphere — The outermost layer of the Earth's atmosphere — the last layer before outer space.

Experimental design — The steps to conduct an experiment. These steps are also called the methods for the experiment. In an experiment, the **experimental group** is compared to the **control group.**

Experimental group — In an experiment, the group that is exposed to a condition that is being tested in order to determine the effects of the condition. For example, in an experiment to determine what would happen to a plant that does not get sunlight, the plants in the experimental group would be deprived of sunlight in order to determine the effects of this condition.

Fermentation — The process of chemical change brought about by conversion of sugar into ethyl alcohol and carbon dioxide by **yeast** enzymes. Fermentation is used to make preserved foods such as pickles and sauerkraut.

Food processing — The process of changing a whole food in order to preserve the food and make it into another form for people to eat. Some examples are adding salt, water, and herbs to tomatoes, then cooking them at a high temperature to make tomato sauce, or removing the bran and endosperm from wheat and grinding it to make white flour.

Food production — The process of growing plants or raising animals for the purpose of human consumption. Many different methods are used for production, such as industrial or organic farming methods.

Food system — The system in place for food production, transportation, processing, packaging, marketing, and sales.

Freezing — A method of food preservation that works by lowering the food's temperature below 32°F or 0°C to stop the growth of microorganisms. Many vegetables are **blanched** before freezing.

Geosphere — The solid part of planet Earth. It includes the various layers of the Earth's interior, rocks, minerals, landforms, and the processes that shape Earth's surface.

Germ — The part of a seed that will become the new plant when the seed is planted. The germ has a high oil and vitamin content.

Goods — Items or products that may be bought and sold in the marketplace.

Greenhouse effect — The process by which **greenhouse gases** absorb heat that leaves Earth and re-emit it back in the direction of the planet, which keeps Earth warmer.

Greenhouse gases — Gases in the Earth's atmosphere such as carbon dioxide, methane, and water vapor.

Gyres — Circulating currents in the oceans that bring warm waters away from the equator and cold waters toward the equator.

Hydrosphere — The subsystem of Earth that includes all the liquid water in oceans, lakes, and streams.

Hypothesis — An informed or educated prediction, or a guess, about the results of an experiment.

Inflorescence — A flower head at the top of the stem. **Panicles** and **spikes** are two types of grass inflorescence.

Inhibit — To prevent or restrain an action from being taken, such as inhibiting the growth of microorganisms.

Lithosphere — The Earth's crust and upper mantle. The lithosphere is made up of constantly moving sections called plates.

Mesosphere — The layer of the atmosphere between the **stratosphere** and the **thermosphere** with the coldest temperatures and the highest clouds.

Milling — The processing of wheat or other grains in order to separate the **endosperm, bran,** and **germ,** producing flour. The processes of washing, soaking, grinding, and sifting are used in milling.

Mine shafts — Tunnels used in underground mining to reach coal as much as 1,000 feet under the Earth's surface.

Monsoon — Changes in air-circulation patterns that occur seasonally and bring heavy rains.

Mortar — A bowl-shaped tool made of a hard material such as stone, used to hold grain or another substance that is ground into powder using a **pestle.**

Nondurable goods — **Goods** that are consumed in one use, or under three years.

Nonrenewable resource — A natural resource that cannot be reproduced in a short period of time.

Nonrenewable resources are used at a faster rate by humans than they are produced by nature.

Osmosis — The diffusion of water through a semipermeable membrane so that the concentration of a solvent (a liquid that can dissolve other substances) is equal on both sides of the membrane.

Packaging — The **subsystem** of the **food system** in which processed or whole foods are enclosed in materials for the purpose of protecting the food or preventing spoiling. Food packaging is often made of metal, plastic, glass, or paper.

Panicle — A type of branched **inflorescence,** or flower cluster.

Particulate matter — A term for air pollution in the form of tiny particles suspended in air that we breathe.

Peat — Plant matter that died, decomposed, and was covered by sediment millions of years ago. As the material was put under intense heat and pressure, it gradually turned into coal.

Pedicel — In grasses, a short stalk that supports a **spikelet.**

Pericarp — The outer wall of a fruit. In grasses, the fruit is a dry fruit, commonly called a grain. The pericarp is the outer wall of the grain.

Pestle — A tool used to pound and grind substances held by a **mortar** into a powder.

Photosynthesis — The process by which a plant makes food using energy it absorbs from sunlight. The plant uses the light energy, carbon dioxide from the air, and water from the soil to make sugars, which are the plant's food.

Pickling — A method of food preservation that soaks food in **brine** or vinegar. Pickling stops microorganisms from growing in an airtight container.

Preservative — A chemical substance used to preserve food from microorganism decomposition or fermentation.

Reefer van — An abbreviation for "refrigerator van," a vehicle with a freight area that cools perishable goods during shipment.

Refrigerated car — A car on a freight train that cools the perishable goods inside during travel.

Reservoir — A natural or artificial place where water is collected and stored.

Resources — The materials, physical or human, used to complete a task.

Risk — The chance of suffering loss when making choices.

Ruminants — Livestock animals with four stomachs that release methane, a greenhouse gas, into the environment as a part of normal digestion.

Seed coat — In a seed, a protective outer layer made mostly of fiber, which protects the **embryo.**

Sodium chloride — The scientific term for table salt. The chemical formula of salt is NaCl.

Spike — A type of **inflorescence,** or flower cluster.

Spikelet — A floral structure made up of one or more flowers enclosed by bracts (leaves).

Stratosphere — The layer of the atmosphere that contains the ozone layer and is found between the **troposphere** and the **mesosphere.**

Subsystem — A unit of interacting parts within a larger **system.** A subsystem contributes to the overall function of the larger system.

Surface mining — A method used for mining coal and other materials that lie close to the surface of the Earth. Topsoil and other surface layers are removed so that the coal is exposed, and then mined.

System — A whole that is more than the sum of its parts.

Tank car — A type of car on a freight train that carries liquid freight.

Tanker — (1) An oceangoing ship designed to carry mass quantities of liquid, such as crude oil. (2) A stainless-steel container pulled by a big rig to transport liquid.

Tassel — The male portion of a corn plant; a flowering structure that produces and sheds pollen.

Thermosphere — The atmospheric layer found above the mesosphere, which has the highest temperatures — up to 3,632°F.

Tip cap — A portion of a corn kernel that attaches the kernel to the cob.

Trade-off — The choice of a greater benefit over a lesser one when it is not possible to choose both. Also, the acceptance of the constraints of a solution, in order to benefit from its strengths.

Transportation — A food **subsystem** used to transport foods between farms, processing facilities, warehouses, and food retailers. Transportation may happen by truck, train, barge, or plane.

Troposphere — The level of the atmosphere closest to the Earth's surface. This level contains most of the water vapor, so weather occurs here.

Variable — In an experiment, the condition or factor that is being tested. It is what will be different between the **control group** and the **experimental group.** For example, in an experiment to determine what would happen to a plant that does not get sunlight, the variable would be the amount of light the plants get.

Wind — The natural movement of air caused by differing temperatures.

Yeast — Single-celled fungi that actively produce carbon dioxide at room temperature. Yeast is useful for leavening bread and fermenting food.